MW01069826

CONTINUOUS-TIME SIGNALS

Continuous-Time Signals

by

YURIY SHMALIY
Guanajuato University, Mexico

Springer

A C.I.P. Catalogue record for this book is available from the Library of Congress.

ISBN-10 1-4020-4817-3 (HB)
ISBN-13 978-1-4020-4817-3 (HB)
ISBN-10 1-4020-4818-1 (e-book)
ISBN-13 978-1-4020-4818-0 (e-book)

Published by Springer,
P.O. Box 17, 3300 AA Dordrecht, The Netherlands.

www.springer.com

Printed on acid-free paper

All Rights Reserved
© 2006 Springer
No part of this work may be reproduced, stored in a retrieval system, or transmitted in any form or by any means, electronic, mechanical, photocopying, microfilming, recording or otherwise, without written permission from the Publisher, with the exception of any material supplied specifically for the purpose of being entered and executed on a computer system, for exclusive use by the purchaser of the work.

Printed in the Netherlands.

To my family

Preface

As far back as the 1870s, when two American inventors Elisha Gray and Graham Bell independently designed devices to transmit speech electrically (the telephone), and the 1890s, when the Russian scientist Aleksandr Popov and the Italian engineer Guglielmo Marconi independently demonstrated the equipment to transmit and receive messages wirelessly (the radio), the theory of electrical signal was born. However, the idea of signals has been employed by mankind all through history, whenever any message was transmitted from a far point. Circles on water indicating that some disturbance is present in the area give a vivid example of such messages. The prehistory of electrical signals takes us back to the 1860s, when the British scientist James Clerk Maxwell predicted the possibility of generating electromagnetic waves that would travel at the speed of light, and to the 1880s, when the German physicist Heinrich Hertz demonstrated this radiation (hence the word "radio").

As a time-varying process of any physical state of an object that serves for representation, detection, and transmission of messages, a modern electrical *signal*, in applications, possesses many specific properties including:

- A flow of information, in information theory;
- Disturbance used to convey information and information to be conveyed over a communication system;
- An asynchronous event transmitted between one process and another;
- An electrical transmittance (either input or output) that conveys information;
- Form of a radio wave in relation to the frequency, serving to convey intelligence in communication;
- A mechanism by which a process may be notified by the kernel of an event occurring in the system;
- A detectable impulse by which information is communicated through electronic or optical means, or over wire, cable, microwave, laser beams, etc;
- A data stream that comes from electrical impulses or electromagnetic waves;

- Any electronic visual, audible, or other indication used to convey information;
- The physical activity of the labeled tracer material that is measured by a detector instrument; the signal is the response that is measured for each sample;
- A varying electrical voltage that represents sound.

How to pass through this jungle and understand the properties of signals in an optimum way? Fundamental knowledge may be acquired by learning the continuous-time signals, for which this book offers five major steps:

1. Observe applications of signals in electronic systems, elementary signals, and basic canons of signals description (Chapter 1).
2. Consider the representation of signals in the frequency domain (by Fourier transform) and realize how the spectral density of a single waveform becomes that of its burst and then the spectrum of its train (Chapter 2).
3. Analyze different kinds of amplitude and angular modulations and note a consistency between the spectra of modulating and modulated signals (Chapter 3).
4. Understand the energy and power presentations of signals and their correlation properties (Chapter 4).
5. Observe the bandlimited and analytic signals, methods of their description, transformation (by Hilbert transform), and sampling (Chapter 5).

This book is essentially an extensive revision of my Lectures on Radio Signals given during a couple of decades in Kharkiv Military University, Ukraine, and several relevant courses on Signals and Systems as well as Signal Processing in the Guanajuato University, Mexico, in recent years. Although, it is intended for undergraduate and graduate students, it may also be useful in postgraduate studies.

Salamanca, Mexico *Yuriy S. Shmaliy*

Contents

1

Introduction

Signals and *processes* in electronic systems play a fundamental role to transfer information from one point or space to any other point or space. Their description, transformation, and conversion are basic in *electrical engineering*. Therefore, it becomes possible to optimize systems with highest efficiency both in the time and frequency domains. This is why the theory of signals is fundamental for almost all electrical engineering fields; Mechanical, chemical, physical, biological, and other systems exploit fundamentals of this theory whenever waves and waveforms appear. Our purpose in this chapter is to introduce a concept and necessary fundamental canons of *signals*, thereby giving readers food for learning the following chapters.

1.1 Signals Application in Systems

The word "signal" has appeared from the Latin term *signum* meaning "sign" and occupied a wide semantic scope in various ranges of science and engineering. It is defined as follows:

> ***Signal***: A *signal* is a time-varying process of any physical state of any object, which serves for representation, detection, and transmission of messages.

□

In electrical engineering, time variations of electric currents and voltages in electronic systems, radio waves radiated by a transmitter in space, and noise processes in electronic units are examples of signals. Application of signals in several most critical electronic systems are illustrated below.

1.1.1 Radars

The *radar* is usually called a device for determining the presence and location of an object by measuring the time for the echo of a radio wave to return from it and the direction from which it returns. In other words, it is a measuring instrument in which the echo of a pulse of microwave radiation is used to detect and locate distant objects. A radar pulse-train is a type of amplitude modulation of the radar frequency carrier wave, similar to how carrier waves are modulated in communication systems. In this case, the information signal is quite simple: a single pulse repeated at regular intervals.

Basic operation principle of radars is illustrated in Fig. 1.1. Here transmitter generates radio frequency (RF) impulse signal that is reflected from the target (moving or stationary object) and is returned to receiver. Conventional ("monostatic") radar, in which the illuminator and receiver are on the same platform, is vulnerable to a variety of countermeasures. Bistatic radar, in which the illuminator and receiver are widely separated, can greatly reduce the vulnerability to countermeasures such as jamming and antiradiation weapons, and can increase slow moving target detection and identification capability by "clutter tuning" (receiver maneuvers so that its motion compensates for the motion of the illuminator; creates zero Doppler shift for the

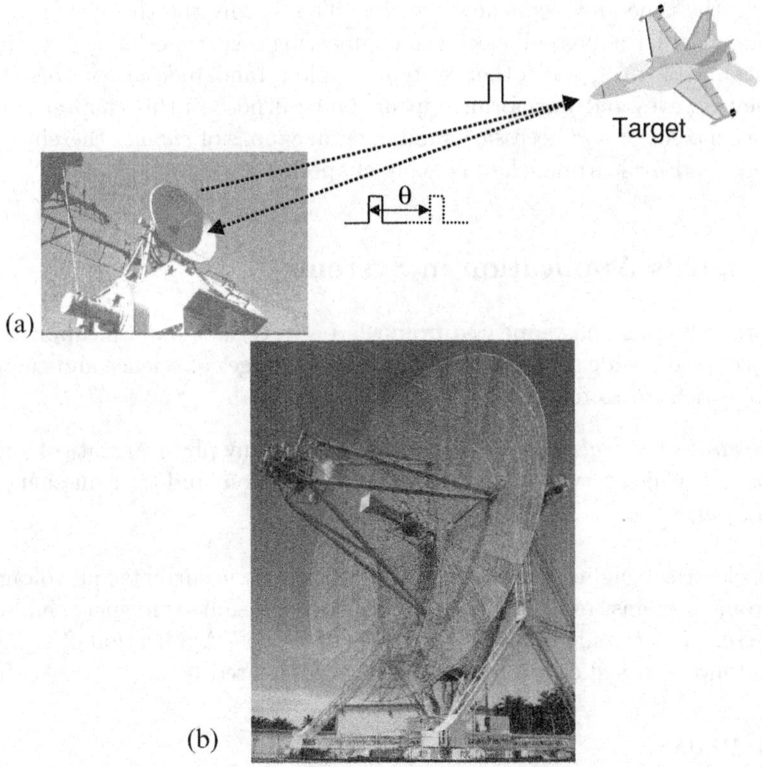

Fig. 1.1 Operation principle of radars: (**a**) pulse radar and (**b**) long-range radar antenna.

area being searched). The transmitter can remain far from battle area, in a "sanctuary." The receiver can remain "quiet."

At the early stage, radars employed simple single pulses to fulfill requirements. With time, for the sake of measuring accuracy, the pulses with frequency modulation and pulse-coded bursts were exploited. The timing and phase coherent problems can be orders of magnitude more severe in bistatic than in monostatic radar, especially when the platforms are moving. The two reference oscillators must remain synchronized and synchronized during a mission so that the receiver knows when the transmitter emits each pulse, so that the phase variations will be small enough to allow a satisfactory image to be formed. Low noise crystal oscillators are required for short-term stability. Atomic frequency standards are often required for long-term stability.

1.1.2 Sonar

Sonar (acronym for *SOund NAvigation and Ranging*) is called a measuring instrument that sends out an acoustic pulse in water and measures distances in terms of the time for the echo of the pulse to return. This device is used primarily for detection and location of underwater objects by reflecting acoustic waves from them, or by interception of acoustic waves from an underwater, surface, or above-surface acoustic source. Note that sonar operates with acoustic waves in the same way that radar and radio direction-finding equipment operate with electromagnetic waves, including use of the Doppler effect, radial component of velocity measurement, and triangulation.

1.1.3 Remote Sensing

Remote sensing is the science — and to some extent, art — of acquiring information about the Earth's surface without actually being in contact with it. This is done by sensing and recording reflected or emitted energy and processing, analyzing, and applying that information. Two kinds of remote sensing are employed. In *active remote sensing*, the object is illuminated by radiation produced by the sensors, such as radar or microwaves (Fig. 1.2a). In *passive remote sensing*, the sensor records energy that is reflected or emitted from the source, such as light from the sun (Fig. 1.2b). This is also the most common type of system.

1.1.4 Communications

Analog and digital communications are likely the most impressive examples of efficient use of signals. In analog communications, an analog method of modulating radio signals is employed so that they can carry information such as voice or data. In digital communications, the carrier signal is modulated digitally by encoding information using a binary code of "0" and "1". Most

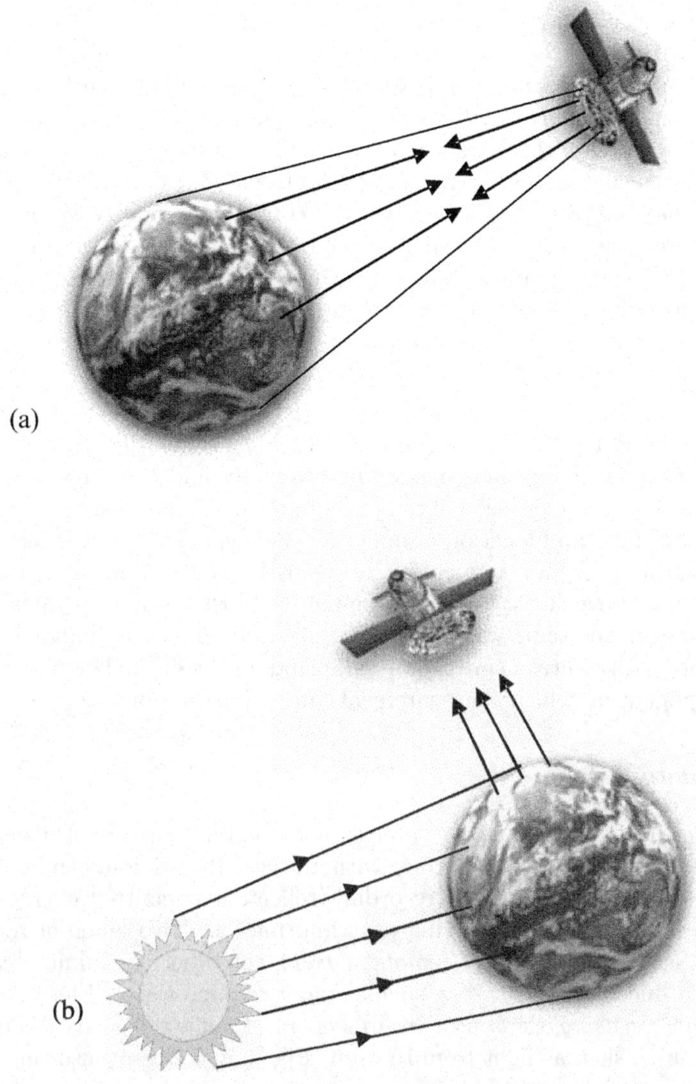

Fig. 1.2 Remote sensing operation principle: (**a**) active and (**b**) passive.

newer wireless phones and networks use digital technology and one of the most striking developments of the past decade has been the decline of public service broadcasting systems everywhere in the world. Figure 1.3 illustrates the basic principle of two-way satellite communications.

To transfer a maximum of information for the shortest possible time duration, different kinds of modulation had been examined for decades at different

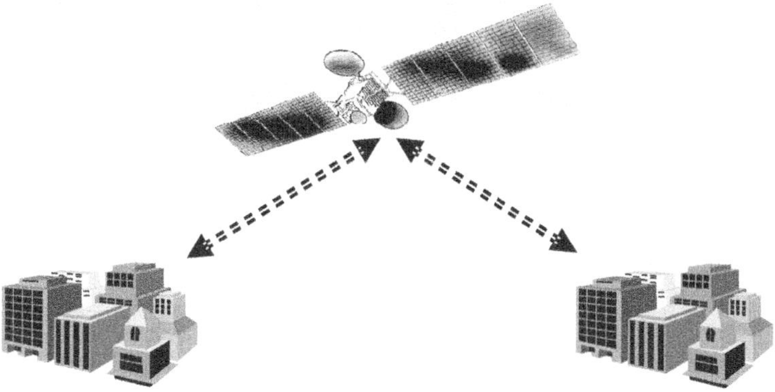

Fig. 1.3 Two-way satellite communications.

carrier frequencies. In digital transmission, either a binary or M-ary keying is used in amplitude, phase, and frequency providing the commercially available resources with a minimum error.

Historically, as the number of users of commercial two-way radios have grown, channel spacing have been narrowed, and higher-frequency spectra have had to be allocated to accommodate the demand. Narrower channel spacings and higher operating frequencies necessitate tighter frequency tolerances for both the transmitters and the receivers. In 1949, when only a few thousand commercial broadcast transmitters were in use, a 500 ppm (ppm $= 10^{-6}$) tolerance was adequate. Today, the millions of cellular telephones (which operate at frequency bands above 800 MHz) must maintain a frequency tolerance of 2.5 ppm. The 896–901 MHz and 935–940 MHz mobile radio bands require frequency tolerances of 0.1 ppm at the base station and 1.5 ppm at the mobile station. The need to accommodate more users will continue to require higher and higher frequency accuracies. For example, NASA concept for a personal satellite communication system would use walkie-talkie-like hand-held terminals, a 30 GHz uplink, a 20 GHz downlink, and a 10 kHz channel spacing. The terminals' frequency accuracy requirement is few parts in 10^{-8}.

1.1.5 Global Positioning System

Navigation systems are used to provide moving objects with information about their positioning. An example is the satellite-based global positioning system (GPS) that consists of (a) a constellation of 24 satellites in orbit 11,000 nmi above the Earth, (b) several on-station (i.e., in-orbit) spares, and (c) a ground-based control segment. Figure 1.4 gives an example of the GPS use in ship navigation. Each space vehicular (SV) transmits two microwave carrier signals (Fig. 1.5). The L1 frequency (1575.42 MHz) carries the navigation message and the standard positioning service (SPS) code signals. The L2 frequency (1227.60 MHz) is used to measure the ionospheric delay by precise

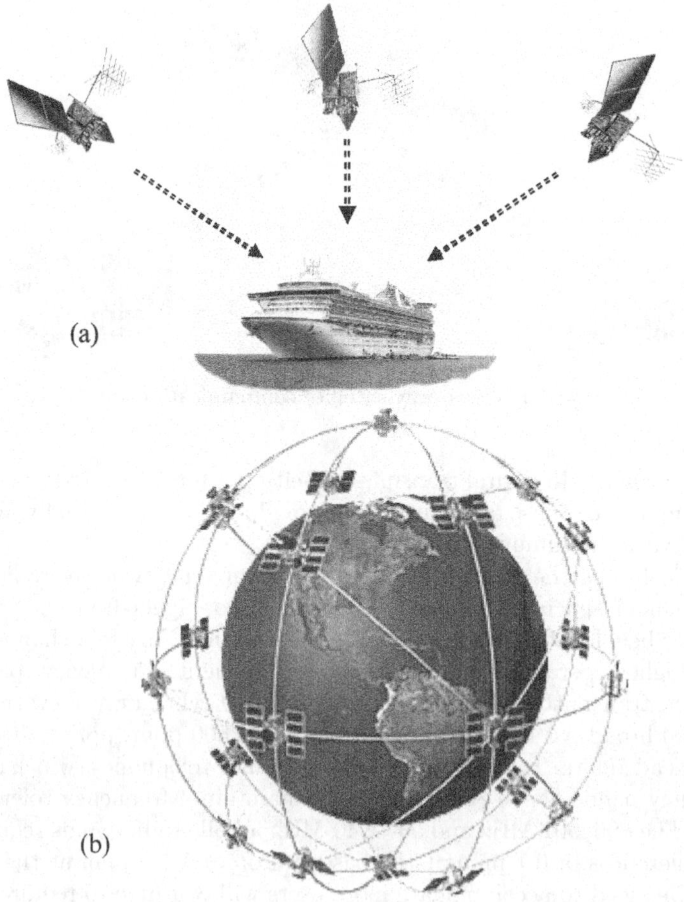

Fig. 1.4 GPS system: (**a**) ship navigation and (**b**) GPS constellation.

positioning service (PPS) equipped receivers. Three binary codes shift the L1 and/or L2 carrier phase:

- The coarse acquisition (C/A) code modulates the L1 carrier phase. The C/A code is a repeating 1 MHz pseudorandom noise (PRN) code. This noise-like code modulates the L1 carrier signal, "spreading" the spectrum over a 1 MHz bandwidth. The C/A code repeats every 1023 bits (one millisecond). There is a different C/A code PRN for each SV. GPS satellites are often identified by their PRN number, the unique identifier for each PRN code. The C/A code that modulates the L1 carrier is the basis for the civil SPS.
- The precision (P) code modulates both the L1 and the L2 carrier phases. The P code is a very long (7 days) 10 MHz PRN code. In the antispoofing (AS) mode of operation, the P code is encrypted into the Y code. The

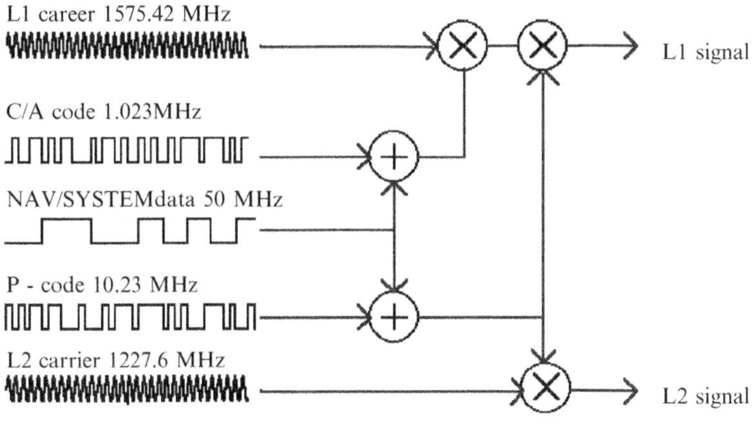

Fig. 1.5 GPS satellite signals.

encrypted Y code requires a classified AS module for each receiver channel and is for use only by authorized users with cryptographic keys. The P(Y) code is the basis for the PPS.

- The navigation message also modulates the L1-C/A code signal. The navigation message is a 50 Hz signal consisting of data bits that describe the GPS satellite orbits, clock corrections, and other system parameters.

Any navigation system operates in time. Therefore, to obtain extremely accurate 3-D (latitude, longitude, and elevation) global navigation (position determination), precise time (time signals) must also be disseminated. These signals are used in what is called timekeeping.

Historically, navigation has been a principal motivator in man's search for better clocks. Even in ancient times, one could measure latitude by observing the stars' position. However, to determine longitude, the problem became one of timing. This is why GPS-derived position determination is based on the arrival times, at an appropriate receiver, of precisely timed signals from the satellites that are above the user's radio horizon. On the whole, in the GPS, atomic clocks in the satellites and quartz oscillators in the receivers provide nanosecond-level accuracies. The resulting (worldwide) navigational accuracies are about 10 m and some nanoseconds. Accordingly, GPS has emerged as the leading methodology for synchronization not only for communication but also for transport, navigation, commercial two-way radio, space exploration, military requirements, Doppler radar systems, science, etc.

1.2 Signals Classification

Classification of signals may be done for a large number of factors that mostly depend on their applications in systems.

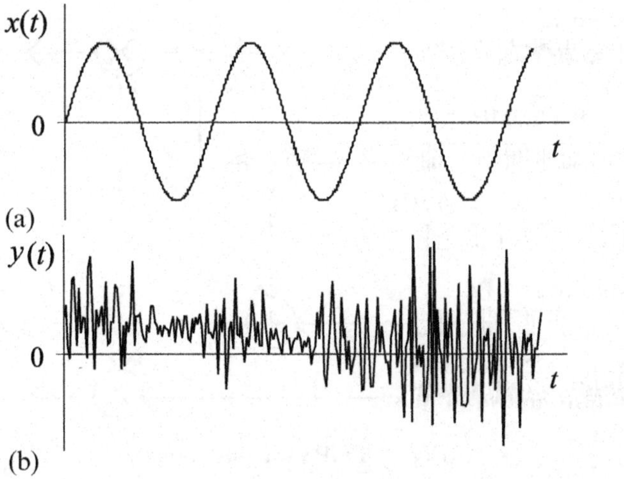

Fig. 1.6 Example of signals: (**a**) deterministic and (**b**) random.

1.2.1 Regularity

Most commonly, the signals are separated into two big classes: *deterministic* (Fig. 1.6a) (*regular* or *systematic* in which a random amount is insignificant) and *random* (*noisy*) (Fig. 1.6b).

- *Deterministic signals* are precisely determined at an arbitrary time instant; their simulation implies searching for proper analytic functions to describe them explicitly or with highest accuracy.
- *Random signal* cannot be described analytically at an arbitrary time instant owing to its stochastic nature; such signals cannot be described in deterministic functions or by their assemblage and are subject to the probability theory and mathematical statistics.

It is important to remember that a recognition of signals as deterministic and random is conditional in a sense. Indeed, in our life there are no deterministic physical processes at all, at least because of noise that exists everywhere. The question is, however, how large is this noise? If it is negligible, as compared to the signal value, then the signal is assumed to be deterministic. If not, the signal is random, and stochastic methods would be in order for its description.

1.2.2 Causality

The signals produced by physical devices or systems are called *causal.* It is assumed that such a signal exists only at or after the time the signal generator is turned on. Therefore, the casual signal $y(t)$ satisfies $y(t) = x(t)$, if $t \geqslant 0$ and

$y(t) = 0$, if $t < 0$. Signals that are not causal are called *noncausal*. Noncausal signals representation is very often used as a mathematical idealization of real signals, supposing that $x(t)$ exists with $-\infty \leq t \leq \infty$.

1.2.3 Periodicity

Both deterministic and random signals may be either *periodic* (Fig. 1.7a) or *single* (*nonperiodic*) (Fig. 1.7b).

- *Periodic signals* (Fig. 1.7a) reiterate their values through the equal time duration T called a period of repetition. For such signals the following equality holds true:
$$x(t) = x(t \pm nT) \tag{1.1}$$
where $x(t)$ is a signal and $n = 0, 1, 2, \ldots$. It seems obvious that simulation of (1.1) implies that a signal may be described only on the time interval T and then repeated n times with period T.
- *Single signals* or *nonperiodic signals* (Fig. 1.7b) do not exhibit repetitions on the unlimited time interval and therefore an equality (1.1) cannot be applied.
- *Impulse signals.* A special class of signals unites the *impulse* signals. A single impulse signal is the one that exists only during a short time. Impulse signals may also be periodic. Two types of impulse signals are usually distinguished:
  - *Video pulse signal*, also called *waveform*, is an impulse signal $x(t)$ without a carrier (Fig. 1.8a).
  - *Radio frequency (RF) pulse* signal $y(t)$ is a video pulse signal $x(t)$ filled with the carrier signal $z(t)$ (Fig. 1.8b).

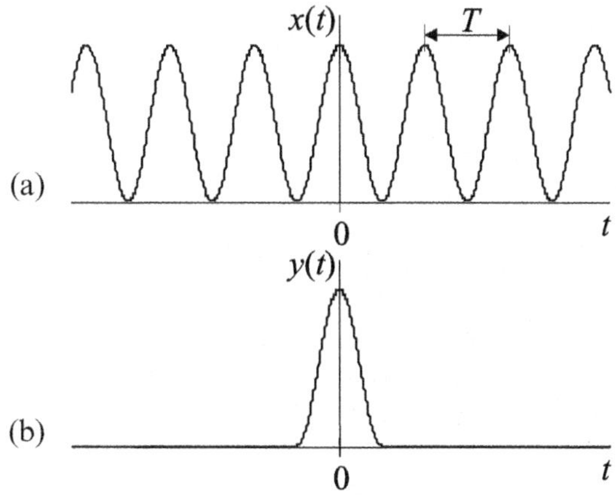

Fig. 1.7 Example of signals: (**a**) $x(t)$ is periodic and (**b**) $y(t)$ is nonperiodic.

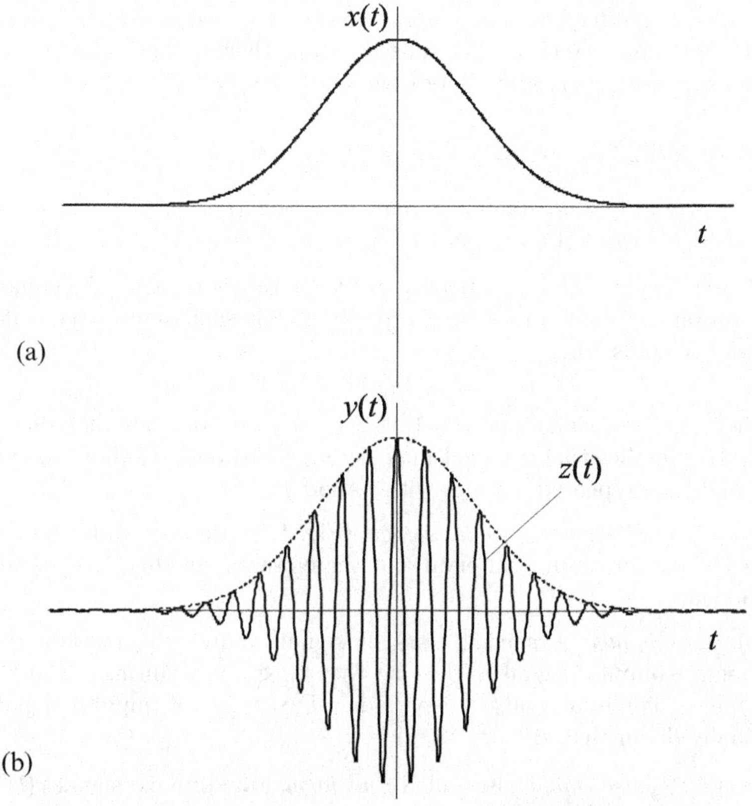

Fig. 1.8 Impulse signals: (**a**) video pulse and (**b**) radio pulse.

1.2.4 Dimensionality

Both periodic and single signals may depend on different factors and exist in the same timescale. Accordingly, they may be *one-dimensional* and *multidimensional*:

- *One-dimensional (scalar) signal* is a function of one or more variables whose range is 1-D. A scalar signal is represented in the time domain by means of only one function. Examples are shown in Figs. 1.6 and 1.7. A physical example is an electric current in an electronic unit.
- *Multidimensional (vector) signal* is a vector function, whose range is 3-D or, in general, N-dimensional (N-d). A vector signal is combined with an assemblage of 1-D signals. An N-d signal is modelled as a vector of dimensions $N \times 1$

$$\mathbf{x} \equiv \mathbf{x}(t) = [x_1(t), x_2(t), \ldots, x_N(t)]^T \tag{1.2}$$

where an integer N is said to be its *order* or *dimensionality*. An example of a multidimensional signal is several voltages on the output of a multipole. An example of 2-D signals is an electronic image of the USA and Mexico obtained by NASA with a satellite remote sensing at some instant t_1 (Fig. 1.9a). An example of 3-D signals is fixed at some instant t_2, a cosine wave attenuated with a Gaussian[1] envelope in the orthogonal directions (Fig. 1.9b).

1.2.5 Presentation Form

Regarding the form of presentation, all signals may be distinguished to fall within three classes:

- An *analog signal* or *continuous-time signal* is a signal $x(t)$, which value may be determined (measured) at an arbitrary time instant (Fig. 1.10a).
- *Discrete-time* signal is a signal $x(t_n)$, where n is an integer that represents an analog signal by discrete values at some time instants t_n, usually with a constant sample time $\Delta = t_{n+1} - t_n$ (Fig. 1.10b).
- *Digital* signal is a signal $x[n]$, which is represented by discrete values at discrete points n with a digital code (binary, as a role) (Fig. 1.10c). Therefore, basically, $x[n] \neq x(t_n)$ and the quantization error depends on the resolution of the analog-to-digital converter.

1.2.6 Characteristics

Every signal may be explicitly described either in the time domain (by time functions) or in the frequency domain (by spectral characteristics). Signals presentations in the time and frequency domains are interchangeable to mean that any signal described in the time domain may be translated to the frequency domain and come back to the time domain without errors. The following characteristics are usually used to describe signals:

- In the time domain: *effective duration*, *covariance function*, *peak amplitude*, *period of repetition*, *speed of change*, *correlation time*, *time duration*, etc.
- In the frequency domain:
  - Spectrum of periodic signals is represented by the Fourier[2] series with the *magnitude spectrum* and *phase spectrum*.

[1] Johann Carl Friedrich Gauss, German mathematician, 30 April 1777–23 February 1855.

[2] Jean Baptiste Joseph Fourier, French mathematician, 21 March 1768–16 May 1830.

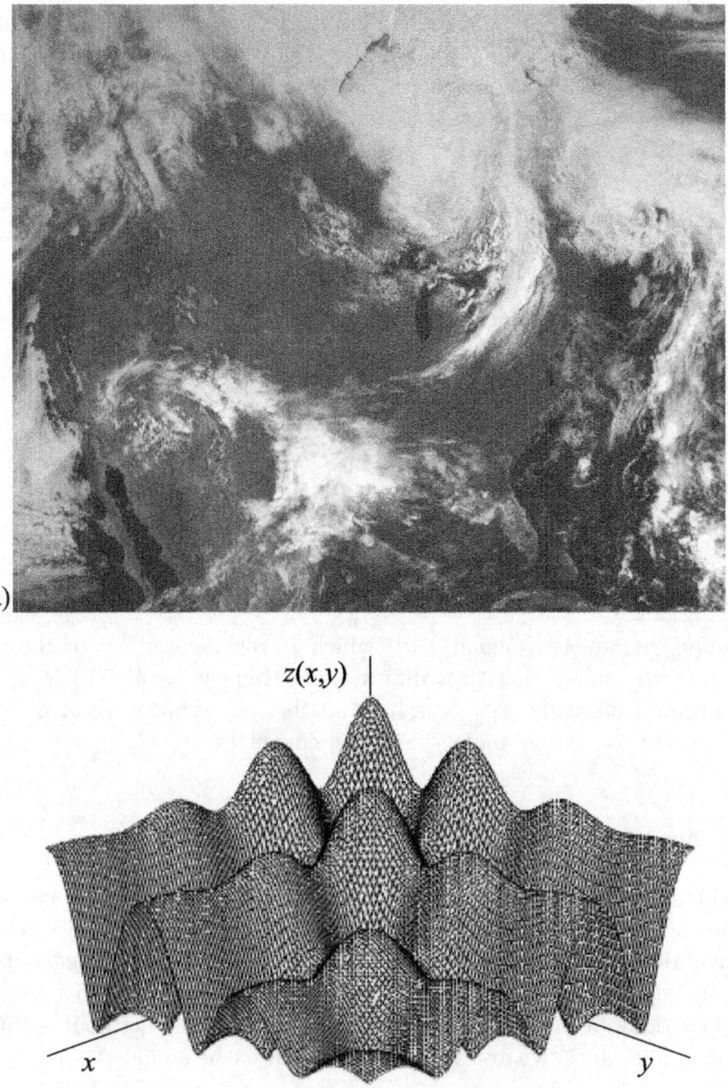

(a)

(b)

Fig. 1.9 Multidimensional signals: (**a**) 2-D satellite electronic image and (**b**) 3-D Gaussian radio pulse.

- Spectral density of nonperiodic signals is represented by the Fourier transform with the *magnitude spectral density* and *phase spectral density*.
- Both the spectrum and spectral density are characterized with the *signal energy*, *signal power*, *spectral width*, *spectral shape*, etc.

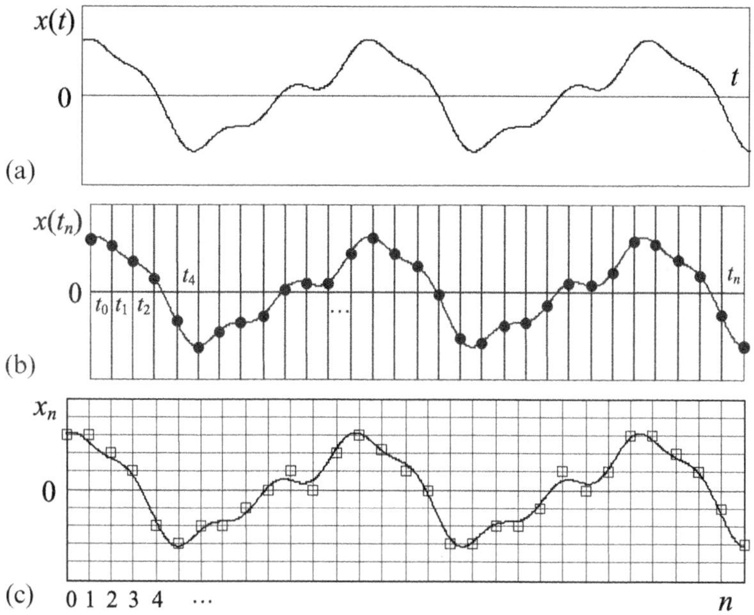

Fig. 1.10 Types of signals: (**a**) continuous-time, (**b**) discrete-time, and (**c**) digital.

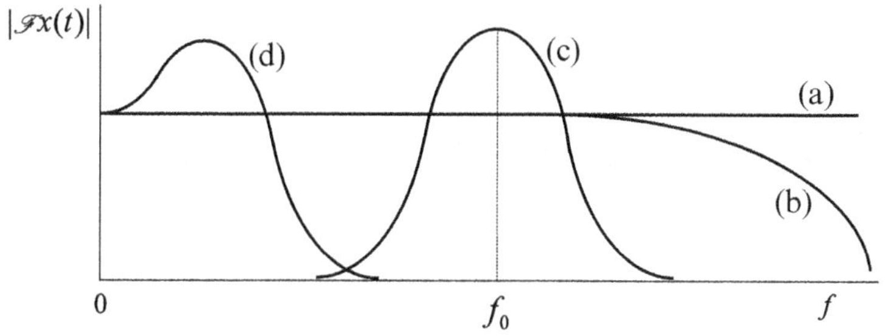

Fig. 1.11 Types of signals: (a) broadband, (b) bandlimited, (c) narrowband, and (d) baseband.

1.2.7 Spectral Width

In the frequency domain, all signals may be classified as follows:

- A *broadband* signal is the one, which spectrum is distributed over a wide range of frequencies as it is shown in Fig. 1.11a.
- A *bandlimited* signal is limited in the frequency domain with some maximum frequency as it is shown in Fig. 1.11b.
- A *narrowband* signal has a spectrum that is localized about a frequency f_0 that is illustrated in Fig. 1.11c.

- A *baseband* signal has a spectral contents in a narrow range close to zero (Fig. 1.11d). Accordingly, a spectrum beginning at 0 Hz and extending contiguously over an increasing frequency range is called a *baseband spectrum.*

1.2.8 Power and Energy

Every signal bears some energy and has some power. However, not each signal may be described in both terms. An example is a constant noncausal signal that has infinite energy.

An instantaneous power of a real signal $x(t)$ is defined by

$$P_x(t) = x^2(t). \tag{1.3}$$

In applications, however, it is much more important to know the signal energy or average power over some time bounds $\pm T$. Accordingly, two types of signals are recognized:

- *Energy signal* or *finite energy signal* is a signal, which energy

$$E_x = \|x\|_2^2 = \lim_{T\to\infty} \int_{-T}^{T} P_x(t)\mathrm{d}t = \lim_{T\to\infty} \int_{-T}^{T} x^2(t)\mathrm{d}t < \infty \tag{1.4}$$

  is finite. The quantity $\|x\|_2$ used in (1.4) is known as the L_2-norm of $x(t)$.

- *Power signal* or *finite power signal* is a signal which average power

$$P_x = \langle x^2(t) \rangle = \lim_{T\to\infty} \frac{1}{2T} \int_{-T}^{T} x^2(t)\mathrm{d}t < \infty \tag{1.5}$$

  is finite. If $x(t)$ is a periodic signal with period T then the limit in (1.5) is omitted.

Example 1.1. Given a harmonic noncausal signal $x(t) = A_0 \cos \omega_0 t$, which energy is infinite, $E_x = A_0^2 \int_{-\infty}^{\infty} \cos^2 \omega_0 t\mathrm{d}t = A_0^2 \infty$. Thus, it is not an energy signal. However, its average power is finite, $P_x(t) = \frac{A_0^2}{2T} \int_{-T}^{T} \cos^2 \omega_0 t\mathrm{d}t = \frac{1}{2}A_0^2 < \infty$. Hence, it is a power signal.

□

1.2.9 Orthogonality

In the correlation analysis and transforms of signals, *orthogonal signals* play an important role.

- Two real signals $x(t)$ and $y(t)$ are said to be *orthogonal*, $x(t) \perp y(t)$, on the interval $[a, b]$ if their inner (scalar) product (and so the joint energy) is zero:

$$\langle x, y \rangle = \int_a^b x(t)y(t)\mathrm{d}t = 0\,. \tag{1.6}$$

- Two same signals are called *orthonormal* if

$$\langle x, y \rangle = \begin{cases} 1, & x(t) = y(t) \\ 0, & \text{otherwise} \end{cases}.$$

In other words, if a function (signal) $x(t)$ has a zero projection on some other function (signal) $y(t)$, then their joint area is zero and they are orthogonal. Such an important property allows avoiding large computational burden in the multidimensional analysis.

Example 1.2. Given three signals:

$$\begin{aligned} x(t) &= A_0 \cos \omega_0 t\,, \\ y(t) &= A_0 \sin \omega_0 t\,, \\ z(t) &= A_0 \cos(\omega_0 t + \pi/4)\,. \end{aligned}$$

It follows, by (1.6), that two first signals are orthogonal and that no other pair of these signals satisfies (1.6).

□

We have already classified the signals with many characteristics. Even so, this list is not exhaustive and may be extended respecting some new methods of signals generation, transmitting, formation, and receiving. Notwithstanding this fact, the above given classification is sufficient for an overwhelming majority of applied problems.

1.3 Basic Signals

Mathematical modeling of signals very often requires its presentation by simple elementary signals, which properties in the time and frequency domains are well studied. Indeed, if we want to describe, for example, a rectangular pulse-train, then a linear combination of gained and shifted elementary unit-step functions will certainly be the best choice. We may also want to describe some continuous function that may be combined with elementary harmonic functions in what is known as the Fourier series. So, basic elementary functions play an important role in the signals theory.

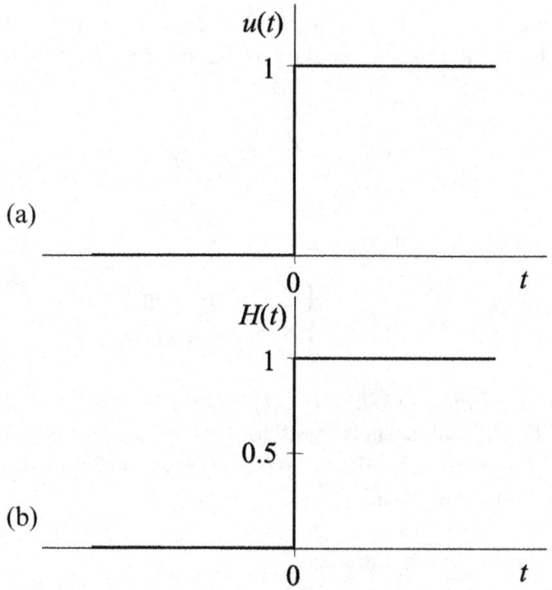

Fig. 1.12 Unit step: (**a**) Unit-step function and (**b**) Heaviside unit-step function.

1.3.1 Unit Step

A *unit-step function* (Fig. 1.12a) is defined by

$$u(t) = \begin{cases} 1, & t \geqslant 0 \\ 0, & t < 0 \end{cases} \tag{1.7}$$

and is usually used in signals to model rectangular waveforms and in systems to define the step response.

The other presentation of a unit step was given by Heaviside[3] in a conventionally continuous form. The *Heaviside unit-step function* (Fig. 1.12b) is performed as

$$H(t) = \begin{cases} 1, & t > 0 \\ 0.5, & t = 0 \\ 0, & t < 0 \end{cases} \tag{1.8}$$

and may be modeled by the function

$$v(t,\xi) = \begin{cases} 1, & t > \xi \\ 0.5(\frac{t}{\xi}+1), & -\xi \leqslant t \leqslant \xi \\ 0, & t < -\xi \end{cases}, \tag{1.9}$$

[3] Oliver Heaviside, English physicist, 18 May 1850–3 February 1925.

once $H(t) = \lim_{\xi \to 0} v(t, \xi)$. This is not the only way to model the unit step. The following function may also be useful:

$$v(t, n) = \frac{1}{1 + e^{-nt}} . \tag{1.10}$$

It follows from (1.10) that tending n toward infinity makes the function to be more and more close to the Heaviside step function, so that one may suppose that $H(t) = \lim_{n \to \infty} v(t, n)$.

Example 1.3. Given a rectangular impulse signal (Fig. 1.13a). By (1.7), it is described to be $x(t) = 8.5[u(t-1) - u(t-3)]$.

□

Example 1.4. Given a truncated ramp impulse signal (Fig. 1.13b). By (1.7) and (1.9), we go to the model $x(t) = 8.5[v(t-2, 1) - u(t-3)]$.

□

Example 1.5. Given an arbitrary continuous signal (Fig. 1.13c). By (1.7), this signal is described as

$$x(t) = \sum_{i=-\infty}^{\infty} x(iT)[u(t - iT) - u(t - iT - T)] ,$$

where a sample time T should be chosen to be small enough to make the approximation error negligible.

□

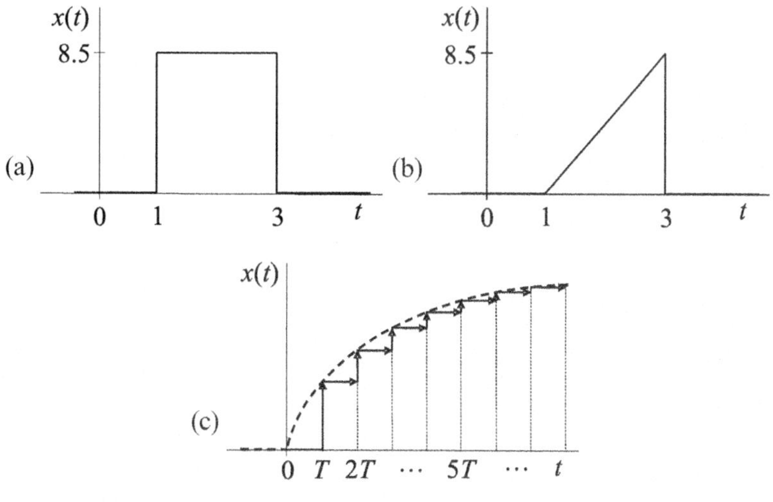

Fig. 1.13 Signals: (**a**) rectangular pulse, (**b**) ramp pulse, and (**c**) arbitrary signal.

1.3.2 Dirac Delta Function

The *Dirac*[4] *delta function*, often referred to as the *unit impulse*, *impulse symbol*, *Dirac impulse*, or *delta function*, is the function that defines the idea of a unit impulse, having the fundamental properties

$$\delta(x) = \begin{cases} \infty, & x = 0 \\ 0, & x \neq 0 \end{cases}, \tag{1.11}$$

$$\int_{-\infty}^{\infty} \delta(x)\mathrm{d}x = 1\,. \tag{1.12}$$

Mathematically, $\delta(t)$ may be defined by the derivative of the unit-step function,

$$\delta(t) = \frac{\mathrm{d}u(t)}{\mathrm{d}t}\,. \tag{1.13}$$

In an equivalent sense, one may also specify the unit step by integrating the delta function,

$$u(t) = \int_{-\infty}^{t} \delta(t)\mathrm{d}t\,. \tag{1.14}$$

The fundamental properties of the delta function, (1.11) and (1.12), are also satisfied if to use the following definition:

$$\delta(t) = \lim_{\xi \to 0} \frac{\mathrm{d}H(t,\xi)}{\mathrm{d}t}\,. \tag{1.15}$$

Therefore, the unit impulse is very often considered as a rectangular pulse of the amplitude $1/2\xi$ (Fig. 1.14). Following (1.11), it needs to set $\xi = 0$ in (1.15) and Fig. 1.14a, and thus the delta function is not physically realizable.

The *Kronecker*[5] *impulse* (or *symbol*) is a discrete-time counterpart of the delta function; however, it is physically realizable, as $\xi \neq 0$ in the discrete scale. Both the delta function (Fig. 1.14b) in the continuous time and the Kronecker impulse in the discrete time are used as test functions to specify the system's impulse response.

The following properties of $\delta(t)$ are of importance.

1.3.2.1 Sifting

This property is also called *sampling property* or *filtering property* . Since the delta function is zero everywhere except zero, the following relations hold true:

$$x(t)\delta(t) = x(0)\delta(t) \quad \text{and} \quad x(t)\delta(t-\theta) = x(\theta)\delta(t-\theta)\,,$$

[4] Paul Adrien Maurice Dirac, English mathematician, 8 August 1902–20 October 1984.

[5] Leopold Kronecker, German mathematician, 7 December 1823–29 December 1891.

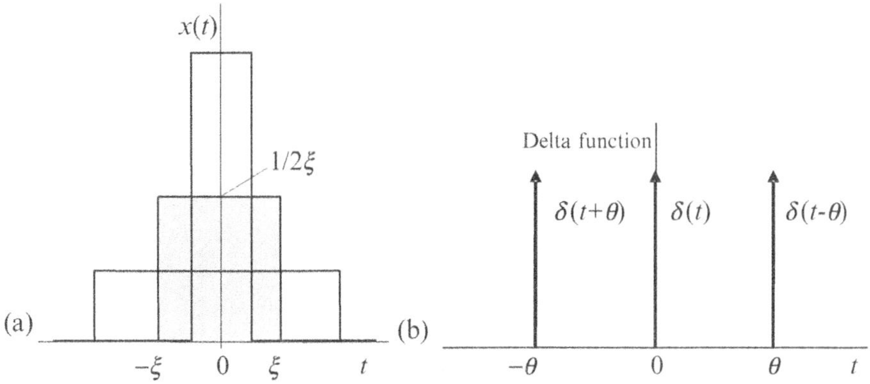

Fig. 1.14 Unit impulse: (**a**) rectangular model and (**b**) positions, by a time-shift $\pm\theta$.

allowing us to write

$$\int_{-\infty}^{\infty} x(t)\delta(t-\theta)\mathrm{d}t = \int_{-\infty}^{\infty} x(\theta)\delta(t-\theta)\mathrm{d}t$$

$$= x(\theta)\int_{-\infty}^{\infty} \delta(t-\theta)\mathrm{d}t = x(\theta)\,. \tag{1.16}$$

So, if to multiply any continuous-time function with the delta function and integrate this product in time, then the result will be the value of the function exactly at the point where the delta function exists. In a case of $\theta = 0$, (1.16) thus degenerates to

$$\int_{-\infty}^{\infty} x(t)\delta(t)\mathrm{d}t = x(0)\,. \tag{1.17}$$

Alternatively, the sifting property also claims that

$$\int_{a}^{b} x(t)\delta(t)\mathrm{d}t = \begin{cases} x(0), & a < 0 < b \\ 0, & a < b < 0 \text{ or } 0 < a < b\,. \\ x(0)\delta(0), & a = 0 \qquad \text{or } b = 0 \end{cases} \tag{1.18}$$

It is important to remember that both (1.17) and (1.18) are symbolic expressions and should not be considered as an ordinary Riemann[6] integral. Therefore, $\delta(t)$ is often called a *generalized function* and $x(t)$ is then said to be a *testing function.*

[6] Georg Friedrich Bernhard Riemann, German mathematician, 17 September 1826–20 July 1866.

Example 1.6. By (1.17), the following double integral is easily calculated to be

$$\int_{-a}^{a}\int_{-\infty}^{\infty}\delta(t)\mathrm{e}^{-j\omega(t+\theta)}\mathrm{d}t\mathrm{d}\omega=\int_{-a}^{a}\mathrm{e}^{-j\omega\theta}\mathrm{d}\omega=2a\frac{\sin a\theta}{a\theta}\,.$$

□

1.3.2.2 Symmetry

It follows from the definition of the delta function that

$$\delta(-t)=\delta(t)\,. \tag{1.19}$$

This symmetry property gives, in turn, the following useful relation:

$$\int_{-\infty}^{0}\delta(t)\mathrm{d}t=\int_{0}^{\infty}\delta(t)\mathrm{d}t=\frac{1}{2}\,. \tag{1.20}$$

Equivalently, (1.20) may also be rewritten as

$$\int_{\theta-\xi}^{\theta}\delta(t-\theta)\mathrm{d}t=\int_{\theta}^{\theta+\xi}\delta(t-\theta)\mathrm{d}t=\frac{1}{2}\,. \tag{1.21}$$

Example 1.7. The following signal is integrated in the positive range. By the symmetry and sifting properties, the closed-form solution becomes

$$\int_{0}^{a}\int_{0}^{\infty}\delta(t)\mathrm{e}^{-j\omega(t+\theta)}\mathrm{d}t\mathrm{d}\omega=\frac{1}{2}\int_{0}^{a}\mathrm{e}^{-j\omega\theta}\mathrm{d}\omega=\frac{1}{2j\theta}\left(1-\mathrm{e}^{ja\theta}\right).$$

□

1.3.2.3 Timescaling

The following timescaling property is useful:

$$\delta(at)=\frac{1}{|a|}\delta(t)\,. \tag{1.22}$$

Example 1.8. The following signal is integrated with $\theta>0$. By changing the variables and using the timescaling property (1.22), we go to

$$\begin{aligned}\int_{-a}^{a}\int_{-\infty}^{\infty}\delta\left(\frac{t}{\theta}\right)\mathrm{e}^{-j\omega(t+\theta)}\mathrm{d}t\mathrm{d}\omega&=|\theta|\int_{-a}^{a}\int_{-\infty}^{\infty}\delta(t)\mathrm{e}^{-j\omega(t+\theta)}\mathrm{d}t\mathrm{d}\omega\\&=2a|\theta|\frac{\sin a\theta}{a\theta}\,.\end{aligned}$$

□

1.3.2.4 Alternative Definitions

$\delta(t)$ is defined as a suitably chosen conventional function having unity area over an infinitesimal time interval. An example is the rectangular pulse as shown in Fig. 1.14. Many other usual functions demonstrate properties of the delta function in limits. Several such functions are shown below:

$$\delta(t-\theta) = \frac{1}{\pi}\lim_{\alpha\to\infty}\frac{\sin\alpha(t-\theta)}{t-\theta} \tag{1.23}$$

$$= \lim_{\alpha\to 0}\frac{1}{\alpha\sqrt{2\pi}}\mathrm{e}^{-\frac{(t-\theta)^2}{2\alpha^2}} \tag{1.24}$$

$$= \frac{1}{\pi}\lim_{\alpha\to\infty}\frac{1-\cos\alpha(t-\theta)}{\alpha(t-\theta)^2} \tag{1.25}$$

$$= \frac{1}{\pi}\lim_{\alpha\to\infty}\frac{\sin^2\alpha(t-\theta)}{\alpha(t-\theta)^2} \tag{1.26}$$

$$= \frac{1}{\pi}\lim_{\alpha\to\infty}\frac{\alpha}{\alpha^2(t-\theta)^2+1} \tag{1.27}$$

$$= \frac{1}{\pi^2(t-\theta)}\lim_{\alpha\to 0}\int\limits_{(t-\theta)-\alpha}^{(t-\theta)+\alpha}\frac{\mathrm{d}y}{y}\,. \tag{1.28}$$

The following formal relation also holds true:

$$\delta(t-\theta) = \frac{1}{2}\frac{\partial^2}{\partial t^2}|t-\theta|\,. \tag{1.29}$$

1.3.2.5 Derivative

The derivative of the delta function can be defined by a relation

$$\int\limits_{-\infty}^{\infty} x(t)\frac{\mathrm{d}\delta(t)}{\mathrm{d}t}\mathrm{d}t = -\left.\frac{\mathrm{d}x(t)}{\mathrm{d}t}\right|_{t=0}, \tag{1.30}$$

where $x(t)$ is continuous at $t=0$ and vanishes beyond some fixed interval to mean that the time derivative $\mathrm{d}x(t)/\mathrm{d}t$ exists at $t=0$.

1.3.2.6 Fourier Transforms

The Fourier transform of the delta function is given by

$$\int\limits_{-\infty}^{\infty}\delta(t-\theta)\mathrm{e}^{-j\omega t}\mathrm{d}t = \mathrm{e}^{-j\omega\theta}\,, \tag{1.31}$$

At $\theta = 0$, the transform (1.31) becomes

$$\int_{-\infty}^{\infty} \delta(t) \mathrm{e}^{-j\omega t} \mathrm{d}t = 1 \tag{1.32}$$

that leads to the important conclusion: the spectral contents of the delta function are uniform over all frequencies.

It follows straightforwardly from (1.31) that the Fourier transform of a half sum of two delta functions shifted on $\pm\theta$ is a harmonic wave. Indeed,

$$\begin{aligned} &\frac{1}{2} \int_{-\infty}^{\infty} [\delta(t+\theta) + \delta(t-\theta)] \mathrm{e}^{-j\omega t} \mathrm{d}t \\ &= \frac{1}{2} [\mathrm{e}^{j\omega\theta} + \mathrm{e}^{-j\omega\theta}] = \cos \omega\theta \,. \end{aligned} \tag{1.33}$$

By taking the inverse Fourier transform of (1.31) and (1.33), respectively, we go to,

$$\delta(t-\theta) = \frac{1}{2\pi} \int_{-\infty}^{\infty} \mathrm{e}^{j\omega(t-\theta)} \mathrm{d}\omega = \frac{1}{\pi} \int_{0}^{\infty} \cos \omega(t-\theta) \mathrm{d}\omega , \tag{1.34}$$

$$\begin{aligned} \frac{1}{2}[\delta(t+\theta) + \delta(t-\theta)] &= \frac{1}{2\pi} \int_{-\infty}^{\infty} \cos \omega\theta e^{j\omega t} \mathrm{d}\omega \\ &= \frac{1}{\pi} \int_{0}^{\infty} \cos \omega\theta \cos \omega t \mathrm{d}\omega \,. \end{aligned} \tag{1.35}$$

1.3.2.7 Delta Spectrum

The delta-shaped spectrum appears as a product of the integration

$$\begin{aligned} \delta(f - f_0) &= \int_{-\infty}^{\infty} \mathrm{e}^{\pm 2\pi j(f-f_0)t} \mathrm{d}t = \int_{-\infty}^{\infty} \mathrm{e}^{\pm j(\omega-\omega_0)t} \mathrm{d}t \\ &= 2\pi \int_{-\infty}^{\infty} \mathrm{e}^{\pm 2\pi j(\omega-\omega_0)z} \mathrm{d}z = 2\pi\delta(\omega - \omega_0) , \end{aligned} \tag{1.36}$$

where an auxiliary variable is $z = t/2\pi$, f_0 is a regular shift frequency in Hz, and ω_0 is an angular shift frequency in rad/s, $\omega_0 = 2\pi f_0$. This relation establishes an important correspondence between two delta functions expressed in terms of regular and angular frequencies, i.e.,

$$\delta(f - f_0) = 2\pi\delta(\omega - \omega_0) . \tag{1.37}$$

1.3.3 Exponential Signal

The following exponential models are used in signals descriptions.

1.3.3.1 Complex Exponential Signal

A *complex exponential signal*

$$x(t) = e^{j\omega_0 t} \tag{1.38}$$

is basic in the complex (symbolic) harmonic presentation of signals. It is usually associated with Euler's[7] formula

$$x(t) = e^{j\omega_0 t} = \cos\omega_0 t + j\sin\omega_0 t\,. \tag{1.39}$$

In accordance with (1.39), $x(t)$ is a complex signal, which real part is a cosine function $\cos\omega_0 t$ and imaginary part is a sine function $\sin\omega_0 t$. It is also periodic with the *period of repetition*

$$T = \frac{2\pi}{\omega_0} = \frac{1}{f_0}. \tag{1.40}$$

1.3.3.2 Generalized Exponential Signal

A *general complex exponential signal* has the combined power of the exponential function, namely it is performed by

$$x(t) = e^{(\alpha+j\omega_0)t} = e^{\alpha t}(\cos\omega_0 t + j\sin\omega_0 t)\,. \tag{1.41}$$

This signal has a real part $e^{\alpha t}\cos\omega_0 t$ and imaginary part $e^{\alpha t}\sin\omega_0 t$, which are exponentially decreasing (if $\alpha < 0$) or increasing (if $\alpha > 0$) harmonic functions (Fig. 1.15).

Example 1.9. An oscillatory system is combined with the inductance L, capacitor C, and resistor R included in parallel. The system is described with the second-order ordinarily differential equation (ODE)

$$C\frac{d^2}{dt^2}v(t) + \frac{1}{R}\frac{d}{dt}v(t) + \frac{1}{L}v(t) = 0,$$

where $v(t)$ is a system voltage. For the known initial conditions, a solution of this equation is a real part of the general complex exponential signal (1.41) taking the form of

$$v(t) = V_0 e^{-\frac{t}{2\tau_0}}\cos\omega_0 t\,,$$

where V_0 is a peak value, $\tau_0 = RC$ is a system time constant, and $\omega_0 = \sqrt{\frac{1}{LC} - \frac{1}{4C^2R^2}}$ is the fundamental angular frequency. Since $\alpha = -1/2\tau_0 < 0$, then oscillations attenuate, as in Fig. 1.15b starting at $t = 0$.

□

[7] Leonhard Euler, Switzerland-born mathematician, 15 April 1707–18 September 1783.

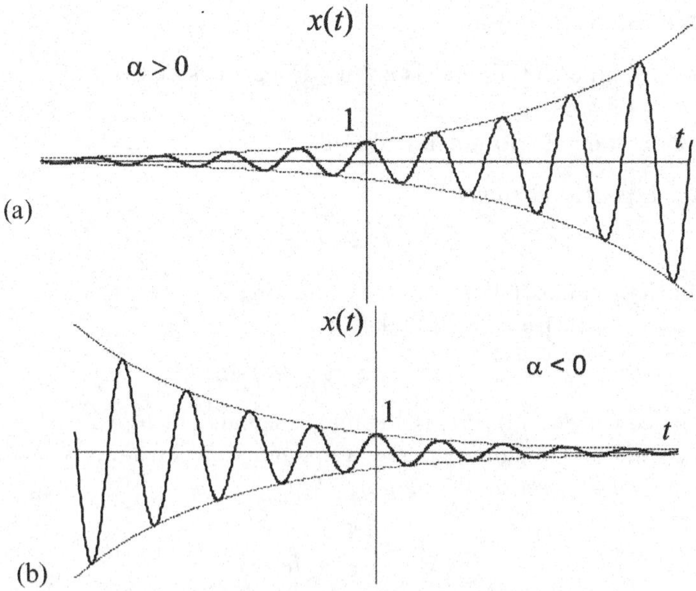

Fig. 1.15 Exponentially gained harmonic signal $e^{\alpha t}\cos\omega_0 t$: (**a**) $\alpha > 0$ and (**b**) $\alpha < 0$. A signal becomes $\cos\omega_0 t$ if $\alpha = 0$.

1.3.3.3 Real Exponential Signal

A *real exponential signal* is a degenerate version of (1.41) with $\omega_0 = 0$,

$$x(t) = \mathrm{e}^{\alpha t}\,. \tag{1.42}$$

Since the harmonic content is absent here, the function is performed either by the increasing ($\alpha > 0$) or by the decaying ($\alpha < 0$) positive-valued envelope of oscillations shown in Fig. 1.15a and b, respectively.

Example 1.10. An *RC* system is described with the ODE

$$RC\frac{\mathrm{d}}{\mathrm{d}t}v(t) + v(t) = 0\,.$$

For the known initial conditions, a solution of this ODE regarding the system signal $v(t)$ is a real exponential function (1.42),

$$v(t) = V_0\mathrm{e}^{-\frac{t}{2\tau_0}}\,,$$

where $\tau_c = RC$. Since $\alpha = -1/2\tau_0 < 0$, then the signal (voltage) reduces with time starting at $t = 0$.

□

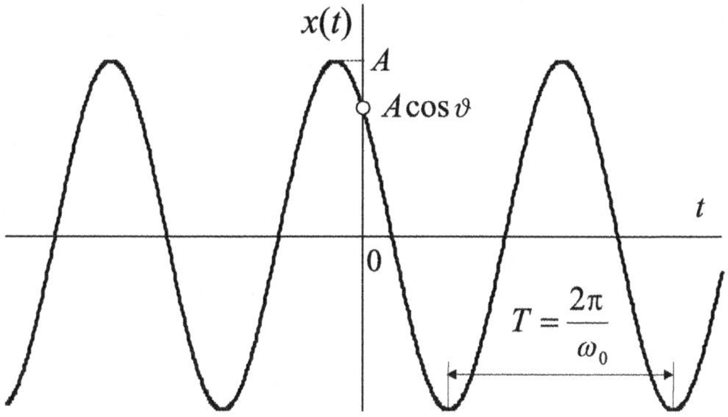

Fig. 1.16 Harmonic signal.

1.3.4 Harmonic Signal

A continuous-time harmonic (sine or cosine) function is fundamental in the spectral presentation of signals. Its cosine version is given by

$$x(t) = A\cos(\omega_0 t + \vartheta), \tag{1.43}$$

where A is a constant real amplitude and ϑ is a constant phase. An example of this function is shown in Fig. 1.16. The function is periodic with period T (1.40) and the reciprocal of T is called the *fundamental frequency*

$$f_0 = \frac{\omega_0}{2\pi} = \frac{1}{T}, \tag{1.44}$$

where ω_0 is a *fundamental angular frequency.*

Euler's formula gives alternative presentations for the cosine and sine functions, respectively,

$$A\cos(\omega_0 t + \vartheta) = A\,\mathrm{Re}\{\mathrm{e}^{j(\omega_0 t+\vartheta)}\}, \tag{1.45}$$

$$A\sin(\omega_0 t + \vartheta) = A\,\mathrm{Im}\{\mathrm{e}^{j(\omega_0 t+\vartheta)}\}, \tag{1.46}$$

which are useful in the symbolic harmonic analysis of signals.

Example 1.11. An LC system is represented with the ODE

$$LC\frac{\mathrm{d}^2}{\mathrm{d}t^2}v(t) + v(t) = 0,$$

which solution for the known initial conditions is a harmonic signal (1.43) performed as $v(t) = V_0\cos\omega_0 t$, where $\omega_0 = 1/\sqrt{LC}$. It follows that the peak amplitude of this voltage remains constant with time.

□

1.4 Methods of Signals Presentation and Description

We now know that signals may exist in different waveforms, lengths, and dimensions and that there are useful elementary signals, which properties are well studied. In different applications to systems and signal processing (SP), signals are represented in various mathematical forms that we outline below.

1.4.1 Generalized Function as a Signal Model

Any real physical signal is finite at every point in the time domain. There is, however, at least one elementary signal $\delta(t)$, which value at $t = 0$ does not exist at all (1.11). To overcome a difficulty with infinity in dealing with such signals, we need to extend the definition of function as a mathematical model of a signal involving the theory of *generalized functions.*

The concept of a generalized function follows from a simple consideration. Let us take a pen and rotate it obtaining different projections on a plate. If a function $f(t)$ represents a "pen" then some other function $\phi(t)$ should help us to rotate it. A "projection" may then be calculated by the integral

$$F[\phi] = \langle f, \phi \rangle = \int_{-\infty}^{\infty} f(t)\phi(t)\mathrm{d}t, \tag{1.47}$$

in which $\phi(t)$ is said to be a *test function.*

It is clear that every function $\phi(t)$ will generate some numerical value of $\langle f, \phi \rangle$. Therefore, (1.47) will specify some *functional* $F[\phi]$ in the space of test functions $\phi(t)$. Given any ordinary function $f(t)$, the functional defined by (1.47) is linear and continuous, provided the following definitions:

- A functional $F[\phi]$ is linear if

$$F[\alpha\phi_1 + \beta\phi_2] = \alpha F[\phi_1] + \beta F[\phi_2], \tag{1.48}$$

  where $\phi_1(t)$, $\phi_2(t)$, α, and β are any real or complex numbers.
- A functional $F[\phi]$ is continuous if whenever a sequence of functions $\phi_n(t)$ converges to a function $\phi(t)$, then the sequence of numbers $F[\mathrm{d}^i\varphi_n(t)/\mathrm{d}t^i]$ converges to $F[\mathrm{d}^i\varphi(t)/\mathrm{d}t^i]$ for all $i = 0, 1, ...$ In other words, the sequence of functions $\phi_n(t)$ converges to a function $\phi(t)$ if $\lim_{n\to\infty} \mathrm{d}^i\phi_n(t)/\mathrm{d}t^i = \mathrm{d}^i\phi(t)/\mathrm{d}t^i$.

If the above conditions are satisfied, we say that, on the space of test functions $\phi(t)$, we have a *generalized function* $f(t)$. The theory of generalized functions was developed by Schwartz[8] who called them *distributions.* We, however, avoid using this term further.

[8] Laurent Schwartz, Jewish mathematician, 5 March 1915–4 July 2002.

All ordinary functions define continuous and linear functionals by the rule of (1.47). The space of continuous linear functionals is much larger than that generated by the ordinary functions. Therefore, a generalized function is defined as any continuous linear functional on the space of test functions. Ordinary functions are called *regular generalized functions.* The rest of generalized functions is called *singular generalized functions.*

An example of singular generalized functions is the Dirac delta function $\delta(t)$, in which case the integral (1.47) with $f(t) = \delta(t)$ is neither the Lebesque[9] nor the Riemann integral. It says that whenever this integral appears, the rule (1.47) gives a solution

$$F[\phi] = \int_{-\infty}^{\infty} \delta(t)\phi(t)\mathrm{d}t = \phi(0) \tag{1.49}$$

leading to a *sifting property* (1.17).

The generalized functions exhibit many properties of classical functions. In particular, they may be differentiated if to note that test functions are finite, i.e., tend toward zero beyond the interval $t_1 \leqslant t \leqslant t_2$. Then the time derivative $f' = \mathrm{d}f(t)/\mathrm{d}t$ of the generalized function $f(t)$ is given by the functional, by differentiating (1.47) by parts,

$$\langle f', \phi\rangle = f(t)\phi(t)|_{-\infty}^{\infty} - \int_{-\infty}^{\infty} f(t)\phi'(t)\mathrm{d}t = -\int_{-\infty}^{\infty} f(t)\phi'(t)\mathrm{d}t = -\langle f, \phi'\rangle. \tag{1.50}$$

Example 1.12. Define the time derivative of the unit-step function (1.7). Consider $u(t)$ to be a generalized function. Then, by (1.50),

$$\langle u', \phi\rangle = -\langle u, \phi'\rangle = -\int_{0}^{\infty} \phi'(t)\mathrm{d}t = \phi(0) = \langle \delta, \phi\rangle\,.$$

Therefore, $\mathrm{d}u(t)/\mathrm{d}t = \delta(t)$ and we go to (1.13). It is important that (1.13) needs to be understood namely in a sense of the theory of the generalized functions. Otherwise, in the classical sense, the time derivative $\mathrm{d}u(t)/\mathrm{d}t$ does not exist at $t = 0$ at all.

□

Example 1.13. Define the time derivative of the Dirac function $\delta(t)$. Following Example 1.12 and (1.50), write

$$\langle \delta', \phi\rangle = -\langle \delta, \phi'\rangle = -\phi'(0)\,.$$

□

As it is seen, generalized functions are a useful tool to determine properties of singular functions. Therefore, the theory of generalized functions fairly occupied an important place in the signals theory.

[9] Henri Léon Lebesgue, French mathematician, 28 June 1875–26 July 1941.

1.4.2 Linear Space of Signals

While operating with signals in space, one faces the problem in their comparison both in shapes or waveforms and magnitudes. The situation may be illustrated with two signal vectors. The first vector is large but projects on some plane at a point. The second vector is short but projects at a line. If one will conclude that the second vector, which projection dominates, is larger, the conclusion will be wrong. We thus have to describe signals properly avoiding such methodological mistakes.

Assume that we have n signals $x_1(t), x_2(t), ..., x_n(t)$ of the same class with some common properties. We then may think that there is some space $\mathbb{R}$ to which these signals belong and write $\mathbb{R} = \{x_1(t), x_2(t), ..., x_n(t)\}$.

Example 1.14. The space $\mathbb{R}$ is a set of various positive-valued electric signals that do not equal zero on $[t_1, t_2]$ and equal to zero beyond this interval.

□

Example 1.15. The space $\mathbb{N}$ is a set of harmonic signals of the type $x(t) = A_n \cos(\omega_n t + \varphi_n)$ with different amplitudes A_n, frequencies ω_n, and phases φ_n.

□

Example 1.16. The space $\mathbb{V}$ is a set of binary vectors such as $\mathbf{v}_1 = [1,\ 0,\ 1]$, $\mathbf{v}_2 = [0,\ 1,\ 1]$, and $\mathbf{v}_3 = [1,\ 1,\ 0]$.

□

Space presentation of signals becomes fruitful when some of its components are expressible through the other ones. In such a case, they say that a set of signals has a certain structure. Of course, selection of signal structures must be motivated by physical reasons and imaginations. For instance, in electric systems, signals may be summed or multiplied with some gain coefficients that outlines a certain structure in linear space.

A set of signals specifies the *real linear space* $\mathbb{R}$ if the following axioms are valid:

- For any signal, $x \in \mathbb{R}$ is real for arbitrary t.
- For any signals $x_1 \in \mathbb{R}$ and $x_2 \in \mathbb{R}$, their sum $y = x_1 + x_2$ exists and also belongs to the space $\mathbb{R}$; that is $y \in \mathbb{R}$.
- For any signal $x \in \mathbb{R}$ and any real coefficient a, the signal $y = ax$ also belongs to the space $\mathbb{R}$; that is $y \in \mathbb{R}$.
- The space $\mathbb{R}$ contains a special empty component $\emptyset$ that specifies a property: $x + \emptyset = x$ for all $x \in \mathbb{R}$.

The above-given set of axioms is not exhaustive. Some other useful postulates are also used in the theory of space presentation of signals and SP. For example, if mathematical models of signals take complex values, then, allowing for complex values and coefficients, we go to the concept of *complex linear space*. If the operators are nonlinear, the space may be specified to be either *real nonlinear space* or *complex nonlinear space*.

1.4.3 Coordinate Basis

In every linear space of signals, we may designate a special subspace to play a role of the coordinate axes. It says that a set of vectors $\{x_1, x_2, x_3, ...\} \in \mathbb{R}$ is *linearly independent* if the equality

$$\sum_i a_i x_i = \emptyset$$

holds true then and only then when all numerical coefficients become zero.

A system of linearly independent vectors forms a *coordinate basis* in linear space. Given an extension of any signal $y(t)$ to

$$y(t) = \sum_i b_i x_i(t),$$

the numerical values b_i are said to be *projections* of a signal $y(t)$ in the coordinate basis. Let us notice that, in contrast to the 3-D space, a number of basis vectors may be finite or even infinite.

Example 1.17. A linear space may be formed by analytic signals, each of which is extended to the Taylor polynomial of the infinite degree:

$$y(t) = \sum_{i=0}^{\infty} \frac{a_i}{i!} t^i .$$

The coordinate basis here is an infinite set of functions $\{x_0 = 1,\ x_1 = t,\ x_2 = t^2,\ ...\}$.

□

1.4.4 Normed Linear Space

There is the other concept to describe a vector length in the linear space. Indeed, comparing different signals of the same class in some space, we ordinarily would like to know how "large" is each vector and how the ith vector is "larger" than the jth one.

In mathematics, the vector length is called the *norm*. Accordingly, a linear space $\mathbb{L}$ is said to be *normed* if every vector $x(t) \in \mathbb{L}$ is specified by its norm $\|x\|$. For the normed space, the following axioms are valid:

- The norm is nonnegative to mean that $\|x\| \geqslant 0$.
- The norm is $\|x\| = 0$ then, and only then, when $x = \emptyset$.

- For any a, the following equality holds true: $\|ax\| = |a| \cdot \|x\|$.
- If $x(t) \in \mathbb{L}$ and $y(t) \in \mathbb{L}$, then the following inequality is valid: $\|x + y\| \leqslant \|x\| + \|y\|$. It is also known as the *triangular inequality.*

Different types of signal norms may be used in applications depending on their physical meanings and geometrical interpretations.

1.4.4.1 Scalar-valued Signals

- The L_1-norm of a signal $x(t)$ is the integral of the absolute value $|x(t)|$ representing its *length* or *total resources*,

$$\|x\|_1 = \int_{-\infty}^{\infty} |x(t)|\mathrm{d}t\,. \tag{1.51}$$

- The L_2-norm of $x(t)$ is defined as the square root of the integral of $x^2(t)$,

$$\|x\|_2 = \sqrt{\int_{-\infty}^{\infty} x^2(t)\mathrm{d}t}\,, \tag{1.52}$$

and if a signal is complex, then $\|x\|_2$ is specified by

$$\|x\|_2 = \sqrt{\int_{-\infty}^{\infty} x(t)x^*(t)\mathrm{d}t}\,, \tag{1.53}$$

where (*) means a complex conjugate value. The L_2-norm is appropriate for electrical signals at least by two reasons:

  - A signal is evaluated in terms of the energy effect, for example, by the amount of warmth induced on a resistance. It then follows that the squared norm may be treated as a *signal energy* (1.54); that is

$$E_x = \|x\|_2^2 = \int_{-\infty}^{\infty} x(t)x^*(t)\mathrm{d}t\,. \tag{1.54}$$

    For instance, suppose that $i(t)$ is a current through a 1 Ω resistor. Then the instantaneous power equals $i^2(t)$ and the total energy equals the integral of this, namely, $\|i\|_2^2$.

  - The energy norm is "insensitive" to changes in the signal waveform. These changes may be substantial but existing in a short time. Therefore, their integral effect may be insignificant.

- The L_p-norm of $x(t)$ is a generalization for both the L_1-norm and the L_2-norm. It is defined as

$$\|x\|_p = \sqrt[p]{\int_{-\infty}^{\infty} |x(t)|^p \mathrm{d}t}\,. \tag{1.55}$$

The necessity to use the L_p-norm refers to the fact that the integrand in (1.55) should be Lebesgue-integrable for the integral to exist. Therefore, this norm is a generalization of the standard Riemann integral to a more general class of signals.

- The L_∞-norm is often called the ∞-norm. It is characterized as the maximum of the absolute value (*peak value*) of $x(t)$,

$$\|x\|_\infty = \max_t |x(t)|\,, \tag{1.56}$$

assuming that the maximum exists. Otherwise, if there is no guarantee that it exists, the correct way to define the L_∞-norm is to calculate it as the least upper bound (*supremum*) of the absolute value,

$$\|x\|_\infty = \sup_t |x(t)|\,. \tag{1.57}$$

- Root mean square (RMS) is calculated by

$$\|x\|_{\mathrm{rms}} = \sqrt{\lim_{T\to\infty} \frac{1}{T} \int_{-T/2}^{T/2} x^2(t)\mathrm{d}t}\,. \tag{1.58}$$

Example 1.18. Given a truncated ramp signal

$$x(t) = \begin{cases} at\,, & \text{if } 0 < t < \theta \\ 0\,, & \text{otherwise} \end{cases}.$$

Its L_2-norm is calculated by (1.52) as $\|x\|_2 = \sqrt{a^2 \int_0^\theta t^2 \mathrm{d}t} = |a|\sqrt{\theta^3/3}$.

□

Example 1.19. Given an RF signal with a rectangular waveform

$$x(t) = \begin{cases} A\cos\omega_0 t + \varphi_0\,, & \text{if } 0 < t < \theta \\ 0\,, & \text{otherwise} \end{cases}.$$

Its L_2-norm is calculated by

$$\|x\|_2 = A\sqrt{\int_0^\theta \cos^2(\omega_0 t + \varphi_0)\mathrm{d}t} = \frac{A}{\sqrt{\omega_0}}\sqrt{\int_0^{\omega_0\theta+\varphi_0} \cos^2 z\mathrm{d}z}$$

$$= \frac{A}{2\sqrt{\omega_0}}\sqrt{2(\omega_0\theta + \varphi_0) + \sin 2(\omega_0\theta + \varphi_0)}\,.$$

It then follows that if a number of oscillations in the pulse is large, $\omega_0\theta \gg 1$ and $\omega_0\theta \gg \varphi_0$, then the L_2-norm is calculated by $\|x\|_2 = A\sqrt{\theta/2}$ without a substantial loss in accuracy.

□

Example 1.20. Given a truncated ramp signal

$$x(t) = \begin{cases} t\,, & \text{if } 0 < t < 1 \\ 0\,, & \text{otherwise} \end{cases}$$

that, to be mathematically rigorous, has no maximum. Instead, we may introduce the least upper bound or supremum defined as the least number N, which satisfies the condition $N \geqslant x(t)$ for all t, i.e., $\sup_t x(t) = \min\{N\colon x(t) \leqslant N\} = 1$.

□

1.4.4.2 Vector Signals

In a like manner, the norms of vector signals may be specified as follows:

- L_1-norm : $\|x\|_1 = \int\limits_{-\infty}^{\infty} \sum\limits_i |x_i(t)|\mathrm{d}t = \int\limits_{-\infty}^{\infty} \|x(t)\|_1\mathrm{d}t = \sum\limits_i \|x_i\|_1$
- L_2-norm : $\|x\|_2 = \sqrt{\int\limits_{-\infty}^{\infty} x^T(t)x(t)\mathrm{d}t}$
- L_∞-norm : $\|x\|_\infty = \sup\limits_t \max\limits_i |x_i(t)| = \sup\limits_t \|x(t)\|_\infty$
- $L_{\infty,p}$-norm : $\|x\|_{\infty,p} = \sup\limits_t \|x(t)\|_p$
- RMS : $\|x\|_{rms} = \sqrt{\lim\limits_{T\to\infty} \frac{1}{T} \int\limits_{-T/2}^{T/2} x^T(t)x(t)\mathrm{d}t}$

1.4.5 Metric Spaces of Signals

We now need to introduce one more fundamental concept that generalizes our imagination about the distance between two points in space. Assume that in the linear space $\mathbb{R}$ we have a pair of vector signals, $x \in \mathbb{R}$ and $y \in \mathbb{R}$. This space is said to be a *metric space* if there is some nonnegative number

$$d(x, y) \equiv \|x - y\| \tag{1.59}$$

called *metric* describing a *distance* between x and y. Any kind of metrics obeys the axioms of the metric space:

- Symmetry: $d(x, y) = d(y, x)$
- Positive definiteness: $d(x, y) \geqslant 0$ for all x and y; note that $d(x, y) = 0$, if $x = y$
- Triangle inequality: $\|x - y\| \leqslant \|x - z\| + \|z - y\|$ or $\mathrm{d}(x, y) \leqslant d(x, z) + d(z, y)$ for all $x, y, z \in \mathbb{R}$

It may be shown that all inner product spaces are metric spaces and all normed linear spaces are metric spaces as well. However, all metric spaces are not normed linear spaces. It may also be shown that the metric allows evaluating the goodness of one signal to approximate the other one.

Example 1.21. Given truncated a sine signal $x(t)$ and ramp signal $y(t)$ (Fig. 1.17),

$$x(t) = \begin{cases} A \sin(\pi t / 2T)\,, & \text{if } 0 < t < T \\ 0\,, & \text{otherwise}\,, \end{cases}$$

$$y(t) = \begin{cases} Bt\,, & \text{if } 0 < t < T \\ 0\,, & \text{otherwise}\,, \end{cases}$$

The square metric with $T = 1$ is determined by

$$d^2(x, y) = \int_0^1 \left(A \sin \frac{\pi t}{2} - Bt \right)^2 \mathrm{d}t = \frac{1}{2} A^2 - \frac{8}{\pi^2} AB + \frac{1}{3} B^2\,.$$

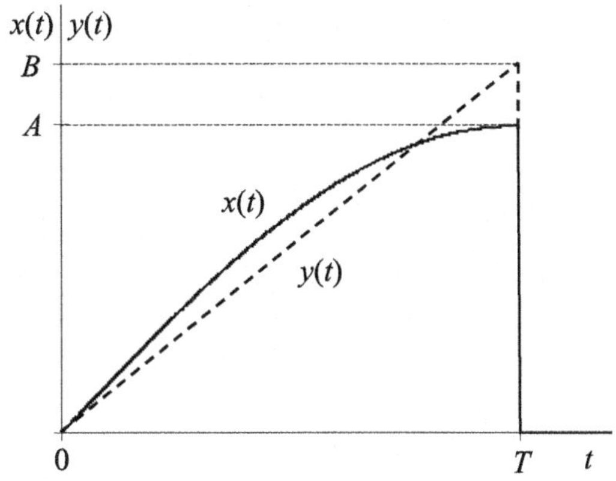

Fig. 1.17 A truncated sinusoid signal $x(t)$ and a ramp signal $y(t)$.

Taking the first derivative with respect to B, we realize that the minimum distance between $x(t)$ and $y(t)$ [the best approximation of $x(t)$ by $y(t)$] is achieved with $B = 12A/\pi^2$. Thus, a minimum distance between two signals is $d_{\min} = A\sqrt{1/2 - 48/\pi^4}$.

□

1.5 Signals Transformation in Orthogonal Bases

In many applications (e.g. in communications), to transmit and receive signals that do not correlate each other are of prime import. Such signals are associated with the *orthogonal functions* and therefore called the *orthogonal signals*. The principle of orthogonality plays a central role in coding, multichannel systems, etc. It is also a basis for the Fourier transforms and many other useful transformations in SP.

Before discussing the orthogonal transformations of signals, we point out that, even though a concept of a linear space, its norm and metric is already given, we still cannot evaluate an angular measure between two vectors. We then need to introduce the *inner* or *scalar product* in the linear space.

1.5.1 Inner Product

Assume two vectors, $\mathbf{x}$ and $\mathbf{y}$, in the rectangular 3-D coordinate space. Then the square of the total value of their sum is calculated as

$$|\mathbf{x}+\mathbf{y}|^2 = |\mathbf{x}|^2 + |\mathbf{y}|^2 + 2\langle \mathbf{x}, \mathbf{y}\rangle , \tag{1.60}$$

where

$$\langle \mathbf{x}, \mathbf{y}\rangle = |\mathbf{x}| \cdot |\mathbf{y}| \cos\psi \tag{1.61}$$

is a scalar called the *inner product* or *scalar product* and ψ is an angle between the vectors.

Assume also two real energy signals, $x(t)$ and $y(t)$, and calculate an energy of their additive sum,

$$E = \int_{-\infty}^{\infty} [x(t) + y(t)]^2 \mathrm{d}t = E_x + E_y + 2\int_{-\infty}^{\infty} x(t)y(t)\mathrm{d}t\,. \tag{1.62}$$

We then arrive, by (1.62), at important conclusions that (1) an additive sum of two signals does not correspond to an additive sum of their energies and (2) there is a term representing a doubled *joint energy* of two signals

$$E_{xy} = \int_{-\infty}^{\infty} x(t)y(t)\mathrm{d}t\,. \tag{1.63}$$

Now, a simple comparison of (1.60) and (1.62) produces the *inner product*,

$$\langle x, y\rangle = \int_{-\infty}^{\infty} x(t)y(t)\mathrm{d}t\,, \tag{1.64}$$

and the angle ψ between two vectors, $\mathbf{x}$ and $\mathbf{y}$,

$$\cos\psi = \frac{\langle x, y\rangle}{\|x\|_2 \cdot \|y\|_2}\,. \tag{1.65}$$

The inner product obeys the following fundamental properties:

- Commutativity (for real signals):

$$\langle x, y\rangle = \langle y, x\rangle\,.$$

□

- Commutativity (for complex signals):

$$\langle x, y\rangle = \langle y, x\rangle^*\,.$$

□

- Nonnegativity:

$$\langle x, y\rangle \geqslant 0\,.$$

If $\psi = \pi/2$, then $\langle x, y\rangle = 0$.

□

- Bilinearity:

$$\langle x, (ay + z)\rangle = a\langle x, y\rangle + \langle x, z\rangle\,,$$

where a is real.

□

- Space presentation. If signals are given, for example, in the 3-D coordinates as $\mathbf{x} = [x_1\, x_2\, x_3]^T$ and $\mathbf{y} = [y_1\, y_2\, y_3]^T$, then

$$\langle x, y\rangle = x_1y_1 + x_2y_2 + x_3y_3 = \mathbf{x}^T\mathbf{y}\,.$$

□

- Cauchy[10]–Schwartz[11] inequality. This inequality is also known as Cauchy–Bunyakovskii[12] inequality and it establishes that the total inner product of two given vectors is less or equal than the product of their norms:

$$|\langle x, y\rangle| \leqslant \|x\| \cdot \|y\|\,.$$

□

---

[10] Augustin Louis Cauchy, French mathematician, 21 August 1789–23 May 1857.

[11] Hermann Amandus Schwarz, German mathematician, 25 January 1843–30 November 1921.

[12] Viktor Yakovlevich Bunyakovskii, Ukrainian-born Russian mathematician, 16 December 1804–12 December 1889.

Every inner product space is a metric space that is given by $d(x, y) = \langle x - y, x - y \rangle$. If this process results in a complete metric space that contains all points of all the converging sequences of vectors signals, it is called the *Hilbert*[13] *space.*

1.5.2 Orthogonal Signals

It is now just a matter of definition to say that two complex energy signals, $x(t)$ and $y(t)$, defined on the time interval $a \leqslant t \leqslant b$ are *orthogonal*, if their joint energy (or the inner product) is zero,

$$E_{xy} = \langle x, y \rangle = \int_a^b x(t) y^*(t) \mathrm{d}t = 0\,. \tag{1.66}$$

Accordingly, a set of signals $u_k(t)$, where k ranges from $-\infty$ to ∞, is an orthogonal basis, provided

- If $i \neq j$, then the signals are mutually orthogonal, i.e.,
$$\langle u_i(t), u_j(t) \rangle = 0\,.$$
- The signals are complete in the sense that the only signal, $x(t)$, which is orthogonal to all $u_k(t)$ is the zero signal; that is, if $\langle x(t), u_k(t) \rangle = 0$ for all k, then $x(t) = 0$.

Figure 1.18 gives examples of sets of orthogonal periodic harmonic functions (a) and nonperiodic Haar[14] functions (b). It may easily be checked out even graphically that the inner product of every pair of different functions in sets is zero.

1.5.3 Generalized Fourier Series

Given an orthogonal basis of signals $u_k(t)$, we can analyze any real signal $x(t)$ in terms of the basis and synthesize the signal vice versa. In such manipulations, the analysis coefficients c_n are called the *generalized Fourier coefficients* and the synthesis equation is called the *generalized Fourier series*, respectively,

$$c_k = \frac{\langle x(t), u_k(t) \rangle}{\langle u_k(t), u_k(t) \rangle} = \frac{1}{\langle u_k(t), u_k(t) \rangle} \int_a^b x(t) u_k^*(t) \mathrm{d}t, \tag{1.67}$$

$$x(t) = \sum_{k=-\infty}^{\infty} c_k u_k(t)\,. \tag{1.68}$$

[13] David Hilbert, German mathematician, 23 January 1862–14 February 1943.
[14] Alfréd Haar, Hungarian mathematician, 11 October 1885–16 March 1933.

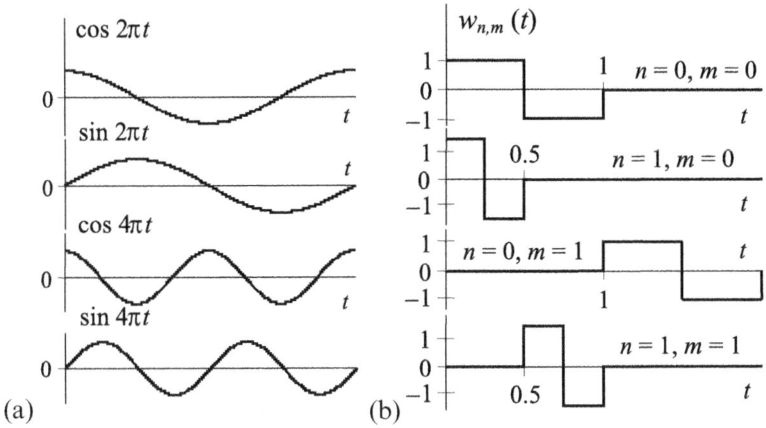

Fig. 1.18 Sets of orthogonal functions: (**a**) periodic harmonic functions and (**b**) nonperiodic Haar functions.

A chance of representing different signals by the generalized Fourier series is of extreme importance in the signals theory and SP. In fact, instead of learning the functional dependence of some signal at an infinite number of points, one has a tool to characterize this signal by a finite set (rigorously, infinite) of coefficients of the generalized Fourier series.

For signals, mathematics offered several useful sets of orthogonal functions that we discuss below.

1.5.4 Fourier Analysis

In the early 1800s, Fourier showed that any periodic function $x(t)$ with period T satisfying the Dirichlet[15] conditions may be extended to the series (1.68) by the orthogonal set of functions $u_k(t) = \mathrm{e}^{-j2\pi f_k t}$, where c_k for each frequency $f_k = k/T$ is calculated by (1.67). This results in the Fourier series

$$c_k = \frac{1}{T}\int_0^T x(t)\mathrm{e}^{-j2\pi f_k t}\mathrm{d}t\,, \tag{1.69}$$

$$x(t) = \sum_{k=-\infty}^{\infty} c_k \mathrm{e}^{j2\pi f_k t}\,, \tag{1.70}$$

for which the complex exponent $\mathrm{e}^{j2\pi f_k t}$ produces a set of orthogonal functions.

The harmonic (cosine and sine) basis has also gained currency in Fourier analysis. If $x(t)$ is symmetrically defined on $-T/2 \leqslant t \leqslant T/2$, then its

[15] Johann Peter Gustav Lejeune Dirichlet, Belgium-born German/French mathematician, 13 February 1805–5 May 1859.

Fourier series will contain only cosine terms. This gives an orthogonal basis of functions $u_k(t) = \cos(2\pi kt/T)$, where $k = 0, 1, 2, \ldots$. If $x(t)$ is asymmetrically defined on $-T/2 \leqslant t \leqslant T/2$, then an orthogonal basis of functions is $u_k(t) = \sin(2\pi kt/T)$, where $k = 0, 1, 2, \ldots$. In a common case, one may also use the combined orthogonal basis of the harmonic functions $u_{2k-1}(t) = \sin(2\pi kt/T)$ and $u_{2k}(t) = \cos(2\pi kt/T)$, where $k = 0, 1, 2, \ldots$.

If to suppose that an absolutely integrable signal $x(t)$ has infinite period $T \to \infty$, then the sum in (1.70) ought to be substituted by integration and the Fourier series goes to the Fourier transform

$$X(j\omega) = \int_{-\infty}^{\infty} x(t)\mathrm{e}^{-j\omega t}\mathrm{d}t\,, \tag{1.71}$$

$$x(t) = \frac{1}{2\pi}\int_{-\infty}^{\infty} X(j\omega)\mathrm{e}^{j\omega t}\mathrm{d}\omega\,, \tag{1.72}$$

where $\omega = 2\pi f$ is an angular frequency. Note that $X(j\omega)$ works here like a set of Fourier coefficients c_k with zero frequency space $f_k - f_{k-1} \to 0$.

Example 1.22. Given a complex harmonic signal $x(t) = \mathrm{e}^{j\omega_0 t}$. Its Fourier transform (1.71) is, by (1.36),

$$X(j\omega) = \int_{-\infty}^{\infty} \mathrm{e}^{-j(\omega-\omega_0)t}\mathrm{d}t = 2\pi\delta(\omega - \omega_0)\,,$$

representing a spectral line (Dirac delta function) at a carrier frequency ω_0. □

In analogous to (1.52), an L_2-norm of the Fourier transform $X(j\omega)$ of a signal $x(t)$ is defined as

$$\|X\|_2 = \sqrt{\frac{1}{2\pi}\int_{-\infty}^{\infty} |X(j\omega)|^2\mathrm{d}\omega} \tag{1.73}$$

and it is stated by the Parseval[16] theorem that

$$\|x\|_2 = \|X\|_2\,. \tag{1.74}$$

[16] Marc-Antoine de Parseval des Chsnes, French mathematician, 27 April 1755–16 August 1836.

1.5.5 Short-time Fourier Transform

Since $x(t)$ in (1.71) and (1.72) is assumed to be known in time from $-\infty$ to ∞, then the Fourier transform becomes low efficient when the frequency content of interest is of a signal localized in time. An example is in speech processing when a signal performance evolves over time and thus a signal is *nonstationary*. To apply the Fourier transform to "short" signals, $x(t)$ first is localized in time by windowing so as to cut off only its well-localized slice. This results in the *short-time Fourier transform* (STFT) also called the *windowed Fourier transform*.

The STFT of a signal $x(t)$ is defined using a weighting function $u(t) = g(t-\theta)\mathrm{e}^{-j\omega t}$ with a *window function* $g(t)$ [rather than a weighting function $u(t) = \mathrm{e}^{-j\omega t}$, as in the Fourier transform] as follows,

$$\mathrm{STFT}(j\omega,\theta) = X(j\omega,\theta) = \int_{-\infty}^{\infty} x(t)g(t-\theta)\mathrm{e}^{-j\omega t}\mathrm{d}t\,, \tag{1.75}$$

where θ is some time shift. The window $g(t)$ may be rectangular or of some other shape. However, the sharp window effectively introduces discontinuities into the function, thereby ruining the decay in the Fourier coefficients. For this reason smoother windows are desirable, which was considered by Gabor[17] in 1946 for the purposes of communication theory. In the STFT, the synthesis problem is solved as follows

$$x(t) = \frac{1}{2\pi\|g\|^2}\int_{-\infty}^{\infty}\int_{-\infty}^{\infty} X(j\omega,\theta)g(t-\theta)\mathrm{e}^{j\omega t}\mathrm{d}\omega\mathrm{d}\theta\,. \tag{1.76}$$

Instead of computing $\mathrm{STFT}(j\omega,\theta)$ for all frequencies $f = \omega/2\pi$ and all time shifts θ, it is very often useful restricting the calculation to $f_k = k/T$ and $\theta = mT$. This forms the orthogonal functions $u_{k,m}(t) = g(t-mT)\mathrm{e}^{j2\pi kt/T}$ leading to $\mathrm{STFT}(j2\pi f_k, mT) = \langle x(t), u_{k,m}(t)\rangle$. The analysis and synthesis transforms are then given by, respectively,

$$\begin{aligned} c_{k,m} &= \frac{\langle x(t)u_{k,m}(t)\rangle}{\langle u_{k,m}(t)u_{k,m}(t)\rangle} = \frac{1}{T}\int_{-\infty}^{\infty} x(t)u_{k,m}^{*}\mathrm{d}t \\ &= \frac{1}{T}\int_{mT}^{(m+1)T} x(t)g(t-mT)\mathrm{e}^{-j2\pi kt/T}\mathrm{d}t\,, \end{aligned} \tag{1.77}$$

[17] Dennis Gabor, Hungarian-British physicist, 5 June 1900–8 February 1979.

$$
\begin{aligned}
x(t) &= \sum_{k=-\infty}^{\infty} \sum_{m=-\infty}^{\infty} c_{k,m} u_{k,m}(t) \\
&= \sum_{k=-\infty}^{\infty} \sum_{m=-\infty}^{\infty} c_{k,m} g(t-mT) e^{j2\pi kt/T} .
\end{aligned} \tag{1.78}
$$

Figure 1.19a gives an example of $x(t)$ representation using a rectangular window of width $\tau = 0.4$ for several values of θ. It follows from this example that the "time resolution" in the STFT is τ, whereas the "frequency resolution" is $1/\tau$.

Example 1.23. Given a complex harmonic signal $x(t) = e^{j\omega_0 t}$. Its STFT (1.75) is, using a rectangular window of width τ,

$$
X(j\omega, \theta) = \int_{(-\tau/2)+\theta}^{(\tau/2)+\theta} e^{-j(\omega-\omega_0)t} dt = \tau \frac{\sin(\omega-\omega_0)\tau/2}{(\omega-\omega_0)\tau/2} e^{-j(\omega-\omega_0)\theta} .
$$

In the limiting case of $\tau \to \infty$, the STFT degenerates to the Fourier transform that, in this example, leads to the Dirac delta-function at a carrier frequency

$$
\lim_{\tau\to\infty} X(j\omega, \theta)|_{\omega=\omega_0} = \lim_{\tau\to\infty} \tau \frac{\sin(\omega-\omega_0)\tau/2}{(\omega-\omega_0)\tau/2}|_{\omega=\omega_0} = \lim_{\tau\to\infty} \tau = \infty .
$$

□

1.5.6 Wavelets

A modification of the Fourier transforms to STFT is not the only way to transform short-time localized nonstationary signals. An alternative set of short-time or "small wave" orthogonal functions is known as *wavelets*. Nowadays, wavelets occupy a wide range of applications in SP, filter banks, image compression, thresholding, denoising, etc. The orthogonal basis is defined here by the function $u(t) = |a|^{-1/2} w[(t-b)/a]$, where $a > 0$ and the factor $a^{-1/2}$ is chosen so that $\|u\|_2 = \|w\|_2$. The *continuous wavelet transform* (CWT) is

$$
X(a,b) = \frac{1}{\sqrt{|a|}} \int_{-\infty}^{\infty} x(t) w^* \left(\frac{t-b}{a} \right) dt . \tag{1.79}
$$

If $w(t)$ decays exponentially with time and $\int_{-\infty}^{\infty} w(t) dt = 0$, then the *inverse continuous wavelet transform* (ICWT) is as follows:

$$
x(t) = \frac{1}{C} \int_{-\infty}^{\infty} \int_{-\infty}^{\infty} X(a,b) \frac{1}{a^2\sqrt{|a|}} w \left(\frac{t-b}{a} \right) da\, db , \tag{1.80}
$$

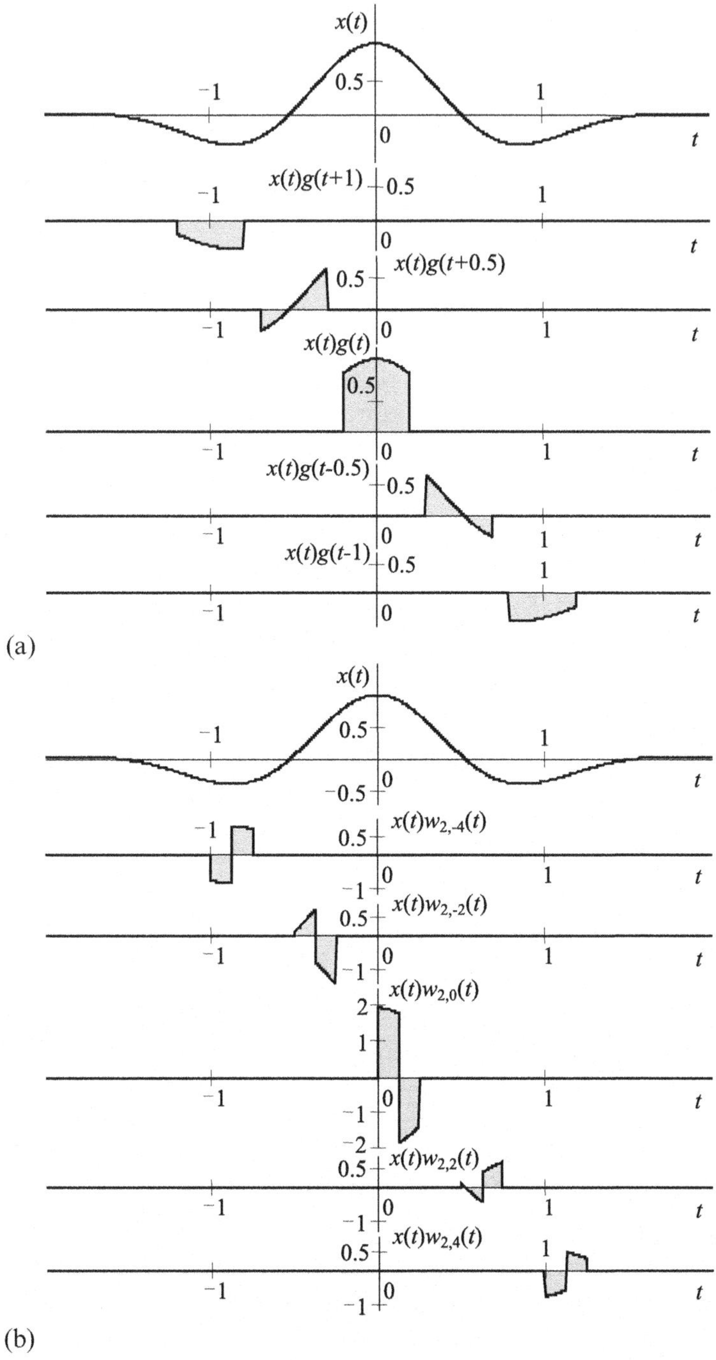

Fig. 1.19 Representation of $x(t)$ using (**a**) rectangular window (STFT) and (**b**) Haar wavelets.

where C is a normalizing constant. Figure 1.19b illustrates a representation of $x(t)$ using several Haar functions (Fig. 1.18b).

A difference between STFT and CWT is as in the following. In the STFT the frequency bands have a fixed width, whereas in the CWT the frequency bands grow and shrink with frequency. This leads to high frequency resolution with CWT at low frequencies and high time resolution at high frequencies.

1.6 Summary

So, we are acquainted now with major fundamentals and cannons of the theory of signals and may start learning more complicated problems. Before continuing, it seems worth to outline the major features of signals and their transforms:

- A deterministic signal is precisely described by analytical functions; a random signal is described in probabilistic terms.
- All real physical signals are causal and the mathematical signals may be noncausal.
- A periodic signal reiterates its values periodically through some time nT; a nonperiodic signal exists uniquely over all the time range.
- A scalar signal is 1-D being represented in the time domain by the only function; a vector signal is multidimensional being combined with an assemblage of scalar signals.
- A continuous-time signal or analog signal is determined exactly at an arbitrary time instant; a discrete-time signal is determined exactly at some fixed time points usually with a constant sample time; a digital signal is represented approximately by some digital code at fixed time points usually with a constant sample time.
- Periodic signals satisfying the Dirichlet conditions may be represented in the frequency domain by the Fourier series through the magnitude and phase spectra.
- Nonperiodic signals satisfying the Dirichlet conditions may be represented in the frequency domain by the Fourier transform through the magnitude and phase spectral densities.
- Periodic signals are typically power signals and nonperiodic signals are energy signals.
- Unit-step function, Dirac delta function, exponential signal, and harmonic function represent elementary signals.
- A concept of generalized functions allows considering the ordinary functions, which do not exist at some points.

- Linear space of signals unite those with common properties; a system of linearly independent functions forms a coordinate basis in linear space.
- The vector length is called the norm; a linear space is normed if every vector in this space is specified with its norm.
- Metric describes a distance between two signals in the metric space.
- The inner product of two signals is an integral over the infinite time of multiplication of these signals.
- Two signals are orthogonal if their inner product (and so joint energy) is zero.
- Any real signal may be analyzed on a basis of given orthogonal functions using the generalized Fourier series.
- In the Fourier analysis, the orthogonal functions are complex exponential or harmonic.
- In the STFT, the orthogonal functions are multiplied with short windows.
- In wavelet transforms, the orthogonal functions are short-time functions.

1.7 Problems

1.1 (Signals application in systems). Explain how a distance to a target is measured with an impulse radar? What should be a pulse waveform to obtain a minimum measurement error? Why not to use a periodic harmonic signal?

1.2. In communications, two modulated (by voice) signals are sent in the same direction at the same time. Can you impose some restrictions for these signals to avoid their interaction in time and space?

1.3. Why precise signals are needed for accurate positioning systems? Explain the necessity of using the coded signals in Fig. 1.5.

1.4 (Signals classification). Given the following signals:

1. $x(t) = A_0 \cos(\omega_0 t + \varphi_0)$
2. $x(t) = A(t) \cos[\omega_0 t + \varphi(t)]$
3. $\mathbf{y}(t) = \mathbf{Bz}(t)$
4. $x(t) = \begin{cases} t, & t > 0 \\ 0.5, & t = 0 \\ 0, & t < 0 \end{cases}$

5. $x(t) = \begin{cases} A_0 \cos[\omega_0 t + \varphi(t)], & t \geqslant 0 \\ 0, & t < 0 \end{cases}$

6. $y(t) = \mathbf{Ax} + \mathbf{Cz}(t)$

7. $x(t) = A_0 e^{j\int_0^t \varphi(t)dt + j\cos\omega_0 t}$

Give graphical representation for these signals.

Which signal is deterministic (regular) and which may be either deterministic or random? Why?

Which signal is causal and which is noncausal? Why?

Which signal is periodic and which is not periodic? Why?

Which signal is 1-D and which is multidimensional? Why?

Which signal is energy signal and which is power signal? Why?

1.5 (Signals orthogonality). Given the following signals. Find the orthogonal pairs among these signals.

1. $x(t) = A_0 \cos(\omega_0 t + \varphi_0)$
2. $x(t) = A(t) \cos[\omega_0 t + \varphi(t)]$
3. $x(t) = A_0 \cos(\omega_0 t + \varphi_0 - \pi/2)$
4. $x(t) = A(t) \sin[\omega_0 t + \varphi(t)]$

1.6 (Signals presentation by elementary functions). Figure 1.20 shows several nonperiodic impulse signals. Describe these pulses analytically, using the elementary functions (1.7) and (1.9).

1.7. Give a graphical presentation for the following signals:

1. $x(t) = 2[u(t) - u(t-2)]$
2. $x(t) = u(t) + u(t-1) - 4u(t-2) + u(t-3) + u(t-4)$
3. $x(t) = [u(t+1) - u(t-1)] \cos(\omega_0 t)$
4. $x(t) = 5[v(t-2,1) - u(t-3)]$
5. $x(t) = 2.5[v(t,1) - v(t-1,1)]$
6. $x(t) = 2u(t) - v(t,1) - u(t-1)$
7. $x(t) = u(\sin \omega_0 t)$
8. $x(t) = v(t-2,1)u(-t+3)$

1.8 (Norms of signals). Given the nonperiodic impulse signals (Fig. 1.20). Determine the following norms of these signals: L_1-norm, L_2-norm, L_∞-norm, and L_{rms}-norm.

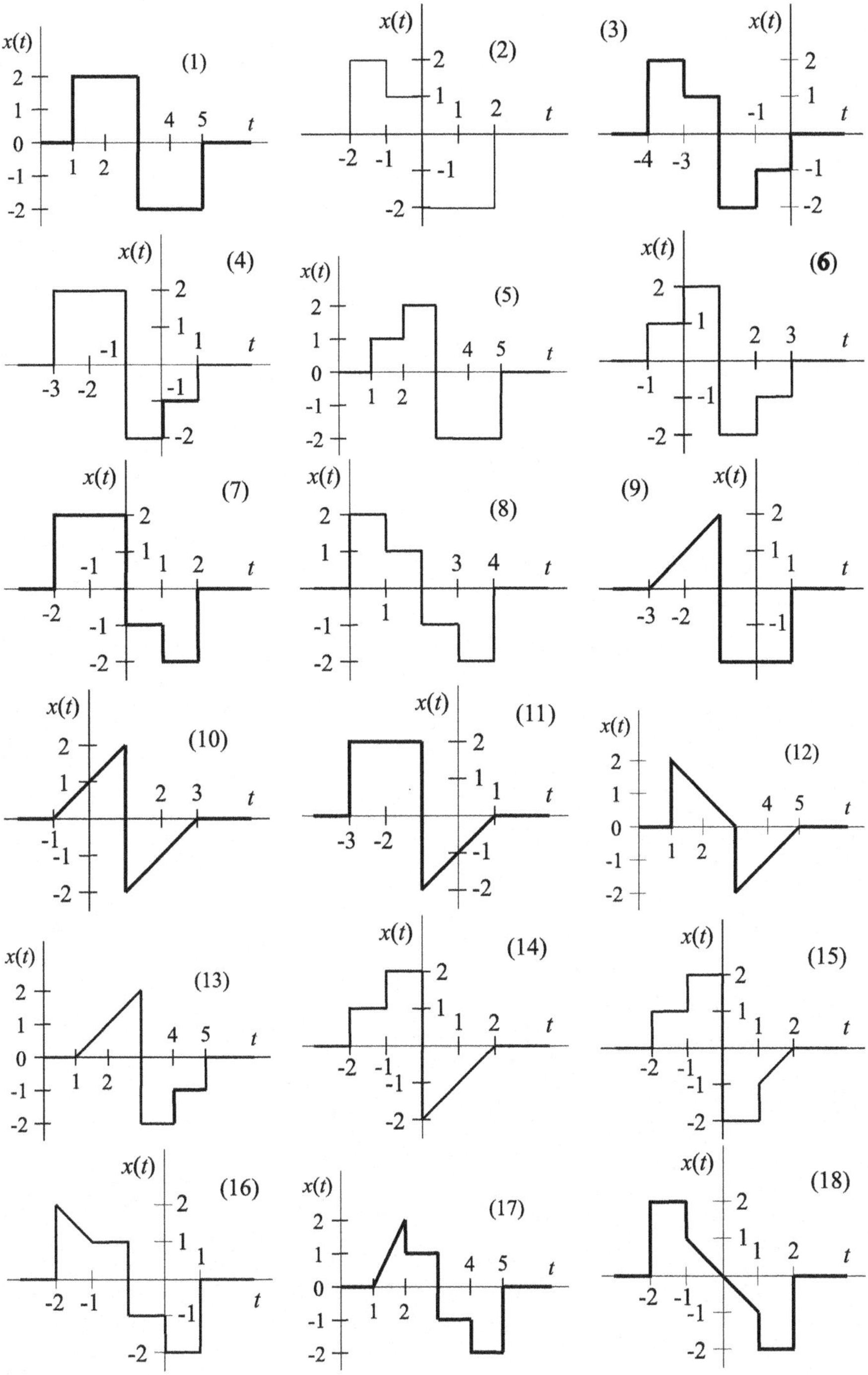

Fig. 1.20 Examples of nonperiodic signals.

1.9 (Metrics of signals). Given the nonperiodic impulse signals (Fig. 1.20). Determine metric between one of these signals and a ramp signal

$$y(t) = \begin{cases} a + bt, & 1 \leqslant t \leqslant 5 \\ 0, & \text{otherwise} \end{cases}$$

Define a and b to obtain a minimum distance between these signals. Give graphical representation for both signals.

1.10 (Signal energy). Given the nonperiodic impulse signals (Fig. 1.20) and a ramp signal

$$y(t) = \begin{cases} a + bt, & -5 \leqslant t \leqslant 5 \\ 0, & \text{otherwise} \end{cases}$$

Determine an energy of an additive sum of these signals. Define the coefficients a and b to obtain a minimum joint energy of two signals. Compare these coefficients with those valid for a minimum distance between these signals.

2

Spectral Presentation of Signals

2.1 Introduction

Signals description in the time domain is natural since it gives a straightforward graphical interpretation. There are, however, many applications for which time plots of signals do not answer on important questions. For example, in wireless systems, it is of high importance to know spectral contents of signals to avoid losing information while transmitting through bandlimited channels.

We thus need to go to the frequency domain. But before, it seems in order to consider a simple example illustrating how a signal waveform influences its spectral width.

Assume we have two harmonic signals $x_1(t) = \cos\omega_0 t$ and $x_2(t) = -0.2\cos 3\omega_0 t$. It is seen (Fig. 2.1) that a sum of these signals $y(t) = x_1(t) + x_2(t)$ may be treated either as a nearly harmonic wave or as a roughly smoothed rectangular pulse-train.

We may also represent $x_1(t)$ and $x_2(t)$ in the frequency domain by the spectral lines with the amplitudes 1 and 0.2 at ω_0 and $3\omega_0$, respectively (Fig. 2.1). What do we see? The signals occupy extremely narrow (theoretically zero) frequency ranges around ω_0 and $3\omega_0$. Therefore, each of them may be transmitted through the channel with a very narrow bandwidth. Just on the contrary, $y(t)$ occupies a wide range of $2\omega_0$ requiring a very wide bandwidth of a channel. Two important conclusions follow instantly:

- A real periodic signal may be represented by a sum of harmonic signals.
- A signal waveform predetermines its spectral content.

Extension of periodic continuous-time waveforms into the series of harmonic waves has been of interest for several centuries. One of the first works is associated with Euler who employed only a cosine series (like in our example) developing the background for the works of d'Alembert,[1] Lagrange,[2]

[1] Jean Le Rond d' Alembert, French mathematician, 17 November 1717–29 October 1783.

[2] Joseph-Louis Lagrange, Italian/French mathematician, 25 January 1736–10 April 1813.

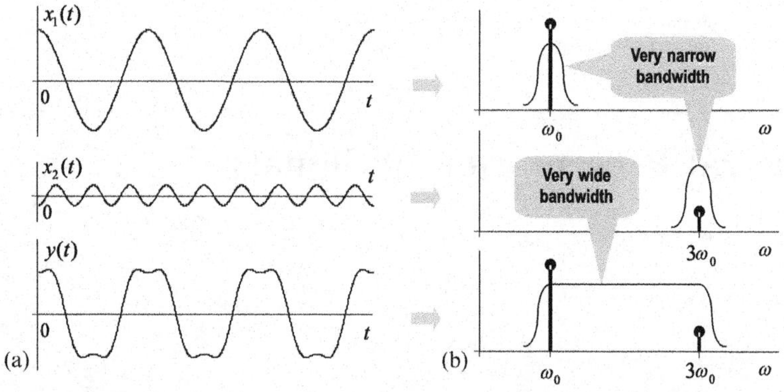

Fig. 2.1 Time (**a**) and spectral (**b**) presentations of periodic waveforms.

and Bernoulli.[3] The next most important steps ahead were done by Joseph Baptiste Fourier who made the major contribution to the problem. His *series* associated with periodic functions and his *transforms* applied to nonperiodic functions were recognized further everywhere as a main tool to represent signals in the frequency domain.

To realize a correspondence between a signal waveform and its spectral content and thus be able to speak free about the waveforms and their spectra, we go in this chapter behind the following logical events. We first examine single pulses and periodic trains. We then consider pulse-bursts and periodic burst-trains. In such a way, understanding interchanges in spectral contents of single waveforms, bursts, and periodic trains becomes the most transparent.

2.2 Presentation of Periodic Signals by Fourier Series

Periodicity is one of the most widely used mathematical idealizations of real processes. It implies that a periodic signal exists within infinite bounds even though no one of the real physical processes can fully satisfy this requirement. A periodic signal is assumed to have a waveform periodically reiterated with a constant *period of repetition* T. The major mathematical tool used in studying periodic signals is the *Fourier series.*

2.2.1 Fourier Series of a Periodic Signal

Any signal $x(t)$ that exhibits strong periodicity with period T may be modeled as

$$x(t) = x(t \pm nT), \tag{2.1}$$

where $n = 0, 1, 2, \ldots$ is arbitrary integer and t is continuous time. A periodic signal has a *fundamental frequency* $F = 1/T$ and may exhaustively be described in the frequency domain by spectral characteristics. A concept of spectral presentation of signals exploits the fact that a time function may be

[3] Daniel Bernoulli, Swiss mathematician, 8 February 1700–17 March 1782.

represented by a sum of harmonic functions with given constant amplitudes and phases related to frequencies kF, $k = 0, 1, 2, \ldots$. Oscillations with frequencies kF are called *harmonics.* The *first harmonic*, $k = 1$, is located at the fundamental frequency F. The *zeroth harmonic*, $k = 0$, is a constant value at zero frequency.

To illustrate the method of substituting a periodic signal by the series of harmonics, we assume that the signal is as in Fig. 2.2 and that it does not equal zero within the time interval

$$t_1 + nT \leqslant t \leqslant t_2 + nT\,. \tag{2.2}$$

Fundamentals of mathematical analysis claim that any periodic function may be extended to the infinite series by the *orthogonal functions* $u_k(t)$. This is supported by the basic property of the orthogonal functions, i.e., the inner product of any pair of unequal orthogonal functions is calculated to be zero, whereas that of the same ones produces some nonzero value $a \neq 0$. With $a = 1$, the functions are said to be *orthonormal* establishing the *orthonormalized basis.* A mathematical tool extending a periodic function to the series on the orthonormalized basis is known as the *generalized Fourier series.* Given an orthogonal basis of signals $u_k(t)$, we can analyze any signal $x(t)$ in terms of the basis and synthesize the signal vice versa (see Chapter 1).

All available orthogonal functions $u_k(t)$ are not used in electrical engineering. To fit a 2π-periodicity of signals, the following orthonormalized basis of harmonic functions may be used:

$$
\begin{aligned}
&u_0 = \frac{1}{\sqrt{T}}\,,\\
&u_1 = \sqrt{\frac{2}{T}}\sin 2\pi\frac{t}{T}\,, \quad u_2 = \sqrt{\frac{2}{T}}\cos 2\pi\frac{t}{T}\,,\\
&u_3 = \sqrt{\frac{2}{T}}\sin 4\pi\frac{t}{T}\,, \quad u_4 = \sqrt{\frac{2}{T}}\cos 4\pi\frac{t}{T}\,,\\
&\quad\vdots\\
&u_{2k-1} = \sqrt{\frac{2}{T}}\sin 2k\pi\frac{t}{T}\,, \quad u_{2k} = \sqrt{\frac{2}{T}}\cos 2k\pi\frac{t}{T}\,,\\
&\quad\vdots
\end{aligned}
\tag{2.3}
$$

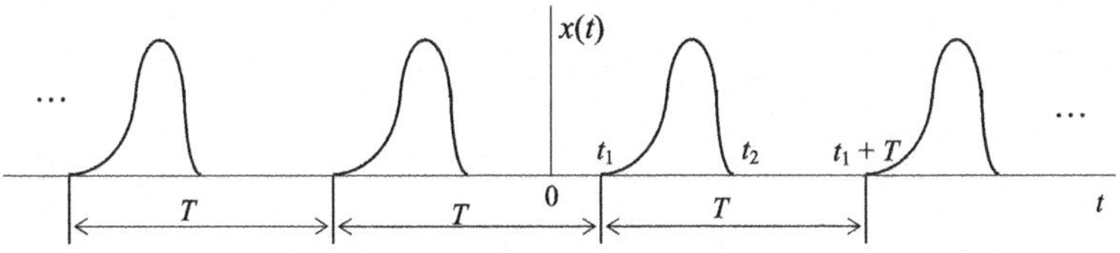

Fig. 2.2 Periodic pulse-train.

As it is seen, every function in (2.3) satisfies the periodicity condition (2.1). Based upon this basis, any periodic function may be extended to the Fourier series if the following Dirichlet conditions are satisfied:

> ***Dirichlet conditions***: Any periodic function may be extended into the Fourier series since this function is defined (absolutely integrable) on the time interval $[-T/2, \ldots, T/2]$, is finite on this interval, and has a finite number of extremes (minima and maxima) and discontinuities on this interval.

□

2.2.1.1 Trigonometric Fourier Series

Satisfied the Dirichlet conditions, an orthogonal decomposition of a periodic signal $x(t)$ (2.1) on a basis (2.3) to the *trigonometric Fourier series* is provided by

$$x(t) = \frac{c_0}{2} + A_1 \cos \Omega t + B_1 \sin \Omega t + A_2 \cos 2\Omega t + B_2 \sin 2\Omega t + \ldots$$
$$+A_k \cos k\Omega t + B_k \sin k\Omega t + \ldots = \frac{c_0}{2} + \sum_{k=1}^{\infty} (A_k \cos k\Omega t + B_k \sin k\Omega t)\,, \tag{2.4}$$

where $\Omega = 2\pi F = 2\pi/T$ is a *natural (angular) fundamental frequency* and $f_k = kF = k/T$ is a frequency of the kth harmonic of a signal $x(t)$. The coefficients A_k, B_k, and c_0 are specified, respectively by,

$$A_k = \frac{2}{T} \int_{-T/2}^{T/2} x(t) \cos k\Omega t \, \mathrm{d}t = \frac{1}{\pi} \int_{-\pi}^{\pi} x(t) \cos k\Omega t \, \mathrm{d}\Omega t\,, \tag{2.5}$$

$$B_k = \frac{2}{T} \int_{-T/2}^{T/2} x(t) \sin k\Omega t \, \mathrm{d}t = \frac{1}{\pi} \int_{-\pi}^{\pi} x(t) \sin k\Omega t \, \mathrm{d}\Omega t\,, \tag{2.6}$$

$$c_0 = \frac{2}{T} \int_{-T/2}^{T/2} x(t) \, \mathrm{d}t = \frac{1}{\pi} \int_{-\pi}^{\pi} x(t) \, \mathrm{d}\Omega t\,. \tag{2.7}$$

2.2.1.2 Harmonic Form of Fourier Series

The orthogonal harmonic basis (2.3) is not unique. We may unite harmonic functions in (2.3) to produce a series of either sine or cosine functions. In the latter case, we go to the *harmonic form of Fourier series*

$$\begin{aligned} x(t) &= \frac{c_0}{2} + c_1 \cos(\Omega t - \Psi_1) + c_2 \cos(2\Omega t - \Psi_2) + \ldots \\ &\quad + c_k \cos(k\Omega t - \Psi_k) + \ldots \\ &= \frac{c_0}{2} + \sum_{k=1}^{\infty} c_k \cos(k\Omega t - \Psi_k) = \frac{c_0}{2} + \sum_{k=1}^{\infty} z_k(t), \end{aligned} \tag{2.8}$$

in which the constant value $c_0/2$ is still associated with zero frequency. The amplitudes and phases of harmonics, respectively,

$$c_k = \sqrt{A_k^2 + B_k^2},$$

$$\tan \Psi_k = \frac{B_k}{A_k},$$

represent the *spectrum* of a signal $x(t)$. In this series, each of the harmonic components

$$z_k(t) = c_k \cos(k\Omega t - \Psi_k) \tag{2.9}$$

may be performed by a vector rotated with an angular frequency $\omega_k = k\Omega = 2\pi kF = 2\pi k/T$ having an amplitude c_k and phase Ψ_k. Figure 2.3 exhibits this vector (a) along with its time function (b).

A special peculiarity of $z_k(t)$ may be noticed by analyzing Fig. 2.3 and (2.9). It follows that $z_k(t)$ is combined with the "fast" component $\cos k\Omega t$

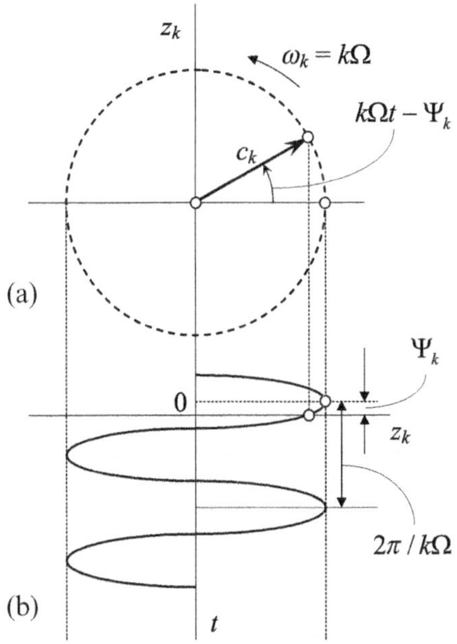

Fig. 2.3 Vector (**a**) and time (**b**) presentations of a signal z_k.

and slowly changed parameters c_k and Ψ_k. Of importance is that c_k and Ψ_k bear information about a signal and may be analyzed separately of the fast function $\cos k\Omega t$. We thus go to the *exponential form of Fourier series* also known as *symbolic form of Fourier series.*

2.2.2 Exponential Form of Fourier Series

Let us apply the Euler formula

$$\cos\gamma = \frac{1}{2}\left(e^{j\gamma} + e^{-j\gamma}\right) \tag{2.10}$$

to the harmonic function $c_k \cos(k\Omega t - \Psi_k)$ and write

$$c_k \cos(k\Omega t - \Psi_k) = \frac{c_k}{2}e^{j(k\Omega t - \Psi_k)} + \frac{c_k}{2}e^{-j(k\Omega t - \Psi_k)}. \tag{2.11}$$

By (2.11), the series (2.8) becomes

$$x(t) = \frac{c_0}{2} + \sum_{k=1}^{\infty}\frac{c_k}{2}e^{j(k\Omega t - \Psi_k)} + \sum_{k=1}^{\infty}\frac{c_k}{2}e^{-j(k\Omega t - \Psi_k)}. \tag{2.12}$$

Several important notations need to be made now. For the harmonic series (2.8) we have

$$c_{-k} = c_k \quad \text{and} \quad \Psi_{-k} = -\Psi_k\,.$$

Moreover, since $B_0 = 0$ in (2.4), we have $\Psi_0 = 0$. This allows us to modify (2.12) as follows: first, having $\Psi_0 = 0$, the constant $c_0/2$ may be merged with the first sum in (2.12); second, we may change the sign of k in the second sum taking into account that $\Psi_{-k} = -\Psi_k$. Thereby, we have two sums, which may also be merged to produce

$$\begin{aligned} x(t) &= \sum_{k=0}^{\infty}\frac{c_k}{2}e^{j(k\Omega t - \Psi_k)} + \sum_{k=-1}^{-\infty}\frac{c_k}{2}e^{-j(-k\Omega t + \Psi_k)} \\ &= \sum_{k=-\infty}^{\infty}\frac{c_k}{2}e^{-j\Psi_k}e^{jk\Omega t}\,. \end{aligned} \tag{2.13}$$

The result (2.13) leads to the synthesis equation known as the *exponential form of Fourier series* and the analysis equation specifying the *spectrum* of a periodic signal $x(t)$, respectively,

$$x(t) = \sum_{k=-\infty}^{\infty} C_k e^{jk\Omega t}\,, \tag{2.14}$$

$$C_k = \frac{1}{T}\int_{-T/2}^{T/2} x(t)e^{-jk\Omega t}\mathrm{d}t\,, \tag{2.15}$$

where the complex amplitudes (vectors)

$$C_k = |C_k|\,\mathrm{e}^{-j\Psi_k} \tag{2.16}$$

represent the *spectrum* of a periodic signal $x(t)$. An assemblage of amplitudes $|C_k| = c_k/2$ is said to be the *magnitude spectrum* and an assemblage of phases Ψ_k the *phase spectrum.*

Example 2.1. Given a complex exponential signal $x(t) = \mathrm{e}^{j\Omega t}$ that is periodic with period $T = 2\pi/\Omega$. Its spectrum C_k is defined, by (2.15), as follows,

$$C_k = \frac{1}{T}\int\limits_{-T/2}^{T/2} \mathrm{e}^{j\Omega t}\mathrm{e}^{-jk\Omega t}\mathrm{d}t = \frac{1}{T}\int\limits_{-T/2}^{T/2} \mathrm{e}^{-j(k-1)\Omega t}\mathrm{d}t$$
$$= \frac{\sin(k-1)\pi}{(k-1)\pi}\,.$$

A simple observation shows that

$$\mathrm{e}^{j\Omega t} \overset{\mathcal{F}}{\Leftrightarrow} C_k = \begin{cases} 1, & k = 1 \\ 0, & k \neq 1 \end{cases}.$$

□

2.2.2.1 An Analysis Problem

An analysis problem (2.15) implies that the spectrum of a periodic signal $x(t)$ may be defined by an assemblage of C_k (2.15), as it is shown in Fig. 2.4.

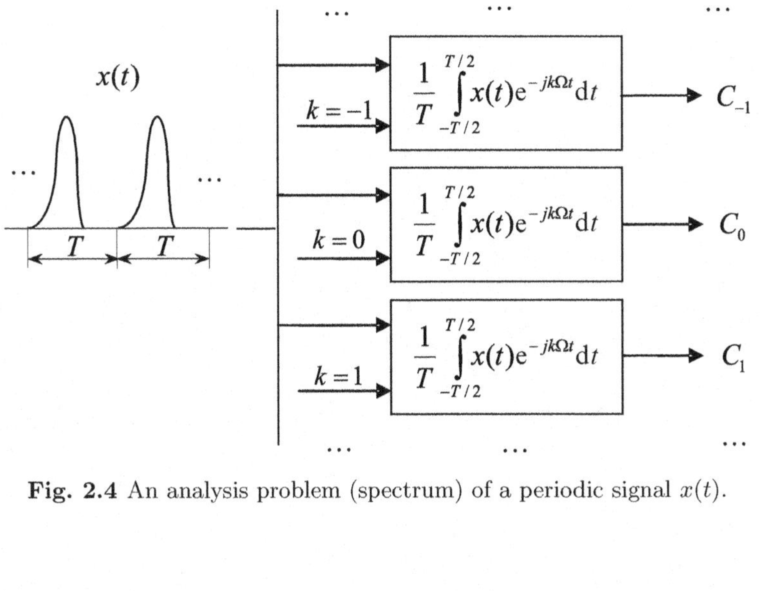

Fig. 2.4 An analysis problem (spectrum) of a periodic signal $x(t)$.

The spectrum is said to be the *double-sided spectrum* or *mathematical spectrum* if it occupies all the frequency range, including negative frequencies. Such a spectrum is represented by the magnitude and phase *lower sidebands* (LSB) and *upper sidebands* (USB). An example of the double-sided spectrum is shown in Fig. 2.5a.

Physically speaking, there are no negative frequencies. Therefore, in applications, the *one-sided spectrums* or *physical spectrums* are used. The *one-sided magnitude spectrum* is produced from the double-sided magnitude spectrum if to get summed the LSBs and USBs. The *one-sided phase spectrum* is represented by the USB of the double-sided spectrum. Figure 2.5b brings an example of the one-sided spectra. Let us notice that the phase spectra are usually represented by the modulo 2π versions that we will exploit further.

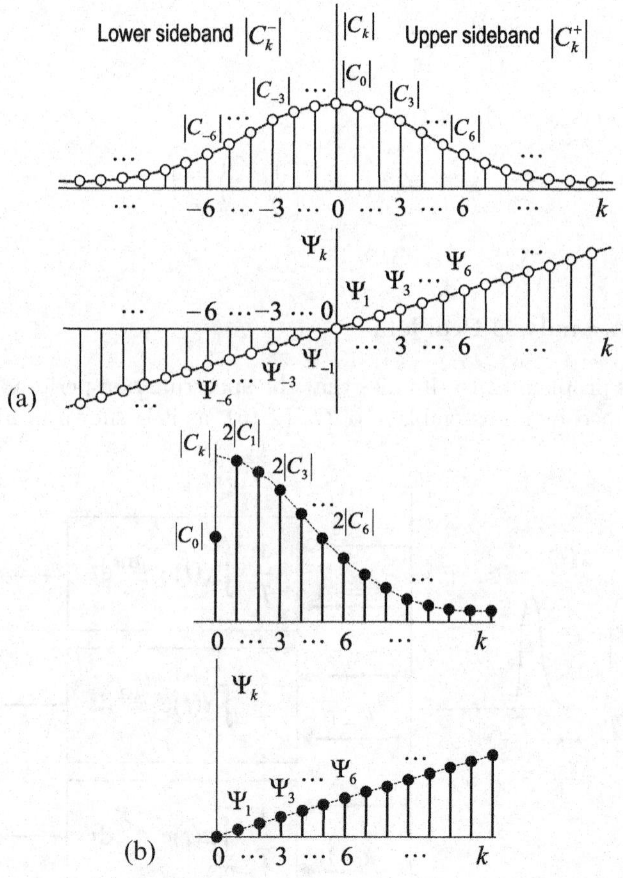

Fig. 2.5 Spectrums of a periodic signal: (**a**) double-sided and (**b**) one-sided.

2.2.2.2 A Synthesis Problem

It follows, by (2.14), that a periodic signal may exhaustively be described by an infinite set of rotated vectors $C_k e^{jk\Omega t}$ representing *harmonics* (or *spectral lines*) if the amplitudes $|C_k|$ and phases Ψ_k are properly predetermined. Restoration of a periodic signal by its spectral components is called a *synthesis problem*. In real applications, a set of vectors cannot be infinite. A finite set produces a synthesis error known as the *Gibbs phenomenon*.

2.2.3 Gibbs Phenomenon

Let us solve a synthesis problem assuming that a set of vectors in (2.14) is finite being restricted with some fixed number $2K$, $K < \infty$. A finite series produces an error that may be evaluated by the measure

$$\varepsilon_K(t) = x(t) - x_K(t) = x(t) - \sum_{k=-K}^{K} C_k e^{jk\Omega t},$$

which mean square value $E[\varepsilon_K^2(t)]$ gives an idea about inaccuracy.

An important point is that $\varepsilon_K(t)$ depends on a signal waveform. In fact, in the above considered simplest example (Fig. 2.1), the synthesis error disappears by accounting only two harmonics, $x_1(t)$ and $x_2(t)$. For more complex waveforms having points of discontinuity, the error $\varepsilon_K(t)$ has a tendency to oscillate that is known as the *Gibbs phenomenon*. To illustrate, we may invoke Fig. 2.1 once again. Let $y(t)$ be synthesizing a rectangular pulse-train. Instantly, we observe that the approximating function oscillates about the assumed rectangular waveform.

More generally, Fig. 2.6 illustrates synthesis of the rectangular pulse-train with an arbitrary set of harmonics. As it is seen, the synthesized pulse-train suffers of the Gibbs phenomenon at the points of discontinuities. It may puzzle the reader, but even with an infinite K, the phenomenon does not fully disappear. All splashes still exist. They just become delta-shaped. We thus conclude that, rigorously, there is some inconsistency between $x(t)$ and its synthesized version even with an infinite K.

2.2.4 Properties of Fourier Series

As well as any other useful transform, the Fourier series demonstrates several important properties that may be used to reduce the burden in the analysis and synthesis problems. We overview below the most widely used of those properties.

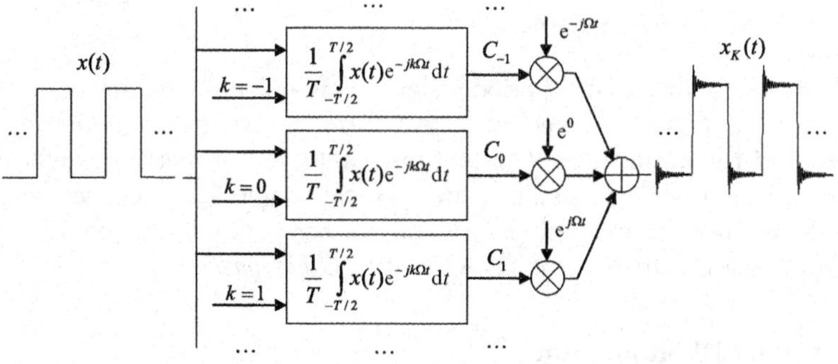

Fig. 2.6 Synthesis problem of the rectangular pulse-train.

2.2.4.1 Time shifting

Suppose $x(t) \overset{\mathcal{F}}{\Leftrightarrow} C_k$ is shifted in time $x(t \pm t_0)$. For a shifted signal, the coefficients C_{1k} are defined, by (2.15), as

$$\begin{aligned} C_{1k} &= \frac{1}{T}\int_T x(t-t_0)\mathrm{e}^{-jk\Omega t}\mathrm{d}t \\ &= \mathrm{e}^{-jk\Omega t_0}\frac{1}{T}\int_T x(t-t_0)\mathrm{e}^{-jk\Omega(t-t_0)}\mathrm{d}(t-t_0) \\ &= \mathrm{e}^{-jk\Omega t_0}\frac{1}{T}\int_T x(\theta)\mathrm{e}^{-jk\Omega\theta}\mathrm{d}\theta = \mathrm{e}^{-jk\Omega t_0}C_k\,. \end{aligned}$$

Thus, the following *time-shift* property holds true:

$$x(t \pm t_0) \overset{\mathcal{F}}{\Leftrightarrow} C_k \mathrm{e}^{\pm jk\Omega t_0}\,. \tag{2.17}$$

Example 2.2. Given a shifted complex exponential signal $\mathrm{e}^{j\Omega(t-t_0)}$. By (2.15) and Example 2.1, its spectrum is readily defined to be

$$\mathrm{e}^{j\Omega(t-t_0)} \overset{\mathcal{F}}{\Leftrightarrow} C_k = \begin{cases} \mathrm{e}^{-j\Omega t_0}, & k=1 \\ 0, & k \neq 1 \end{cases}.$$

Hence, the magnitude spectrum remains unchanged and phase spectrum possesses a shift $-\Omega t_0$.

□

2.2.4.2 Timescaling

Let $x(t)$ be squeezed or stretched in time, i.e., $x(\alpha t)$, $\alpha > 0$. The *time-scale* property claims that such a signal is characterized by C_k with a new period

T/α. Indeed, for $x(\alpha t)$, a synthesis problem (2.14) is solved by

$$x(\alpha t) = \sum_{k=-\infty}^{\infty} C_k e^{jk\Omega \alpha t} = \sum_{k=-\infty}^{\infty} C_k e^{jk\Omega_1 t},$$

thus representing a signal $x(t)$ with a new frequency $\Omega_1 = \alpha\Omega$ or, equivalently, new period $T_1 = T/\alpha = 2\pi/\Omega$.

Example 2.3. Consider a scaled signal $x(\alpha t) = e^{j\Omega \alpha t}$. In a like manner (see Example 2.1), an analysis problem is solved here by

$$C_k = \frac{\sin(k-\alpha)\pi}{(k-\alpha)\pi},$$

meaning that

$$e^{j\Omega\alpha t} \overset{\mathcal{F}}{\Leftrightarrow} C_k = \begin{cases} 1, & k = \alpha \\ 0, & k \neq \alpha \end{cases}.$$

Since the scaling coefficient α may equally be related either to time or to frequency, we may say that $e^{j\Omega\alpha t}$ is represented by the spectrum of an unscaled signal $e^{j\Omega t}$ with a new analysis frequency $\Omega_1 = \alpha\Omega$ or, equivalently, new period $T_1 = T/\alpha$.

□

2.2.4.3 Time differentiation

If a signal $y(t)$ is given as a time derivative of $x(t) \overset{\mathcal{F}}{\Leftrightarrow} C_k$, then, by (2.14), we may write

$$y(t) = \frac{d}{dt}x(t) = \sum_{k=-\infty}^{\infty} C_k \frac{d}{dt} e^{jk\Omega t} = \sum_{k=-\infty}^{\infty} jk\Omega C_k \frac{d}{dt} e^{jk\Omega t}.$$

The *differentiation* property then reads

$$\frac{d}{dt}x(t) \overset{\mathcal{F}}{\Leftrightarrow} jk\Omega C_k. \tag{2.18}$$

Example 2.4. Given a signal $x(t) = e^{j\Omega t}$ and its differentiated version $y(t) = \frac{d}{dt}x(t) = j\Omega e^{j\Omega t}$. It follows straightforwardly from Example 2.1 that

$$j\Omega e^{j\Omega t} \overset{\mathcal{F}}{\Leftrightarrow} C_k = \begin{cases} j\Omega, & k = 1 \\ 0, & k \neq 1 \end{cases}.$$

It coincides with (2.18) if $k = 1$.

□

2.2.4.4 Time integration

If $x(t)$ is integrated in time and the integral is finite-valued, then, for a signal $y(t) = \int_{-\infty}^{t} x(t)\mathrm{d}t$, we may write, by (2.14),

$$y(t) = \int_{-\infty}^{t} x(t)\mathrm{d}t = \sum_{k=-\infty}^{\infty} C_k \int_{-\infty}^{t} \mathrm{e}^{jk\Omega t}\mathrm{d}t = \sum_{k=-\infty}^{\infty} \frac{1}{jk\Omega} C_k \mathrm{e}^{jk\Omega t}\,.$$

The time-integration property may now be performed with

$$\int_{-\infty}^{t} x(t)\mathrm{d}t < \infty \overset{\mathcal{F}}{\Leftrightarrow} \frac{1}{jk\Omega} C_k\,. \tag{2.19}$$

Example 2.5. Consider a signal $x(t) = \mathrm{e}^{j\Omega t}$ and its integrated version $y(t) = \int_{-\infty}^{t} x(t)\mathrm{d}t = \frac{1}{j\Omega}\mathrm{e}^{j\Omega t}$. Without any extra explanations and following Example 2.4, we write

$$\frac{1}{j\Omega}\mathrm{e}^{j\Omega t} \overset{\mathcal{F}}{\Leftrightarrow} C_k = \begin{cases} \frac{1}{j\Omega}, & k = 1 \\ 0, & k \neq 1 \end{cases}.$$

It fits (2.19) with $k = 1$.

□

2.2.4.5 Linearity or Superposition Principle

Consider a set of periodic signals and their transforms:

$$\alpha_1 x_1(t) \overset{\mathcal{F}}{\Leftrightarrow} \alpha_1 C_{1k}\,, \quad \alpha_2 x_2(t) \overset{\mathcal{F}}{\Leftrightarrow} \alpha_2 C_{2k}\,, \quad \ldots, \quad \alpha_i x_i(t) \overset{\mathcal{F}}{\Leftrightarrow} \alpha_i C_{ik}\,.$$

Here C_{ik} is a complex amplitude of the kth harmonic of the ith periodic signal $x_i(t)$, where $i = 1, 2, \ldots$. The linearity property claims that

$$\sum_i \alpha_i x_i(t) \overset{\mathcal{F}}{\Leftrightarrow} \sum_i \alpha_i C_{ik}\,. \tag{2.20}$$

This property was used in Examples 2.4 and 2.5.

2.2.4.6 Conjugation

If $x(t)$ is complex, then

$$x^*(t) \overset{\mathcal{F}}{\Leftrightarrow} C^*_{-k}\,. \tag{2.21}$$

2.2.4.7 Time reversal

If $x(t)$ is given in an inverse time, then

$$x(-t) \overset{\mathcal{F}}{\Leftrightarrow} C_{-k} \,. \tag{2.22}$$

2.2.4.8 Modulation

If $x(t)$ is multiplied with $e^{jK\Omega t}$, the following *modulation* property may be useful,

$$x(t)e^{jK\Omega t} \overset{\mathcal{F}}{\Leftrightarrow} C_{k-K} \,. \tag{2.23}$$

There are some other useful properties of the Fourier series. Among them, the Parseval relation is critical to specify an average power of periodic signals in the time and frequency domains.

2.2.5 Parseval's Relation for Periodic Signals

In applications, it sometimes becomes important to know an average power P_x of a periodic signal $x(t)$. If the spectral coefficients C_k of a periodic signal $x(t)$ are already known, then its average power may be defined using the Parseval theorem that establishes the following: An average power of a periodic signal $x(t)$ is equal to the infinite sum of the square amplitudes $|C_k|$ of its spectrum.

$$P_x = \frac{1}{T}\int\limits_{-T/2}^{T/2} |x(t)|^2 \,\mathrm{d}t = \frac{1}{T}\int\limits_{-T/2}^{T/2} x(t)x^*(t)\mathrm{d}t$$

$$\frac{1}{T}\int\limits_{-T/2}^{T/2} x(t) \sum_{k=-\infty}^{\infty} C^*_{-k} e^{jk\Omega t}\mathrm{d}t = \frac{1}{T}\int\limits_{-T/2}^{T/2} x(t) \sum_{k=\infty}^{-\infty} C^*_k e^{-jk\Omega t}\mathrm{d}t$$

$$= \sum_{k=\infty}^{-\infty} C^*_k \left[\frac{1}{T}\int\limits_{-T/2}^{T/2} x(t)e^{-jk\Omega t}\mathrm{d}t\right] = \sum_{k=\infty}^{-\infty} C_k C^*_k$$

$$= \sum_{k=\infty}^{-\infty} |C_k|^2 = \sum_{k=-\infty}^{\infty} |C_{-k}|^2 = \sum_{k=-\infty}^{\infty} |C_k|^2. \tag{2.24}$$

In other words, power calculated in the time domain equals the power calculated in the frequency domain.

2.3 Presentation of Single Pulses by Fourier Transform

Fourier series presentation of periodic signals may be extended to single (nonperiodic) signals if we suppose that period T goes to infinity. If we set $T \to \infty$ in the pair of transforms (2.14) and (2.15), then the fundamental frequency $F = 1/T$ will tend toward zero, and the spectral lines of a discrete spectrum will be placed along the frequency axis with an extremely small (zero) interval. Because such a spectrum is combined with snugly placed lines it is called a *continuous spectrum*. The mathematical tool to investigate nonperiodic signals and continuous spectra is the Fourier transforms.

2.3.1 Spectral Presentation of Nonperiodic Signals

Let us consider an arbitrary nonperiodic single pulse $x(t)$ that is determined at the time interval $t_1 < t < t_2$ becomes zero if $t \leqslant t_1$ and $t \geqslant t_2$ (Fig. 2.7), and satisfies the Dirichlet conditions. We suppose that t_1 and t_2 can also be arbitrary.

The Dirichlet conditions claim a signal $x(t)$ to be periodic with period T. To satisfy, we isolate the time range $[0, \ldots, T]$ overlapping the interval $[t_1, \ldots, t_2]$ so that $t_1 > 0$ and $t_2 < T$. We now may tend T toward infinity and think that $x(t)$ is still periodic with $T = \infty$. The relations (2.14) and (2.15) then become, respectively,

$$x(t) = \sum_{k=-\infty}^{\infty} C_k e^{jk\Omega t} , \tag{2.25}$$

$$C_k = \frac{1}{T} \int_{t_1}^{t_2} x(t) e^{-jk\Omega t} dt . \tag{2.26}$$

In (2.26), we have accounted the fact that $x(t)$ exists only in the time span $[t_1, \ldots, t_2]$.

Substituting (2.26) into (2.25) yields

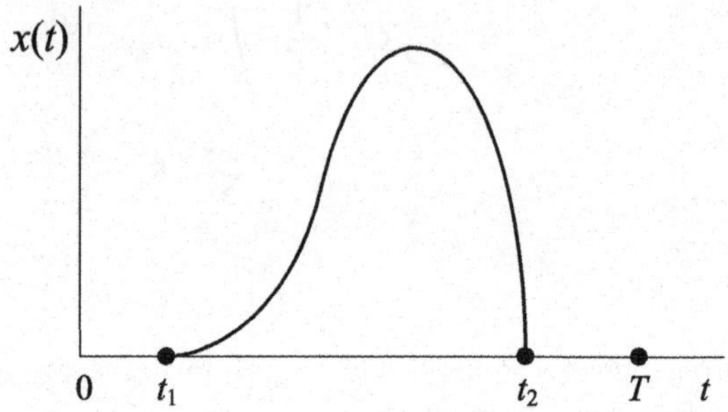

Fig. 2.7 A nonperiodic signal (single pulse).

$$x(t) = \frac{1}{2\pi} \sum_{k=-\infty}^{\infty} \left(\int_{t_1}^{t_2} x(\theta) \mathrm{e}^{-jk\Omega\theta} \mathrm{d}\theta \right) \mathrm{e}^{jk\Omega t} \Omega \,, \quad 0 \leqslant t \leqslant T \,. \tag{2.27}$$

By $T \to \infty$, we have $\Omega = 2\pi/T \to 0$. Thus the frequency $k\Omega$ may be treated as the current angular frequency ω, and hence Ω may be assigned to be $\mathrm{d}\omega$. Having $\Omega = \mathrm{d}\omega$, an infinite sum in (2.27) may be substituted by integration that yields

$$x(t) = \frac{1}{2\pi} \int_{-\infty}^{\infty} \mathrm{e}^{j\omega t} \left[\int_{t_1}^{t_2} x(\theta) \mathrm{e}^{-j\omega\theta} \mathrm{d}\theta \right] \mathrm{d}\omega \,. \tag{2.28}$$

So, the discrete-frequency synthesis problem (2.25) associated with periodic signals has transferred to the continuous-frequency synthesis problem (2.28) associated with single pulses. In (2.28), an inner integral product

$$X(j\omega) = \int_{t_1}^{t_2} x(t) \mathrm{e}^{-j\omega t} \mathrm{d}t \tag{2.29}$$

plays a role of the *spectral characteristic* of a nonperiodic signal $x(t)$.

2.3.2 Direct and Inverse Fourier Transforms

Since a nonperiodic signal $x(t)$ does not exist beyond the time interval $[t_1, \ldots, t_2]$, the integral bounds in the analysis relation (2.29) may be extended to infinity. But, substituting (2.29) with infinite bounds to (2.28) produces a synthesis relation. These relations are called the *direct Fourier transform* and *inverse Fourier transform*, respectively,

$$X(j\omega) = \int_{-\infty}^{\infty} x(t) \mathrm{e}^{-j\omega t} \mathrm{d}t \,. \tag{2.30}$$

$$x(t) = \frac{1}{2\pi} \int_{-\infty}^{\infty} X(j\omega) \, \mathrm{e}^{j\omega t} \mathrm{d}\omega \,. \tag{2.31}$$

Both the direct and the inverse Fourier transforms are fundamental in the theory of signals, systems, and SP.

Example 2.6. A signal is delta-shaped, $x(t) = \delta(t)$. By (2.30) and sifting property of the delta-function, its transform becomes

$$\Delta(j\omega) = \int_{-\infty}^{\infty} \delta(t) \mathrm{e}^{-j\omega t} \mathrm{d}t = \mathrm{e}^{-j\omega 0} = 1 \,,$$

thus

$$\delta(t) \overset{\mathcal{F}}{\Leftrightarrow} 1 .$$

□

The quantity $X(j\omega)$ is called the *spectral density* of a nonperiodic signal $x(t)$. As may easily be observed, (2.15) differs from (2.30) by a gain coefficient $1/T$ and finite bounds. This means that the complex amplitudes of a periodic signal spectrum may be expressed by the spectral density of a relevant single pulse by the relation

$$C_k = \frac{1}{T} X(j2\pi kFt),$$

in which with the frequency ω is taken at discrete points $k\Omega = 2\pi kF = 2\pi k/T$. Thus the spectral density $X(j\omega)$ bears all properties of the complex amplitudes C_k.

Commonly, the spectral density $X(j\omega)$ is a complex function. As such, it may be performed by the real and imaginary spectral components, $A_x(\omega)$ and $B_x(\omega)$, respectively, as follows:

$$X(j\omega) = A_x(\omega) - jB_x(\omega) = |X(j\omega)| \, e^{-j\varphi_x(\omega)} , \tag{2.32}$$

where

$$A_x(\omega) = \int_{-\infty}^{\infty} x(t) \cos \omega t dt , \tag{2.33}$$

$$B_x(\omega) = \int_{-\infty}^{\infty} x(t) \sin \omega t dt . \tag{2.34}$$

The total spectral density $|X(j\omega)| \geqslant 0$ is a positive-valued function called the *magnitude spectral density* or *amplitude spectral density* and the phase spectral characteristic $\varphi_x(\omega)$ is called the *phase spectral density*. The magnitude and phase spectral densities are defined, respectively, by

$$|X(j\omega)| = \sqrt{A_x^2(\omega) + B_x^2(\omega)} , \tag{2.35}$$

$$\tan \varphi_x(\omega) = \frac{B_x(\omega)}{A_x(\omega)} . \tag{2.36}$$

The most common properties of (2.35) and (2.36) are the following:

- $|X(j\omega)|$ is an even function; it is a symmetric function about zero.
- $\varphi_x(\omega)$ is an odd function; it is an antisymmetric function about zero.

The other common property claims that a single pulse and its periodic train have qualitatively equal shapes of their spectral characteristics.

We may now summarize that the spectral density exhaustively describes a nonperiodic signal $x(t)$ in the frequency domain by the magnitude spectral density $|X(j\omega)|$ and phase spectral density $\varphi_x(\omega)$. Both these characteristics are performed as either double-sided (mathematical) or one-sided (physical). In the double-sided presentation, they have the LSBs and USBs (Fig. 2.8a). In the one-sided form, the LSB $|X_-(j\omega)|$ and USB $|X_+(j\omega)|$ of the magnitude spectral density are summed, and the phase spectral density is represented by the USB.

2.3.3 Properties of the Fourier Transform

Typically, in applications, signal waveforms are complex. Therefore, their transformations to the frequency domain and back to the time domain entail certain difficulties. To pass over, it is in order to use properties of the Fourier transform. An additional benefit of these properties is that they efficiently help answering the principal questions of the transforms: *What is going on with the spectral density if a signal waveform undergoes changes? What waveform corresponds to the given spectral density?*

We discuss below the most interesting and widely used properties of the Fourier transform.

2.3.3.1 Time shifting

Given a signal $x(t) \overset{\mathcal{F}}{\Leftrightarrow} X(j\omega)$ and its shifted version $y(t) = x(t - t_0)$ (Fig. 2.9). The spectral density of $y(t)$ is determined, by (2.30), as

$$Y(j\omega) = \int_{-\infty}^{\infty} y(t)e^{-j\omega t}dt = \int_{-\infty}^{\infty} x(t - t_0)e^{-j\omega t}dt\,. \tag{2.37}$$

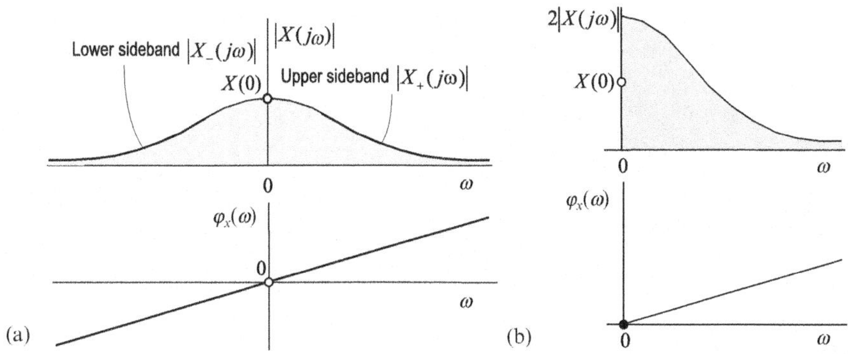

Fig. 2.8 Spectral densities of a nonperiodic signal: (**a**) double-sided and (**b**) one-sided.

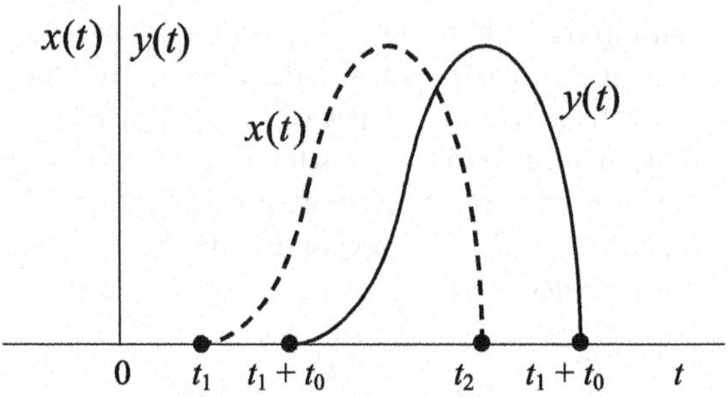

Fig. 2.9 Time shift of a single pulse.

By changing a variable to $\theta = t - t_0$ and writing $t = t_0 + \theta$, we get

$$Y(j\omega) = \mathrm{e}^{-j\omega t_0} \int_{-\infty}^{\infty} x(\theta)\mathrm{e}^{-j\omega\theta}\mathrm{d}\theta\,. \tag{2.38}$$

The integral in the right-hand side of (2.38) is the spectral density of $x(t)$ and thus

$$Y\,(j\omega) = \mathrm{e}^{-j\omega t_0} X\,(j\omega)\,. \tag{2.39}$$

The *time-shifting property* is thus formulated by

$$x(t \pm t_0) \overset{\mathcal{F}}{\Leftrightarrow} \mathrm{e}^{\pm j\omega t_0} X\,(j\omega)\,, \tag{2.40}$$

meaning that any time shift $\pm t_0$ in a signal does not affect its magnitude spectral density and results in an additional phase shift $\pm j\omega t_0$ in its phase spectral density.

Example 2.7. A signal $x(t) = \delta(t) \overset{\mathcal{F}}{\Leftrightarrow} \Delta(j\omega) = 1$ is shifted in time to be $y(t) = \delta(t - t_0)$. By (2.30) and the sifting property of the delta function, the spectral density of $y(t)$ becomes

$$Y(j\omega) = \int_{-\infty}^{\infty} \delta(t - t_0)\mathrm{e}^{-j\omega t}\mathrm{d}t = \mathrm{e}^{-j\omega t_0}\,,$$

thus, $Y(j\omega) = \mathrm{e}^{-j\omega t_0}\Delta(j\omega)$ that is stated by (2.40).

□

2.3.3.2 Time Scalling

A signal $x(t) \overset{\mathcal{F}}{\Leftrightarrow} X(j\omega)$ of a duration τ is squeezed (or stretched) in the time domain to be $y(t) = x(\alpha t)$, where $\alpha > 0$ (Fig. 2.10). The transform of $y(t)$ is

$$Y\,(j\omega) = \int_{-\infty}^{\infty} x\,(\alpha t)\,\mathrm{e}^{-j\omega t}\mathrm{d}t\,. \tag{2.41}$$

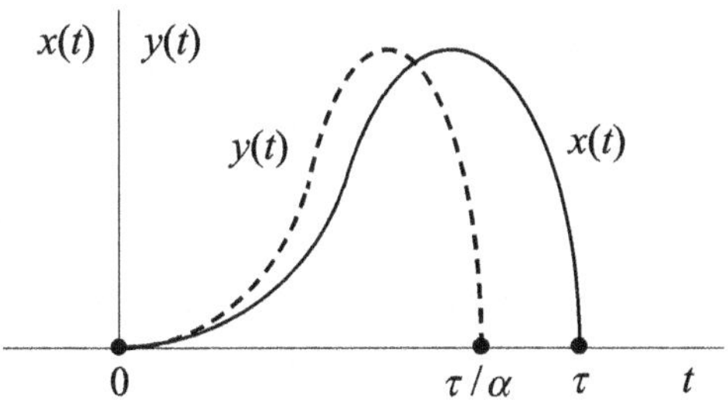

Fig. 2.10 Time scaling of a single pulse.

By a new variable $\theta = \alpha t$ and $t = \theta/\alpha$, (2.41) becomes

$$Y(j\omega) = \frac{1}{\alpha}\int_{-\infty}^{\infty} x(\theta)\, \mathrm{e}^{-j\frac{\omega}{\alpha}\theta}\mathrm{d}\theta\,. \tag{2.42}$$

Since the integral in the right-hand side of (2.42) is a scaled spectral density of $x(t)$, we arrive at

$$Y(j\omega) = \frac{1}{\alpha}X\left(j\frac{\omega}{\alpha}\right) \tag{2.43}$$

that, most generally, leads to the following *scaling theorem* or *similarity theorem*:

$$x(\alpha t) \overset{\mathcal{F}}{\Leftrightarrow} \frac{1}{|\alpha|}X\left(j\frac{\omega}{\alpha}\right)\,. \tag{2.44}$$

So, squeezing ($\alpha > 1$) or stretching ($0 < \alpha < 1$) the origin signal in time results in scaling and gaining its spectral density by the factor of $1/\alpha$.

Example 2.8. A signal $x(t) = \delta(t)$ is scaled with a coefficient $\alpha > 0$ to be $y(t) = \delta(\alpha t)$. By (1.22) and (2.30), the spectral density of $y(t)$ calculates

$$Y(j\omega) = \int_{-\infty}^{\infty} \delta(\alpha t)\mathrm{e}^{-j\omega t}\mathrm{d}t = \frac{1}{|\alpha|}\int_{-\infty}^{\infty} \delta(t)\mathrm{e}^{-j\omega t}\mathrm{d}t = \frac{1}{|\alpha|}$$

that is consistent to (2.44), since $\delta(t) \overset{\mathcal{F}}{\Leftrightarrow} 1$.

□

2.3.3.3 Conjugation

The conjugation theorem claims that if $x(t) \overset{\mathcal{F}}{\Leftrightarrow} X(j\omega)$, then for the conjugate signal $x^*(t)$

$$x^*(t) \overset{\mathcal{F}}{\Leftrightarrow} X^*(-j\omega)\,. \tag{2.45}$$

2.3.3.4 Time reversal

if $x(t) \overset{\mathcal{F}}{\Leftrightarrow} X(j\omega)$, then

$$x(-t) \overset{\mathcal{F}}{\Leftrightarrow} X(-j\omega)\,. \tag{2.46}$$

Example 2.9. A signal is given by $y(t) = \delta(-t)$. Its transform is performed as

$$\begin{aligned} Y(j\omega) &= \int_{-\infty}^{\infty} \delta(-t)\mathrm{e}^{-j\omega t}\mathrm{d}t = -\int_{\infty}^{-\infty} \delta(t)\mathrm{e}^{-j(-\omega)t}\mathrm{d}t \\ &= \int_{-\infty}^{\infty} \delta(t)\mathrm{e}^{-j(-\omega)t}\mathrm{d}t = \mathrm{e}^{-j(-\omega)0} = 1\,. \end{aligned}$$

As in (2.46), the sign of the frequency is changed. However, the spectral density here is unity over all frequencies.

□

2.3.3.5 Differentiation in Time

The differentiation theorem claims that if $x(t) \overset{\mathcal{F}}{\Leftrightarrow} X(j\omega)$, then the time derivative $\mathrm{d}x(t)/\mathrm{d}t$ has the Fourier transform $j\omega X(j\omega)$. Indeed, by integrating the below integral by parts, we arrive at

$$\begin{aligned} \int_{-\infty}^{\infty} \mathrm{e}^{-j\omega t}\frac{\mathrm{d}}{\mathrm{d}t}x(t)\mathrm{d}t &= x(t)\mathrm{e}^{-j\omega t}\Big|_{-\infty}^{\infty} - \int_{-\infty}^{\infty} x(t)\frac{\mathrm{d}}{\mathrm{d}t}\mathrm{e}^{-j\omega t}\mathrm{d}t \\ &= j\omega \int_{-\infty}^{\infty} x(t)\mathrm{e}^{-j\omega t}\mathrm{d}t = j\omega X(j\omega)\,, \end{aligned}$$

since $x(t)$ is supposed to vanish at $t \to \pm\infty$ (Dirichlet condition). Most generally, the differentiation in time theorem results in

$$\frac{\mathrm{d}^n}{\mathrm{d}t^n}x(t) \overset{\mathcal{F}}{\Leftrightarrow} (j\omega)^n X(j\omega)\,. \tag{2.47}$$

Thus, a multiple differentiation of $x(t)$ in time gains its spectral density by $(j\omega)^n$. This property is supported by the duality of operators $\mathrm{d}/\mathrm{d}t \equiv j\omega$.

Example 2.10. A signal is shaped with the time derivative of the delta function, $y(t) = \frac{\mathrm{d}}{\mathrm{d}t}\delta(t)$. Its spectral density, by (1.30), calculates

$$Y(j\omega) = \int_{-\infty}^{\infty} \mathrm{e}^{-j\omega t}\frac{\mathrm{d}}{\mathrm{d}t}\delta(t)\mathrm{d}t = -\frac{\mathrm{d}}{\mathrm{d}t}\,\mathrm{e}^{-j\omega t}\Big|_{t=0} = j\omega\,.$$

Thus, $Y(j\omega) = j\omega\Delta(j\omega)$, where $\Delta(j\omega) = 1$ is the transform of $\delta(t)$.

□

2.3.3.6 Integration

If $x(t) \overset{\mathcal{F}}{\Leftrightarrow} X(j\omega)$, then the transform of $\int_{-\infty}^{t} x(t)\mathrm{d}t$ satisfies the relation

$$\int_{-\infty}^{t} x(t)\mathrm{d}t \overset{\mathcal{F}}{\Leftrightarrow} \frac{1}{j\omega}X(j\omega) + \pi X(0)\delta(\omega)\,. \tag{2.48}$$

Note that if $X(j0) = 0$, then the last term in the right-hand side is zero.

2.3.3.7 Linearity

Given a set of signals,

$$\alpha_1 x_1(t) \overset{\mathcal{F}}{\Leftrightarrow} \alpha_1 X_1(j\omega)\,, \quad a_2 x_2(t) \overset{\mathcal{F}}{\Leftrightarrow} a_2 X_2(j\omega)\,, \ldots, \quad a_i x_i(t) \overset{\mathcal{F}}{\Leftrightarrow} a_i X_i(j\omega)\,, \ldots$$

The addition theorem claims that

$$\sum_i a_i x_i(t) \overset{\mathcal{F}}{\Leftrightarrow} \sum_i a_i X_i(j\omega)\,. \tag{2.49}$$

This property is also known as a *superposition principle* or *addition property* and is critical in linear systems and linear SP.

2.3.3.8 Spectral Density Value at Zero

It follows straightforwardly, by (2.30), that

$$X(j0) = \int_{-\infty}^{\infty} x(t)\mathrm{d}t\,. \tag{2.50}$$

Thus, the value $X(j0)$ is calculated by the area of a relevant nonperiodic signal $x(t)$.

2.3.3.9 Signal Value at Zero

By setting $t = 0$ to (2.31), we get

$$x(0) = \frac{1}{2\pi}\int_{-\infty}^{\infty} X(j\omega)\mathrm{d}\omega = \int_{-\infty}^{\infty} X(j2\pi f)\mathrm{d}f, \tag{2.51}$$

meaning that the value $x(0)$ is provided by the area of the spectral density.

2.3.3.10 Duality

If $x(t)$ has the Fourier transform $X(j\omega)$, then $y(t)$ with the waveform shaped by $X(j\omega)$ will get the Fourier transform $Y(j\omega)$ shaped by $x(t)$. Mathematically, this property results in the following:

$$\text{If } x(t) \overset{\mathcal{F}}{\Leftrightarrow} X(jf)\,, \text{ then } X(t) \overset{\mathcal{F}}{\Leftrightarrow} x(-jf)$$

or

$$\text{If } x(t) \overset{\mathcal{F}}{\Leftrightarrow} X(j\omega)\,, \text{ then } X(t) \overset{\mathcal{F}}{\Leftrightarrow} 2\pi x(-j\omega)\,. \tag{2.52}$$

In fact, if we first write

$$x(t) = \frac{1}{2\pi}\int_{-\infty}^{\infty} X(j\omega)e^{j\omega t}d\omega$$

and then substitute t by ω and ω by t, we will get

$$x(j\omega) = \frac{1}{2\pi}\int_{-\infty}^{\infty} X(t)e^{j\omega t}d\omega\,.$$

Now it just needs to change the sign of frequency and we arrive at

$$2\pi x(-j\omega) = -\int_{\infty}^{-\infty} X(t)e^{-j\omega t}d\omega = \int_{-\infty}^{\infty} X(t)e^{-j\omega t}d\omega\,,$$

meaning that $X(t) \overset{\mathcal{F}}{\Leftrightarrow} 2\pi x(-j\omega)$. Figure 2.11 illustrates this property allowing us to realize a correspondence between the signal waveform and its spectral performance.

Example 2.11. The transform of the delta signal $\delta(t)$ is $\Delta(j\omega) = 1$. Now find the transform of the uniform signal $x(t) = 1$. By (2.30), we get $X(j\omega) = \int_{-\infty}^{\infty} e^{-j\omega t}dt$ that, by (1.36), becomes $X(j\omega) = 2\pi\delta(\omega)$.

□

2.3.3.11 Modulation

The following modulation theorem plays an important role in the modulation theory: if $x(t) \overset{\mathcal{F}}{\Leftrightarrow} X(j\omega)$, then modulating $e^{\pm j\omega_0 t}$ with $x(t)$ results in

$$x(t)e^{\pm j\omega_0 t} \overset{\mathcal{F}}{\Leftrightarrow} X[j(\omega \pm \omega_0)t]\,. \tag{2.53}$$

To prove, one merely may define the transform of $x(t)e^{j\omega_0 t}$, i.e.,

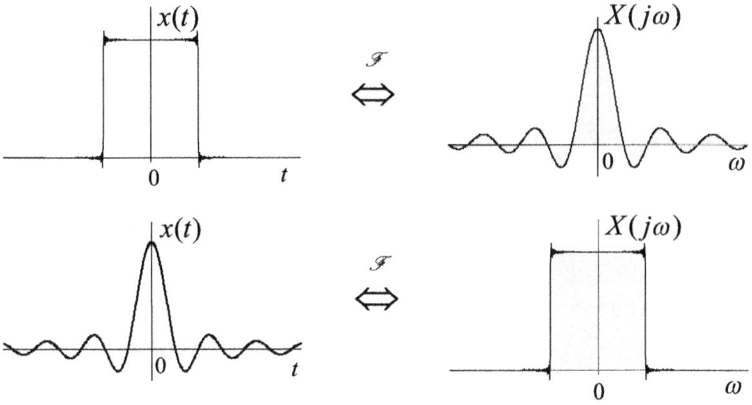

Fig. 2.11 Duality of waveforms and spectral densities.

$$\int_{-\infty}^{\infty} x(t)e^{j\omega_0 t}e^{-j\omega t}dt = \int_{-\infty}^{\infty} x(t)e^{-j(\omega-\omega_0)t}dt = X[j(\omega-\omega_0)] .$$

Note, if a sign of ω_0 is arbitrary, we go to (2.53).

Example 2.12. Consider a signal $x(t) = \delta(t)e^{j\omega_0 t}$. Its transform

$$\int_{-\infty}^{\infty} \delta(t)e^{j\omega_0 t}e^{-j\omega t}dt = \int_{-\infty}^{\infty} \delta(t)e^{-j(\omega-\omega_0)t}dt = 1$$

is unity over all frequencies irrespective of the shift ω_0.

□

2.3.3.12 Multiplication

Consider a signal $y(t) = x_1(t)x_2(t)$. Its spectral density is specified by

$$Y(j\omega) = \int_{-\infty}^{\infty} x_1(t)x_2(t)e^{-j\omega t}dt , \tag{2.54}$$

where $x_1(t)$ and $x_2(t)$ are given by the transforms, respectively,

$$x_1(t) = \frac{1}{2\pi}\int_{-\infty}^{\infty} X_1(j\omega)\, e^{j\omega t}d\omega, \tag{2.55}$$

$$x_2(t) = \frac{1}{2\pi}\int_{-\infty}^{\infty} X_2(j\omega)\, e^{j\omega t}d\omega. \tag{2.56}$$

Substituting (2.56) into (2.54) yields

$$
\begin{aligned}
Y(j\omega) &= \frac{1}{2\pi}\int_{-\infty}^{\infty} x_1(t)\mathrm{e}^{-j\omega t}\int_{-\infty}^{\infty} X_2(j\omega')\,\mathrm{e}^{j\omega' t}\mathrm{d}\omega'\mathrm{d}t \\
&= \frac{1}{2\pi}\int_{-\infty}^{\infty} X_2(j\omega')\left[\int_{-\infty}^{\infty} x_1(t)\mathrm{e}^{-j(\omega-\omega')t}\mathrm{d}t\right]\mathrm{d}\omega'. \qquad (2.57)
\end{aligned}
$$

We notice that the integral in brackets is the spectral density of a $x_1(t)$ at $\omega - \omega'$, so that it may be substituted by $X_1[j(\omega - \omega')]$. This leads to the relation

$$
Y(j\omega) = \frac{1}{2\pi}\int_{-\infty}^{\infty} X_2(j\omega')\,X_1[j(\omega-\omega')]\mathrm{d}\omega' \qquad (2.58)
$$

that, in its right-hand side, consists of a convolution of spectral densities. We thus conclude that if $x_1(t) \overset{\mathcal{F}}{\Leftrightarrow} X_1(j\omega)$ and $x_2(t) \overset{\mathcal{F}}{\Leftrightarrow} X_2(j\omega)$, then the transform of $2\pi x_1(t)x_2(t)$ is a convolution of $X_1(j\omega)$ and $X_2(j\omega)$, e.g.,

$$
x_1(t)x_2(t) \overset{\mathcal{F}}{\Leftrightarrow} \frac{1}{2\pi}X_1(j\omega) * X_2(j\omega). \qquad (2.59)
$$

Example 2.13. Given two signals, $x_1(t) = 1$ and $x_2(t) = \delta(t)$. The transform of the product of $y(t) = x_1(t)x_2(t)$ is defined by

$$
Y(j\omega) = \int_{-\infty}^{\infty} [1 \times \delta(t)]\mathrm{e}^{-j\omega t}\mathrm{d}t = 1\,.
$$

On the other hand, $Y(j\omega)$ may be calculated, by (2.59) and known transforms, $1 \overset{\mathcal{F}}{\Leftrightarrow} 2\pi\delta(\omega)$ and $\delta(t) \overset{\mathcal{F}}{\Leftrightarrow} 1$, as follows:

$$
Y(j\omega) = \frac{1}{2\pi}\left[2\pi\int_{-\infty}^{\infty}\delta(\nu)\mathrm{d}\nu\right] = 1
$$

and we arrive at the same result.

□

2.3.3.13 Convolution

Let $x_1(t) \overset{\mathcal{F}}{\Leftrightarrow} X_1(j\omega)$ and $x_2(t) \overset{\mathcal{F}}{\Leftrightarrow} X_2(j\omega)$. It may be shown that the transform of the convolution of signals, $x_1(t) * x_2(t)$, is provided by a multiplication of their spectral densities, i.e.,

$$x_1(t) * x_2(t) \overset{\mathcal{F}}{\Leftrightarrow} X_1(j\omega) X_2(j\omega) . \tag{2.60}$$

To prove (2.60), we first write

$$x_1(t) * x_2(t) = \int_{-\infty}^{\infty} x_1(\theta) x_2(t-\theta) \mathrm{d}\theta$$

and substitute $x_2(t-\theta)$ with

$$x_2(t-\theta) = \frac{1}{2\pi} \int_{-\infty}^{\infty} X_2(j\omega) \mathrm{e}^{j\omega(t-\theta)} \mathrm{d}\omega$$

that leads to

$$\begin{aligned} x_1(t) * x_2(t) &= \frac{1}{2\pi} \int_{-\infty}^{\infty} x_1(\theta) \int_{-\infty}^{\infty} X_2(j\omega) \mathrm{e}^{j\omega(t-\theta)} \mathrm{d}\omega \mathrm{d}\theta \\ &= \frac{1}{2\pi} \int_{-\infty}^{\infty} X_2(j\omega) \mathrm{e}^{j\omega t} \left[\int_{-\infty}^{\infty} x_1(\theta) \mathrm{e}^{-j\omega\theta} \mathrm{d}\theta \right] \mathrm{d}\omega \\ &= \frac{1}{2\pi} \int_{-\infty}^{\infty} X_1(j\omega) X_2(j\omega) \mathrm{e}^{j\omega t} \mathrm{d}\omega \end{aligned}$$

and we arrive at (2.60).

There are some other properties of the Fourier transform that may be of importance in applications. Among them, the Rayleigh theorem establishes a correspondence between the energy of a nonperiodic signal in the time and frequency domains.

2.3.4 Rayleigh's Theorem

There are a number of applications, which are sensitive not only to the signal waveform and spectral density but also to its energy. An example is radars in which a distance is evaluated for the maximum joint energy of the transmitted and received pulses.

By definition, the instantaneous power of a signal $x(t)$ is specified by $P_x(t) = |x(t)|^2 = x(t)x^*(t)$. The total signal energy is calculated by integrating $P_x(t)$ over time,

$$E_x = \int_{-\infty}^{\infty} P_x(t) \mathrm{d}t = \int_{-\infty}^{\infty} |x(t)|^2 \mathrm{d}t . \tag{2.61}$$

Rayleigh's[4] theorem establishes that the integral of the square absolute value of a nonperiodic function $x(t)$ is equal to the integral of the square magnitude of its Fourier transform, i.e.,

$$\int_{-\infty}^{\infty} |x(t)|^2 \, \mathrm{d}t = \frac{1}{2\pi} \int_{-\infty}^{\infty} |X(j\omega)|^2 \, \mathrm{d}\omega = \frac{1}{\pi} \int_{0}^{\infty} |X(j\omega)|^2 \, \mathrm{d}\omega \, . \tag{2.62}$$

Indeed, if we apply the multiplication rule (2.59) to $|x(t)|^2 = x(t)x^*(t)$, we go step by step from (2.61) to (2.62):

$$E_x = \int_{-\infty}^{\infty} |x(t)|^2 \, \mathrm{d}t = \int_{-\infty}^{\infty} x(t)x^*(t)\mathrm{e}^{-j0t}\mathrm{d}t = X(j0) * X^*(-j0)$$

$$= \frac{1}{2\pi} \int_{-\infty}^{\infty} X(j\omega')\, X\left[j\,(0 - \omega')\right]\mathrm{d}\omega' = \frac{1}{2\pi} \int_{-\infty}^{\infty} X(j\omega)\, X^*(j\omega)\mathrm{d}\omega$$

$$= \frac{1}{2\pi} \int_{-\infty}^{\infty} |X(j\omega)|^2 \mathrm{d}\omega = \frac{1}{\pi} \int_{0}^{\infty} |X(j\omega)|^2 \mathrm{d}\omega \, . \tag{2.63}$$

The property (2.62) is also known as the *Parseval relation* for nonperiodic signals and *Plancherel*[5] *identity*.

2.4 Spectral Densities of Simple Single Pulses

In the theory of signals, systems, and SP, nonperiodic (single) waveforms are fundamental to solve many applied problems. Therefore, knowledge about their spectral properties is of high importance. Single pulses may form *pulse-bursts* also called *nonperiodic pulse-trains* and *periodic pulse sequences* called *periodic pulse-trains*. For each of these signals, the pulse waveform is usually selected to attain maximum efficiency in the applied problem. We will consider spectral densities of the most common single waveforms.

Before applying the Fourier transform, to go to the frequency domain, and learning spectral properties of single waveforms, one needs to remember that the transform will exist if $x(t)$ has finite energy or the following integral is finite,

[4] John William Strutt (Lord Rayleigh), English mathematician, 12 November 1842–30 June 1919.

[5] Michel Plancherel, Swiss mathematician, 16 January 1885–4 March 1967.

$$\int_{-\infty}^{\infty} |x(t)|^2 \mathrm{d}t < \infty\,.$$

The Dirichlet conditions give an alternative: the transform will exist if a signal $x(t)$ is absolutely integrable, i.e.,

$$\int_{-\infty}^{\infty} |x(t)| \mathrm{d}t < \infty\,. \tag{2.64}$$

It may be shown that both these conditions are sufficient, but unnecessary for the Fourier transform to exist. Among them, the Dirichlet conditions (2.64) have gained wider currency in applications.

2.4.1 Truncated Exponential Pulse

A real truncated decaying exponential waveform is one of the most interesting models, since in the limiting cases it degenerates to either a delta-shaped function or a unit-step function. Figure 2.12 illustrates this waveform and we notice that since the slope at zero is infinite, then it is rather an idealization of some real pulse.

The signal is described by

$$x(t) = \begin{cases} A\mathrm{e}^{-\alpha t}, & t \geqslant 0 \\ 0, & t < 0 \end{cases} = A\mathrm{e}^{-\alpha t}u(t)\,, \tag{2.65}$$

where A is an amplitude at zero and $\alpha > 0$ is a positive-valued constant.

It may be shown that, by $\alpha > 0$, the function (2.65) satisfies the Dirichlet condition (2.64). We may thus solve an analysis problem (2.30) that defines the spectral density by

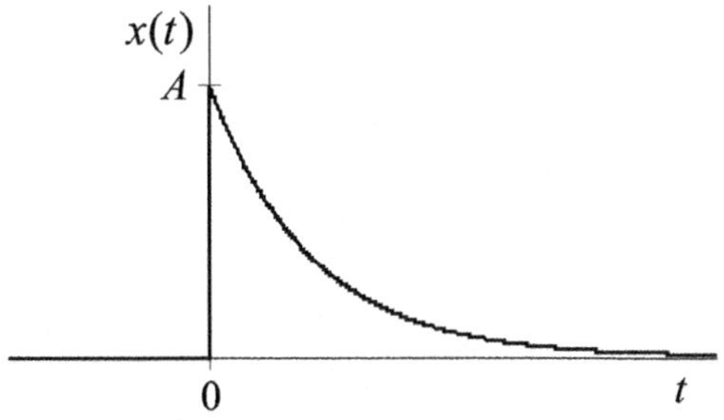

Fig. 2.12 Truncated decaying exponential waveform.

$$X(j\omega) = A\int_{0}^{\infty} \mathrm{e}^{-\alpha t}\mathrm{e}^{-j\omega t}\mathrm{d}t = A\int_{0}^{\infty} \mathrm{e}^{-(\alpha+j\omega)t}\mathrm{d}t$$

$$= -\frac{A}{\alpha + j\omega}\mathrm{e}^{-(\alpha+j\omega)t}\Big|_{0}^{\infty} = \frac{A}{\alpha + j\omega} \quad (2.66)$$

and determines the magnitude and phase spectral densities, respectively, by

$$|X(j\omega)| = \frac{A}{\sqrt{\alpha^2 + \omega^2}}, \quad (2.67)$$

$$\varphi_x(\omega) = -\arctan\frac{\omega}{\alpha}. \quad (2.68)$$

As it is seen, a signal (2.65) has a baseband spectral density (2.66) with its maximum $X(j0) = A/\alpha$ at zero that is a reciprocal of α. Accordingly, if A is constant, then, by $\alpha \to 0$, the maximum tends toward infinity and, by $\alpha \to 0$, it goes to zero. Figure 2.13 shows evolution of $|X(j\omega)|$ and $\varphi_x(\omega)$ if α takes three different values, $\alpha_1 < \alpha < \alpha_2$.

One may also conclude that increasing α broadens the magnitude spectral density and reduces the slope of the phase spectral density. Contrary, reducing α results in narrowing the magnitude density and a larger slope of the phase density. We may thus distinguish two limiting waveforms.

2.4.1.1 Transition to Unit-step Pulse

A limiting value $\alpha \to 0$ degenerates the real truncated exponential pulse (2.65) to the unit-step pulse (Fig. 2.14)

$$x(t) = \begin{cases} A, & t \geqslant 0 \\ 0, & t < 0 \end{cases} = Au(t). \quad (2.69)$$

It may be shown that (2.69) does not satisfy (2.64). However, the problem with integrating may be passed over if to consider (2.69) as a limiting case of (2.65) with $\alpha \to 0$,

$$x(t) = \begin{cases} \lim\limits_{\alpha\to 0} A\mathrm{e}^{-\alpha t} & , t \geqslant 0 \\ 0, & t < 0 \end{cases}. \quad (2.70)$$

The spectral density of (2.70) may then be defined by (2.66) if we tend α toward zero,

$$X(j\omega) = \lim_{\alpha\to 0}\frac{A}{\alpha + j\omega}. \quad (2.71)$$

To calculate (2.71) properly, we exploit a decomposition $1/(\alpha + j\omega) = \alpha/(\alpha^2+\omega^2) - j\omega/(\alpha^2+\omega^2)$, use an identity $\int_{-\infty}^{\infty} a/(a^2+x^2)\mathrm{d}x = \pi$, and go to

$$X(j\omega) = A\pi\delta(\omega) + \frac{A}{j\omega}, \quad (2.72)$$

where $\delta(\omega)$ is the Dirac delta function. We notice that the first term in the right-hand side of (2.72) indicates that the spectral density goes toward

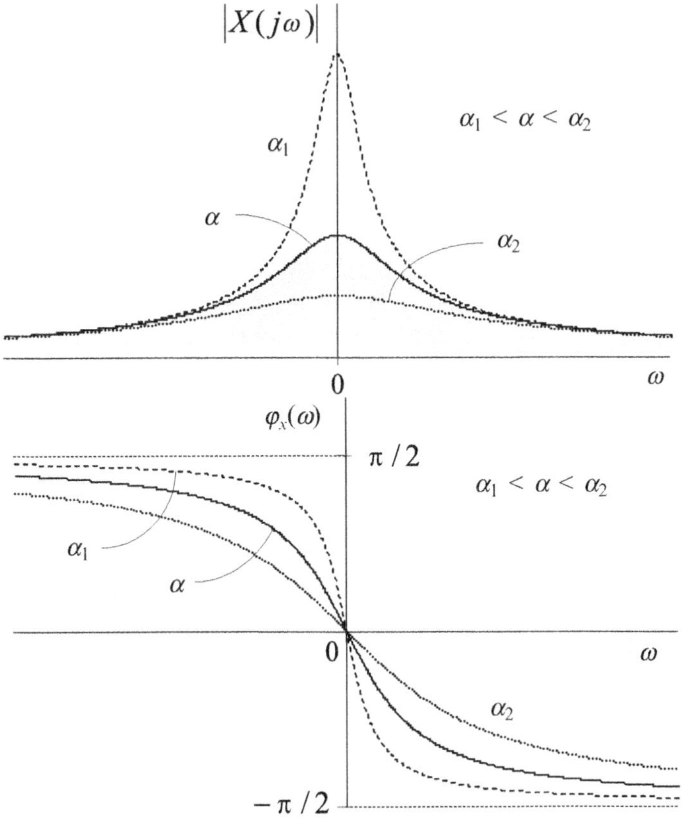

Fig. 2.13 Spectral densities of a real truncated decaying exponential pulse: (**a**) magnitude and (**b**) phase.

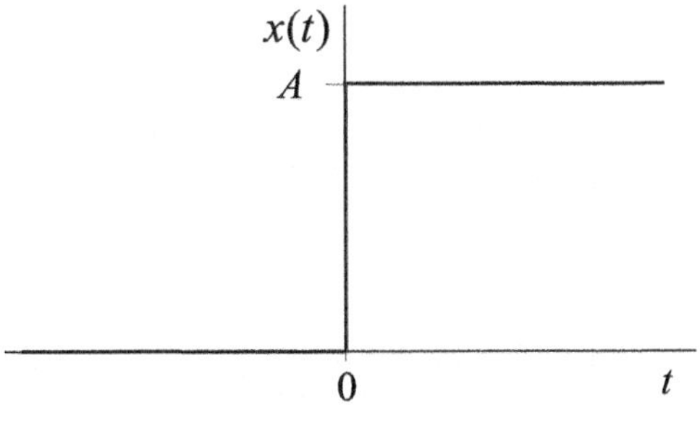

Fig. 2.14 Gained unit-step pulse.

infinity at zero that is also established by the second term. Therefore, very often, this first term is avoided and the magnitude and phase spectral densities are written, respectively, as

$$|X(j\omega)| = \frac{A}{\omega}, \tag{2.73}$$

$$\varphi_x(\omega) = -\lim_{\alpha\to 0} \arctan\left(\frac{\omega}{\alpha}\right) = \begin{cases} \frac{\pi}{2}, & \omega < 0 \\ 0, & \omega = 0 \\ -\frac{\pi}{2}, & \omega > 0 \end{cases} \tag{2.74}$$

Figure 2.15 illustrates both (2.73) and (2.74) and we may now outline the properties of $u(t)$ in the frequency domain:

- Its magnitude spectral density tends toward infinity at $\omega = 0$ and diminishes to zero asymptotically as a reciprocal of ω.
- Its phase spectral density takes the values of $\pi/2$ and $-\pi/2$ with $\omega < 0$ and $\omega > 0$, respectively, and has a discontinuity at zero.

2.4.1.2 Transition to Delta-shaped Pulse

The limiting case of $\alpha \to \infty$ degenerates (2.65) to the delta-shaped pulse with a constant amplitude A,

$$x(t) = \begin{cases} \lim\limits_{\alpha\to\infty} A\mathrm{e}^{-\alpha t}, & t \geqslant 0 \\ 0, & t < 0 \end{cases} = \begin{cases} A, & t = 0 \\ 0, & \text{otherwise} \end{cases}. \tag{2.75}$$

If the amplitude A is chosen such that

$$\int_{-\infty}^{\infty} x(t)\mathrm{d}t = 1\,,$$

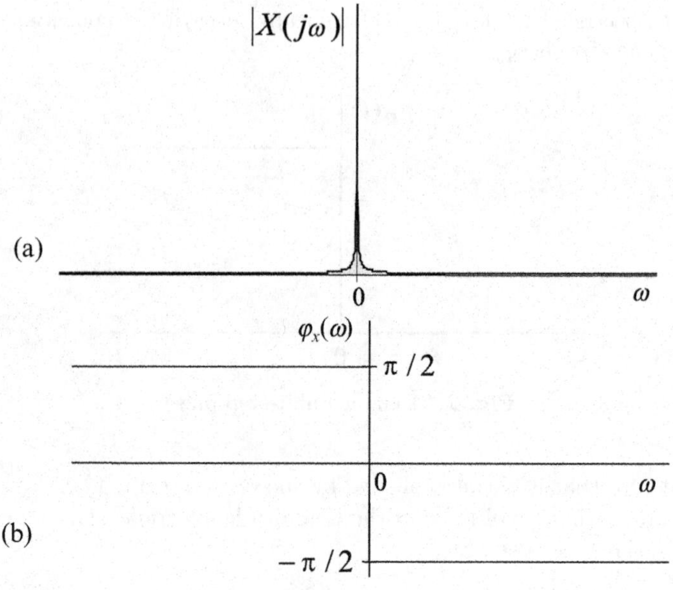

Fig. 2.15 Spectral densities of a unit-step pulse: (**a**) magnitude and (**b**) phase.

so A tends toward infinity at $t = 0$, then (2.75) becomes delta function. Accordingly, its spectral density becomes uniform with unit amplitude, $X(j\omega) = |X(j\omega)| = 1$, and the phase density is zero, $\varphi_x(\omega) = 0$.

2.4.2 Rectangular Pulse

A rectangular pulse waveform has appeared to be one of the most successful mathematical idealizations of real impulse signals. In particular, it represents an elementary bit-pulse in digital communications and simulates well operation of pulse radars.

An analysis problem may be solved here straightforwardly by using the direct Fourier transform. A shorter way implies using the properties of linearity and time shifting of the transform and presenting a rectangular pulse $y(t)$ of duration τ by a sum of two weighted unit-step functions, $x_1(t)$ and $x_2(t)$, as it is shown in Fig. 2.16. Thus, we perform $y(t)$ by

$$y(t) = x_1(t) + x_2(t) = Au\left(t + \frac{\tau}{2}\right) - Au\left(t - \frac{\tau}{2}\right). \tag{2.76}$$

By the linearity property, the spectral density of $y(t)$ is defined as

$$Y(j\omega) = X_1(j\omega) + X_2(j\omega), \tag{2.77}$$

where $X_1(j\omega) \overset{\mathcal{F}}{\Leftrightarrow} x_1(t)$ and $X_2(j\omega) \overset{\mathcal{F}}{\Leftrightarrow} x_2(t)$. To determine $X_1(j\omega)$ and $X_2(j\omega)$, we employ (2.71), apply the time-shifting property, and go to

$$x_1(t) = Au\left(t + \frac{\tau}{2}\right) \overset{\mathcal{F}}{\Leftrightarrow} X_1(j\omega) = \frac{A}{j\omega}e^{j\omega\frac{\tau}{2}}, \tag{2.78}$$

$$x_2(t) = -Au\left(t - \frac{\tau}{2}\right) \overset{\mathcal{F}}{\Leftrightarrow} X_2(j\omega) = -\frac{A}{j\omega}e^{-j\omega\frac{\tau}{2}}. \tag{2.79}$$

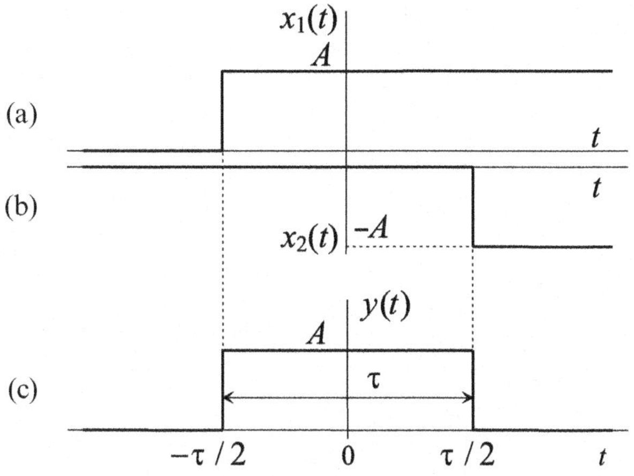

Fig. 2.16 A rectangular pulse (**c**) composed by two step functions (**a**) and (**b**).

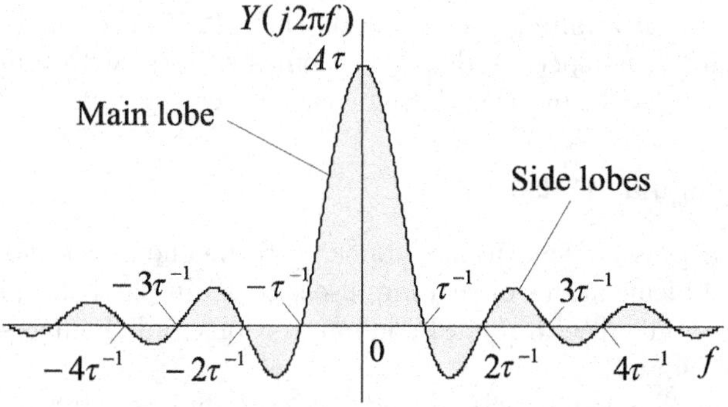

Fig. 2.17 Spectral density of a rectangular pulse.

By (2.78) and (2.79), the spectral density (2.77) becomes

$$\begin{aligned} Y(j\omega) &= \frac{A}{j\omega}e^{j\omega\frac{\tau}{2}} - \frac{A}{j\omega}e^{-j\omega\frac{\tau}{2}} = \frac{2A}{\omega}\frac{e^{j\omega\tau/2} - e^{-j\omega\tau/2}}{2j} \\ &= A\tau\frac{\sin(\omega\tau/2)}{\omega\tau/2}\,. \end{aligned} \tag{2.80}$$

It is seen that (2.80) is a real function symmetric about zero (Fig. 2.17) exhibiting all the properties featured to the sinc-function. In fact, $Y(j\omega)$ attains a maximum at zero, $Y(j0) = A\tau$, and oscillates with frequency about zero. It becomes identically zero at

$$f = \frac{\omega}{2\pi} = \pm\frac{m}{\tau}\,, \qquad m = 1, 2, ...$$

and its envelope diminishes toward zero asymptotically by increasing f. It has a main positive lobe at zero and the side lobes (to the left and right of the main lobe) that may be either positive or negative.

The magnitude spectral density associated with (2.80) is specified by

$$|Y(j\omega)| = A\tau\left|\frac{\sin(\omega\tau/2)}{\omega\tau/2}\right| \tag{2.81}$$

and the phase spectral density by

$$\varphi_y(\omega) = \begin{cases} 0 \pm 2\pi n, & \frac{\sin(\omega\tau/2)}{\omega\tau/2} \geqslant 0 \\ \pm\pi \pm 2\pi n, & \frac{\sin(\omega\tau/2)}{\omega\tau/2} < 0 \end{cases}, \quad n = 0, 1, ... \tag{2.82}$$

Both (2.81) and (2.82) are sketched in Fig. 2.18, where the phase is shown for $n = 0$ and we let $+\pi$ for negative frequencies and $-\pi$ for positive. Two limiting cases may now be recognized.

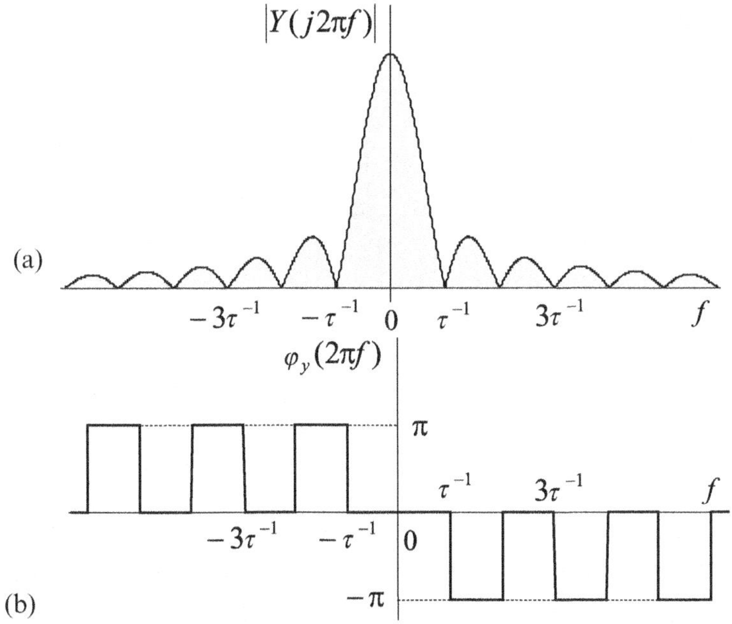

Fig. 2.18 Spectral density of a rectangular pulse: (**a**) magnitude and (**b**) phase.

2.4.2.1 Transition to Delta-pulse

In the limiting cases of $\tau \to 0$ and $A = 1/\tau$, the rectangular pulse $y(t)$ possesses the properties of the delta function $\delta(t)$. Figure 2.19 illustrates this degenerate version of a rectangular pulse. Inherently, by $\tau \to 0$ and $A = 1/\tau$, the spectral density (2.80) becomes uniform with a unit amplitude,

$$Y(j\omega) = \lim_{\tau \to 0} \frac{\sin(\omega\tau/2)}{\omega\tau/2} = 1\,,$$

so that

$$\delta(t) \overset{\mathcal{F}}{\Leftrightarrow} 1 \tag{2.83}$$

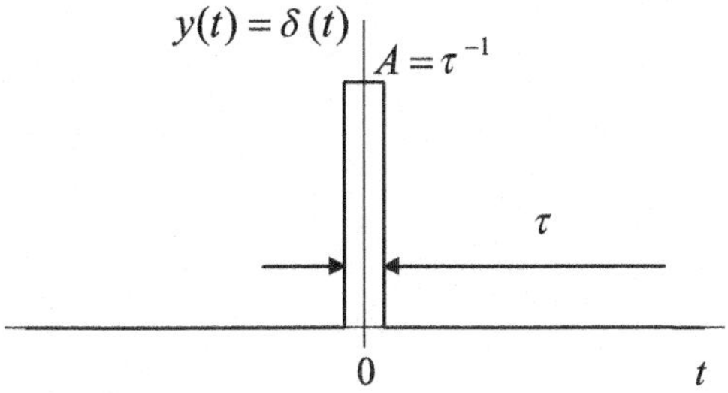

Fig. 2.19 Delta-shaped rectangular pulse.

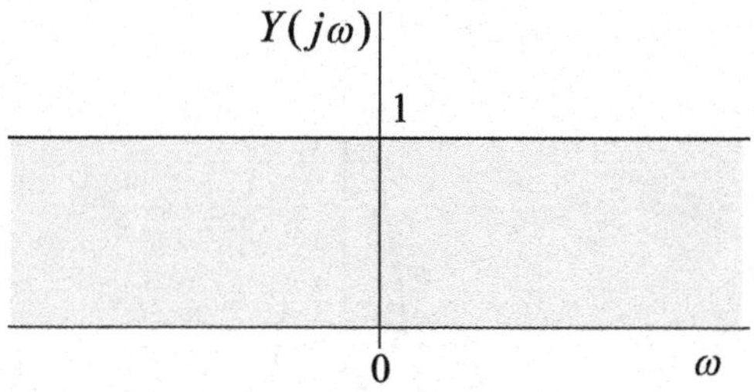

Fig. 2.20 Spectral density of a delta-pulse.

as shown in Fig. 2.20. We then deduce that the *shorter the pulse, the wider its spectral density*, and that in the limiting case we have a uniform unit spectral density (Fig. 2.20) corresponding to the unit pulse (Fig. 2.19).

2.4.2.2 Transition to the Uniform Signal

In the other limiting case, one may suppose that $\tau \to \infty$ and thus the rectangular pulse should degenerate to the uniform signal $y(t) = A$. To define the relevant spectral density, we let $\tau \to \infty$ in (2.80) and use one of the forms of the delta function (1.23) that gives, by $\theta = 0$,

$$\delta(t) = \lim_{\alpha \to \infty} \frac{\alpha}{\pi} \frac{\sin(\alpha t)}{\alpha t} .$$

Applying this relation to (2.80), we go to

$$\lim_{\tau \to \infty} A\tau \frac{\sin(\omega\tau/2)}{\omega\tau/2} = \pi A \delta\left(\frac{\omega}{2}\right) ,$$

change a variable from $\omega/2$ to ω, and arrive at

$$A \overset{\mathcal{F}}{\Leftrightarrow} 2\pi A \delta(\omega) \tag{2.84}$$

that, by $A = 1$, leads to

$$1 \overset{\mathcal{F}}{\Leftrightarrow} 2\pi \delta(\omega)$$

and allows us to think that *the wider the pulse, the shorter its spectral density*, and thus in the limiting case we have a constant signal with the delta-shaped spectral density (2.84).

2.4.3 Triangular Pulse

A symmetric triangular pulse is the other waveform of importance. Unlike the rectangular pulse, the triangular one fits better some physical process, once its slopes are finite. Its function satisfies the Dirichlet condition (2.64) and the Fourier transform thus exists.

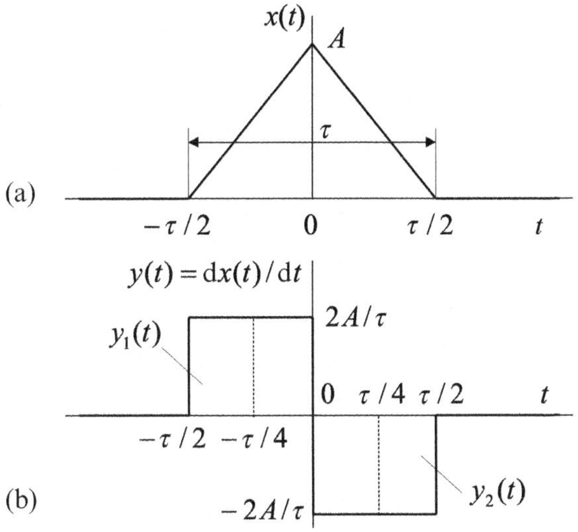

Fig. 2.21 Symmetric triangle pulse: (**a**) waveform and (**b**) time derivative.

Figure 2.21a shows this pulse of duration τ described by

$$x(t) = \frac{2A}{\tau}\left(\frac{\tau}{2} - |t|\right)\left[u\left(t+\frac{\tau}{2}\right) - u\left(t-\frac{\tau}{2}\right)\right]. \quad (2.85)$$

If to take the first time derivative, the triangular waveform becomes rectangular (Fig. 2.21b), and one can easily solve an analysis problem invoking the integration property of the transform. The first time derivative of (2.85) is

$$y(t) = \frac{\mathrm{d}x(t)}{\mathrm{d}t} = y_1(t) + y_2(t), \quad (2.86)$$

where

$$y_1(t) = \frac{2A}{\tau}u\left(t+\frac{\tau}{2}\right) - \frac{2A}{\tau}u(t),$$
$$y_2(t) = -\frac{2A}{\tau}u(t) + \frac{2A}{\tau}u\left(t-\frac{\tau}{2}\right).$$

The spectral densities of $y_1(t)$ and $y_2(t)$ may now be written straightforwardly using (2.80) and the time-shifting property of the transform, respectively,

$$Y_1(j\omega) = \frac{2A}{\tau}\frac{\tau}{2}\frac{\sin(\omega\tau/4)}{\omega\tau/4}\mathrm{e}^{j\frac{\omega\tau}{4}} = A\frac{\sin(\omega\tau/4)}{\omega\tau/4}\mathrm{e}^{j\frac{\omega\tau}{4}}, \quad (2.87)$$

$$Y_2(j\omega) = -\frac{2A}{\tau}\frac{\tau}{2}\frac{\sin(\omega\tau/4)}{\omega\tau/4}\mathrm{e}^{-j\frac{\omega\tau}{4}} = -A\frac{\sin(\omega\tau/4)}{\omega\tau/4}\mathrm{e}^{-j\frac{\omega\tau}{4}}. \quad (2.88)$$

The spectral density of $y(t)$ then becomes

$$Y(j\omega) = Y_1(j\omega) + Y_2(j\omega)$$

$$= A\frac{\sin(\omega\tau/4)}{\omega\tau/4}\left(e^{j\frac{\omega\tau}{4}} - e^{-j\frac{\omega\tau}{4}}\right) = 2Aj\frac{\sin^2(\omega\tau/4)}{\omega\tau/4}$$

$$= j\omega\frac{A\tau}{2}\frac{\sin^2(\omega\tau/4)}{(\omega\tau/4)^2}\,. \tag{2.89}$$

To transfer from (2.89) to the spectral density of $x(t)$, we use the integration property (2.48), note that $Y(j0) = 0$, divide (2.89) with $j\omega$, and finally arrive at

$$X(j\omega) = \frac{A\tau}{2}\frac{\sin^2(\omega\tau/4)}{(\omega\tau/4)^2}\,. \tag{2.90}$$

As it is seen, the spectral density (2.90) is real and positive-valued. Thus, it is also the magnitude spectral density (Fig. 2.22a). Accordingly, the phase spectral density is zero (Fig. 2.22b). An analysis of the density (2.90) shows the following:

- It has a maximum at zero, $X(j0) = A\tau/2$, and becomes zeros at the frequencies

$$f = \frac{2n}{\tau}\,, \quad n = 1, 2, ...$$

- Its side lobes attenuate with frequency by square of a sinc-function.
- By $\tau \to \infty$, it becomes delta-shaped and, by $\tau \to 0$, uniform.

2.4.4 Sinc-shaped Pulse

Thanks to the duality property of the transform, a sinc-pulse is used in applications whenever a rectangular spectral density is needed. Most commonly, its function is described as

$$x(t) = A\frac{\sin(\alpha t)}{\alpha t}\,, \tag{2.91}$$

having a maximum A at zero, $t = 0$, and tending to zero at

$$t = \pm\frac{n\pi}{\alpha}\,, \quad n = 1, 2, ...$$

Figure 2.23 shows this waveform and we realize that the function exists in infinite time bounds and the pulse is thus noncausal.

It is now time to recall that the rectangular pulse (Fig. 2.16) has a sinc-shaped spectral density (Fig. 2.17). Thus, by the duality property, the spectral

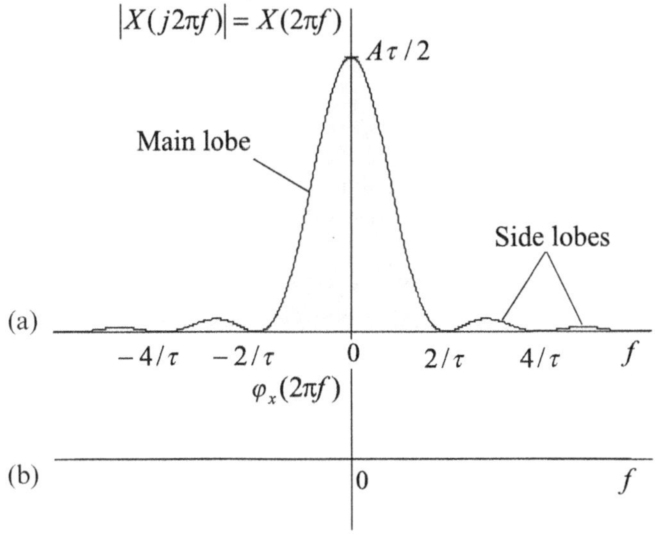

Fig. 2.22 Spectral density of a symmetric triangular pulse: (**a**) magnitude and (**b**) zero phase.

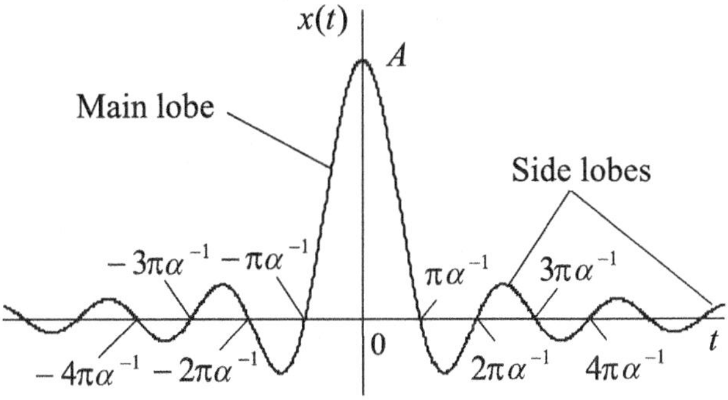

Fig. 2.23 Sinc-shaped pulse.

density of a sinc-pulse (Fig. 2.23) ought to be rectangular with some amplitude B and spectral width $\Delta\omega = 2\pi\Delta f$.

To determine B, we recall that $X(j0)$ is defined by integrating $x(t)$ (2.50). In doing so, we invoke an identity $\int_{-\infty}^{\infty} \sin ax/x \, \mathrm{d}x = \pi$ and arrive at

$$B = X(j0) = A \int_{-\infty}^{\infty} \frac{\sin(\alpha t)}{\alpha t} \mathrm{d}t = A\frac{\pi}{\alpha}. \tag{2.92}$$

Using the other property (2.51) and the value (2.92), we go to

$$x(0) = \frac{1}{2\pi} \int_{-\infty}^{\infty} X(j\omega)\mathrm{d}\omega = \frac{1}{2\pi} \int_{-\Delta\omega/2}^{\Delta\omega/2} B\mathrm{d}\omega = \frac{B}{2\pi}\Delta\omega = \frac{A}{2\alpha}\Delta\omega$$

that, by an equality $x(0) = A = \frac{A}{2\alpha}\Delta\omega$, allows us to define $\Delta\omega = 2\alpha$. The transform of (2.91) is thus given by

$$X(j\omega) = \begin{cases} A\pi/\alpha, & |\omega| < \alpha \\ 0, & |\omega| > \alpha \end{cases} . \tag{2.93}$$

So, the spectral density of a noncausal sinc-pulse is real and positive-valued. It is also the magnitude spectral density and hence the phase is zero with frequency.

One can now wonder if α may be expressed by the pulse duration τ. For the noncausal pulse, not since $\tau \to \infty$ and α merely becomes zero. Frequently, however, they predetermine τ by a distance between two neighboring zero points of the main lobe that gives $\tau = 2\pi/\alpha$. For this value, the coefficient α and spectral width $\Delta\omega$ are defined, respectively, by

$$\alpha = \frac{2\pi}{\tau} \quad \text{and} \quad \Delta\omega = \frac{4\pi}{\tau} . \tag{2.94}$$

The case of $\tau = 2\pi/\alpha$ is illustrated in Fig. 2.24 and we see that the Gibbs phenomenon accompanies the spectral density, since the integration range is finite, $\tau < \infty$. We thus come to the conclusion that the spectral density of a finite sinc-pulse is not positive-valued and thus its phase is not zero over the frequencies.

2.4.5 Gaussian Pulse

So far, we examined nonperiodic signals, which waveforms and transforms have different shapes. There is, however, a noble pulse that has equal presentations in time and frequency. This pulse is Gaussian, which is specified by

$$x(t) = A\mathrm{e}^{-\alpha^2 t^2} , \tag{2.95}$$

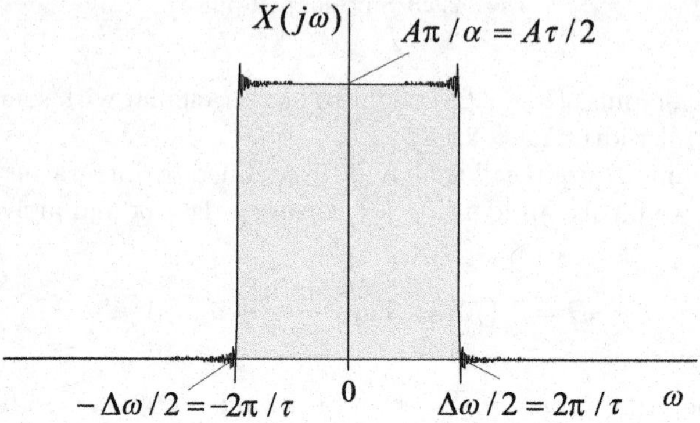

Fig. 2.24 Spectral density of a sinc-shaped pulse for $\tau = 2\pi/\alpha < \infty$.

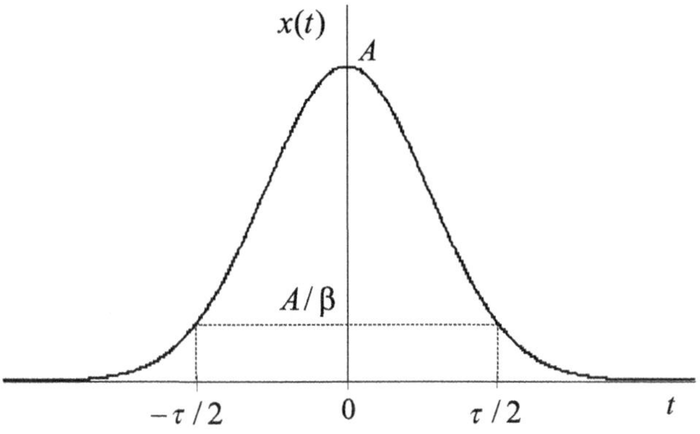

Fig. 2.25 Gaussian pulse.

where α is an arbitrary attenuation coefficient and A is constant (Fig. 2.25). Like a sinc-pulse, the Gaussian one exists in infinite time bounds and is thus noncausal, even though attenuates with time faster.

In solving an analysis problem, we apply (2.30) to (2.95), make the transformations, and first go to

$$X(j\omega) = A\int_{-\infty}^{\infty} e^{-\alpha^2t^2}e^{-j\omega t}dt = A\int_{-\infty}^{\infty} e^{-\alpha^2t^2-j\omega t-\left(\frac{j\omega}{2\alpha}\right)^2+\left(\frac{j\omega}{2\alpha}\right)^2}dt$$

$$= Ae^{-\frac{\omega^2}{4\alpha^2}}\int_{-\infty}^{\infty} e^{-\left(\alpha t+j\frac{\omega}{2\alpha}\right)^2}dt\,.$$

By an identity $\int_{-\infty}^{\infty} e^{-z^2}dz = \sqrt{\pi}$, the function becomes

$$X(j\omega) = \frac{A\sqrt{\pi}}{\alpha}e^{-\frac{\omega^2}{4\alpha^2}}\,. \tag{2.96}$$

Even a quick comparison of (2.95) and (2.96) shows that the functions are both Gaussian. This means that (2.96) also occupies an infinite range in the frequency domain. For $\alpha > 0$, the function (2.96) is real and positive-valued; therefore, it is also the magnitude spectral density. The phase is zero over all frequencies.

2.4.5.1 Pulse Duration

Again we wonder if duration τ of a Gaussian pulse may somehow be specified to express α by τ. To predetermine τ, we may allow for some level A/β, $\beta > 1$, corresponding to the assumed pulse time bounds $\pm\tau/2$ (Fig. 2.25). Thereafter, an equality of (2.95) with $t = \tau/2$ and A/β,

$$Ae^{-\left(\frac{\alpha\tau}{2}\right)^2} = \frac{A}{\beta},$$

will produce

$$\tau = 2\frac{\sqrt{\ln\beta}}{\alpha} \quad \text{or} \quad \alpha = 2\frac{\sqrt{\ln\beta}}{\tau}.$$

So, if some β is allowed by practical reasons, the Gaussian pulse will be specified with the above given τ and α. We notice that the spectral density of such a bounded pulse will not be absolutely Gaussian, but rather a near Gaussian.

2.4.5.2 Spectral Bounds

In a like manner, the spectral density of a Gaussian pulse may also be bounded, as it is shown in Fig. 2.26. If we let $\omega = \Delta\omega/2$ in (2.96) for the level $A\sqrt{\pi}/\alpha\gamma$, where $\gamma > 1$, then an equality

$$\frac{A\sqrt{\pi}}{\alpha}e^{-\frac{\Delta\omega^2}{16\alpha^2}} = \frac{A\sqrt{\pi}}{\alpha\gamma}$$

will produce a spectral bound (Fig. 2.26)

$$\frac{\Delta\omega}{2} = 2\alpha\sqrt{\ln\gamma}.$$

Again we notice that for the bounded spectral density, the time waveform will not be purely Gaussian.

Example 2.14. The Gaussian pulse is performed with $x(t) = Ae^{-[\alpha(t-t_0)]^2}$, $A = 2$ V, $t_0 = 10^{-2}$ s, and $\alpha = 2\,\mathrm{s}^{-1}$. By (2.40), its spectral density calculates $X(j\omega) = \sqrt{\pi}e^{-\frac{\omega^2}{16}}e^{-j10^{-2}\omega}$. Accordingly, the amplitude and phase spectral

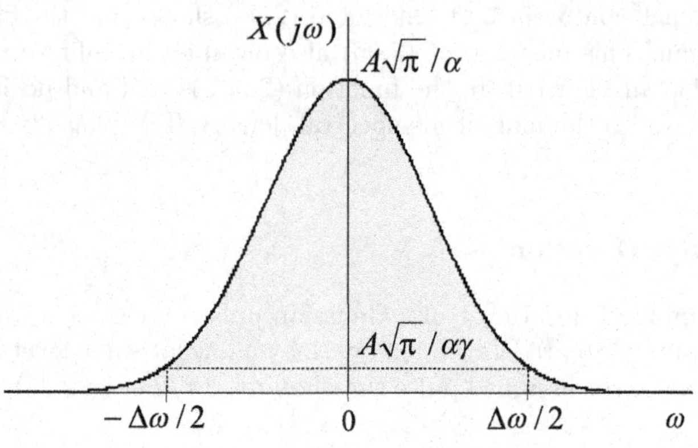

Fig. 2.26 Spectral density of a Gaussian pulse.

characteristics are, $|X(j\omega)| = \sqrt{\pi}e^{-\frac{\omega^2}{16}}$ and $\varphi_x(\omega) = -10^{-2}\omega$, respectively. Since the phase is shifted, then its phase spectral density is not zero, unlike the symmetric pulse case.

□

2.5 Spectral Densities of Complex Single Pulses

We now continue on in learning complex single pulses, which waveforms are often combined by simple and elementary pulses. The methodology, on the whole, remains the same. Before solving an analysis problem, it is recommended, first, to compose the pulse with others, which spectral densities are known. Then use the properties of the transform and go to the required spectral density. If a decomposition is not available, an application of the direct Fourier transform and its properties has no alternative.

2.5.1 Complex Rectangular Pulse

Consider a complex pulse $y(t)$ composed with two rectangular pulses $x_1(t)$ and $x_2(t)$ of the same duration $\tau/2$ and different amplitudes, B and A, respectively (Fig. 2.27), where

$$y(t) = x_1(t) + x_2(t)\,, \tag{2.97}$$

$$x_1(t) = Bu\left(t + \frac{\tau}{2}\right) - Bu(t)\,, \tag{2.98}$$

$$x_2(t) = Au(t) - Au\left(t - \frac{\tau}{2}\right)\,. \tag{2.99}$$

Using (2.80) and employing the time-shift property of the transform, we perform the spectral densities of $x_1(t)$ and $x_2(t)$, respectively, by

$$X_1(j\omega) = \frac{B\tau}{2}\frac{\sin(\omega\tau/4)}{\omega\tau/4}e^{j\frac{\omega\tau}{4}}\,, \tag{2.100}$$

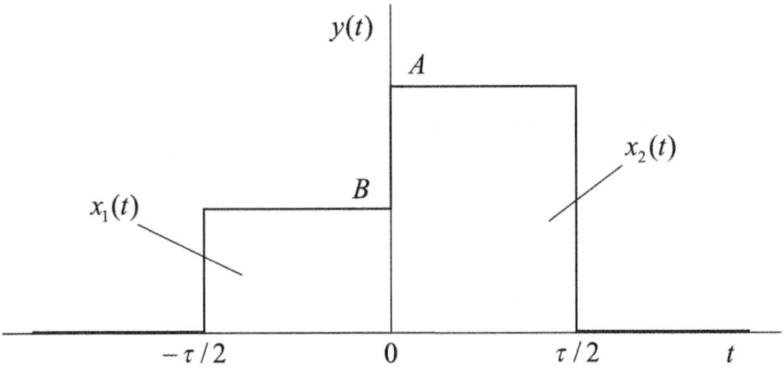

Fig. 2.27 A complex rectangular pulse.

$$X_2(j\omega) = \frac{A\tau}{2}\frac{\sin(\omega\tau/4)}{\omega\tau/4}\mathrm{e}^{-j\frac{\omega\tau}{4}}. \tag{2.101}$$

The Fourier transform of $y(t)$ is then written as

$$Y(j\omega) = X_1(j\omega) + X_2(j\omega) = \frac{\tau}{2}\frac{\sin(\omega\tau/4)}{\omega\tau/4}\left(B\mathrm{e}^{j\frac{\omega\tau}{4}} + A\mathrm{e}^{-j\frac{\omega\tau}{4}}\right) \tag{2.102}$$

and the Euler formula, $\mathrm{e}^{jz} = \cos z + j\sin z$, produces the final expression for the spectral density

$$\begin{aligned} Y(j\omega) &= \frac{\tau}{2}\frac{\sin(\omega\tau/4)}{\omega\tau/4}\left[(A+B)\cos\frac{\omega\tau}{4} - j(A-B)\sin\frac{\omega\tau}{4}\right] \\ &= |Y(j\omega)|\,\mathrm{e}^{-j\varphi_y(\omega)}\,, \end{aligned} \tag{2.103}$$

specifying the magnitude and phase responses, respectively, by

$$|Y(j\omega)| = \frac{\tau}{2}\left|\frac{\sin(\omega\tau/4)}{\omega\tau/4}\right|\sqrt{A^2 + B^2 + 2AB\cos\frac{\omega\tau}{2}}\,, \tag{2.104}$$

$$\varphi_y(\omega) = \begin{cases} \arctan\left(\frac{A-B}{A+B}\tan\frac{\omega T}{4}\right), & \frac{\sin(\omega\tau/4)}{\omega\tau/4} \geqslant 0 \\ \arctan\left(\frac{A-B}{A+B}\tan\frac{\omega T}{4}\right) \pm \pi, & \frac{\sin(\omega\tau/4)}{\omega\tau/4} < 0 \end{cases}. \tag{2.105}$$

Both (2.104) and (2.105) allow for tracing a transition of the spectral density from one state to the other if the amplitudes A and B are changed.

In particular, Fig. 2.28 shows what will go on with the magnitude spectral density if to vary B from 0 to A with the fixed $A = 1$ and $\tau = 1$. It is seen that $|Y(j\omega)|$ evolves from $\frac{A\tau}{2}\left|\frac{\sin(\omega\tau/4)}{\omega\tau/4}\right|$ with $B = 0$ to $A\tau\left|\frac{\sin(\omega\tau/2)}{\omega\tau/2}\right|$ when $B = A$.

Depending on the relationship between the amplitudes A and B, several important limiting cases may now be recognized:

- **Even rectangular pulse.** By $A = B > 0$, the pulse evolves to the even rectangular pulse (Fig. 2.16c) with the spectral density (2.80).
- **Odd rectangular pulse.** Supposing $A = -B > 0$, we arrive at the odd rectangular pulse that is orthogonal to the even one (Fig. 2.16c). Its spectral density is given by

$$Y(j\omega) = j\omega\frac{B\tau^2}{4}\frac{\sin^2(\omega\tau/4)}{(\omega\tau/4)^2}.$$

- **Shifted rectangular pulse.** If $A = 0$ and $B \neq 0$ or $A \neq 0$ and $B = 0$, the pulse becomes rectangular and shifted with the spectral density, respectively,

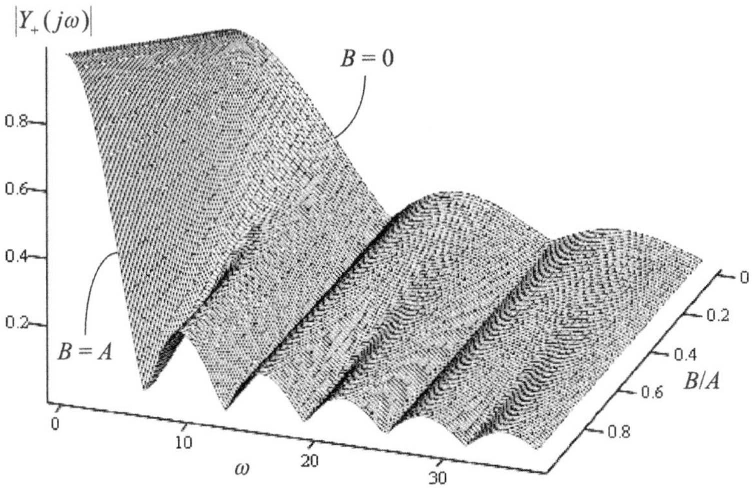

Fig. 2.28 Evolution of the magnitude spectral density (2.104) by changing B with $A = 1$ and $\tau = 1$.

$$Y(j\omega) = \frac{B\tau}{2}\frac{\sin(\omega\tau/4)}{\omega\tau/4}e^{j\frac{\omega\tau}{4}},$$

$$Y(j\omega) = \frac{A\tau}{2}\frac{\sin(\omega\tau/4)}{\omega\tau/4}e^{-j\frac{\omega\tau}{4}}.$$

Example 2.15. Given a complex pulse (Fig. 2.27) having a ratio $A/B = 0.5$. By (2.104), its magnitude spectral density becomes

$$|Y(j\omega)| = \frac{A\tau}{2}\left|\frac{\sin(\omega\tau/4)}{\omega\tau/4}\right|\sqrt{\frac{5}{4}+\cos\left(\frac{\omega\tau}{2}\right)} = Z_1(\omega)Z_2(\omega), \qquad (2.106)$$

where $Z_1(\omega) = \frac{A\tau}{2}\left|\frac{\sin(\omega\tau/4)}{\omega\tau/4}\right|$ and $Z_2(\omega) = \sqrt{\frac{5}{4}+\cos\left(\frac{\omega\tau}{2}\right)}$. Figure 2.29 demonstrates how the pulse asymmetry affects its transform. It follows that extra oscillations in (2.106) are coursed by the multiplier $Z_2(\omega)$ and associated with the pulse asymmetry. It may be shown that $Z_2(\omega)$ becomes constant (frequency-invariant) when $B = 0$ and $B = A$.

□

We notice that the magnitude spectral density of the asymmetric complex rectangular pulse is still symmetric and its phase spectral density acquires the slope caused by asymmetry.

2.5.2 Trapezoidal Pulse

A special feature of the trapezoidal pulse is that it occupies an intermediate place between the rectangular and the triangular pulses possessing their

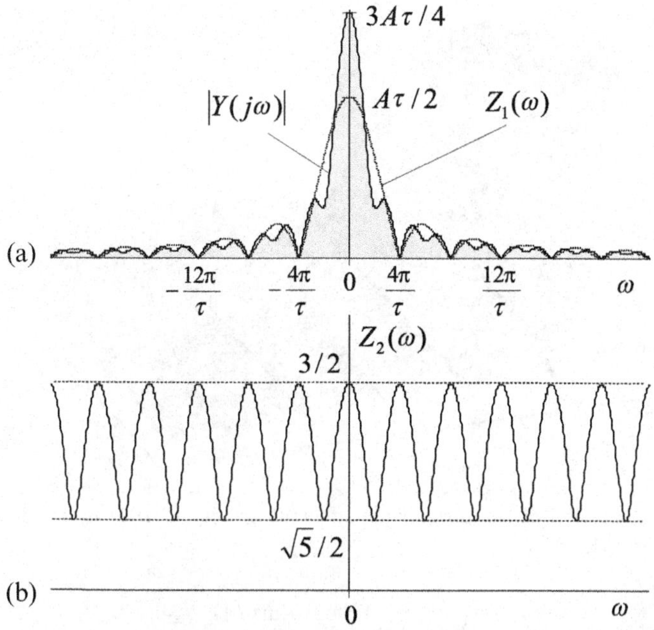

Fig. 2.29 Magnitude spectral density (2.106): **(a)** $|Y(j\omega)|$ and $Z_1(\omega)$, and **(b)** $Z_2(\omega)$.

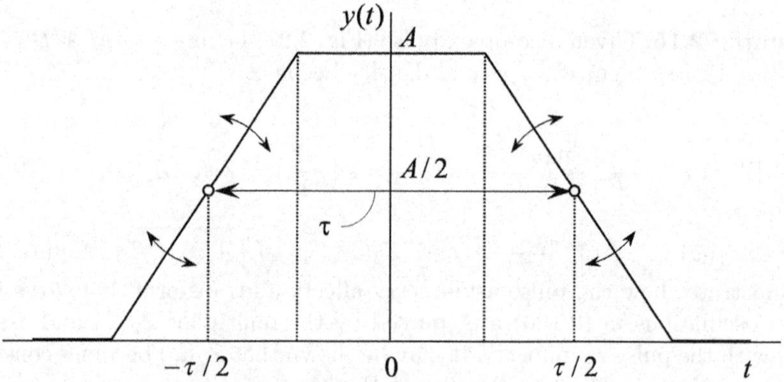

Fig. 2.30 Trapezoidal pulse.

waveforms in particular cases. Yet, it simulates the rectangular pulse with finite slopes of the sides that fits better practical needs.

To solve an analysis problem properly, we fix the amplitude A and suppose that duration τ is constant for the given level $A/2$ (Fig. 2.30). We also allow for the sides to be synchronously rotated about the fixed points (circles) to change the waveform from triangular to rectangular.

A differentiation of this pulse leads to Fig. 2.31. The Fourier transform of the differentiated pulse is specified by the sum of two shifted rectangular

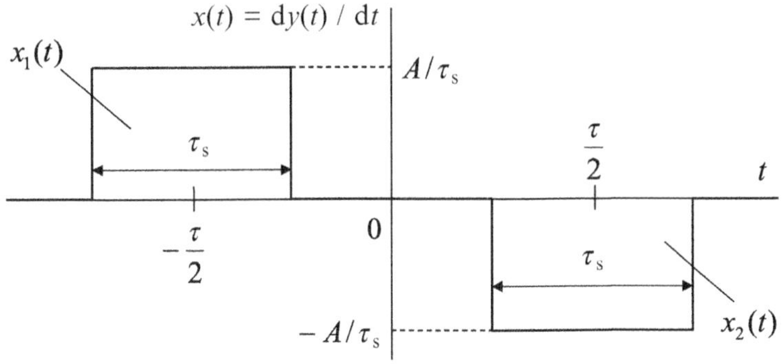

Fig. 2.31 Differentiated trapezoidal pulse.

pulses, which duration changes from 0 to τ and thus the amplitudes range from infinity to $\pm A/\tau$, respectively. The integration property of the transform leads finally to the required spectral density of $y(t)$,

$$\begin{aligned} Y(j\omega) &= \frac{1}{j\omega}\left[X_1(j\omega) + X_2(j\omega)\right] \\ &= \frac{A}{j\omega}\frac{\sin(\omega\tau_s/2)}{\omega\tau_s/2}\left(e^{j\frac{\omega\tau}{2}} - e^{-j\frac{\omega\tau}{2}}\right) \\ &= A\tau\frac{\sin(\omega\tau/2)}{\omega\tau/2}\frac{\sin(\omega\tau_s/2)}{\omega\tau_s/2}\,. \end{aligned} \tag{2.107}$$

It is seen that (2.107) corresponds to the rectangular pulse (2.80) if τ_s reaches zero and it degenerates to the spectral density of a triangular pulse (2.90) of duration τ when $\tau_s = \tau$. A transition between these two waveforms is illustrated in Fig. 2.32. One may observe (see Example 2.16) that there is an intermediate value $\tau_s/\tau = \sqrt{2}/2$ that suppresses the side lobes even more than in the triangular pulse.

Example 2.16. Given a trapezoidal pulse (Fig. 2.30) with the ratio $\tau_s/\tau = 1/\sqrt{2}$. By (2.107), its spectral density becomes

$$Y(j\omega) = A\tau\frac{\sin(\omega\tau/2)}{\omega\tau/2}\frac{\sin\left(\omega\tau/2\sqrt{2}\right)}{\omega\tau/2\sqrt{2}} \tag{2.108}$$

leading to the following magnitude and phase spectral densities, respectively,

$$|Y(j\omega)| = A\tau\left|\frac{\sin(\omega\tau/2)}{\omega\tau/2}\frac{\sin\left(\omega\tau/2\sqrt{2}\right)}{\omega\tau/2\sqrt{2}}\right|, \tag{2.109}$$

$$\varphi_y(\omega) = \begin{cases} \pm 2n\pi, & \frac{\sin(\omega\tau/2)}{\omega\tau/2}\frac{\sin\left(\omega\tau/2\sqrt{2}\right)}{\omega\tau/2\sqrt{2}} < 0 \\ 0, & \frac{\sin(\omega\tau/2)}{\omega\tau/2}\frac{\sin\left(\omega\tau/2\sqrt{2}\right)}{\omega\tau/2\sqrt{2}} \geqslant 0 \end{cases}. \tag{2.110}$$

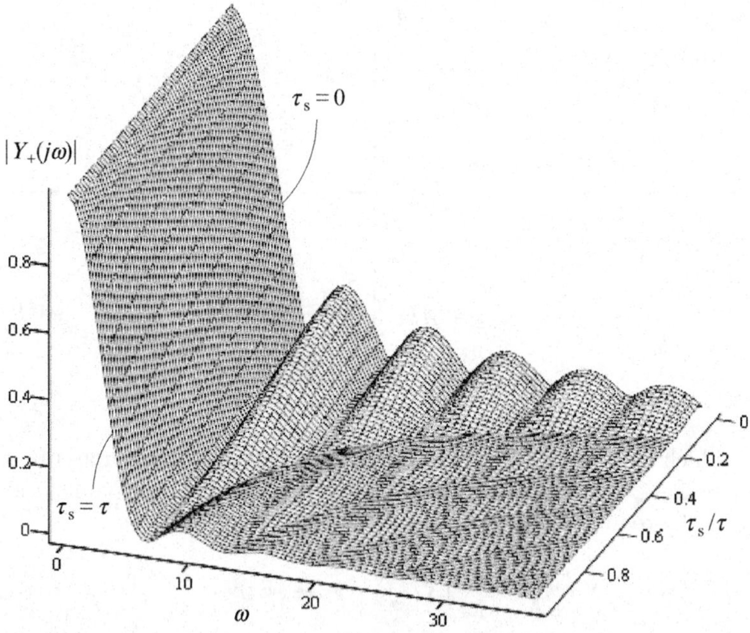

Fig. 2.32 Evolution of the magnitude spectral density (2.107) by changing τ_s with $A = 1$ and $\tau = 1$.

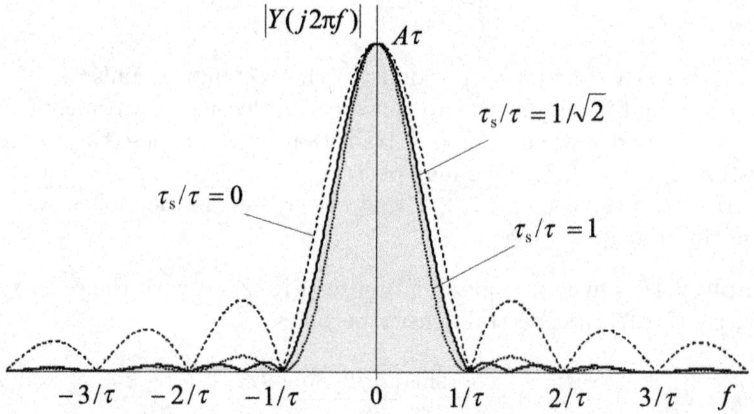

Fig. 2.33 Magnitude spectral density of the trapezoidal pulse with $\tau_s/\tau = 1/\sqrt{2}$: $\tau_s/\tau = 0$ fits the rectangular pulse and $\tau_s/\tau = 1$ corresponds to the triangular pulse.

Figure 2.33 illustrates (2.109) along with the densities associated with $\tau_s/\tau = 0$ (rectangular pulse) and $\tau_s/\tau = 1$ (triangular pulse). It follows that having $\tau_s/\tau = 1/\sqrt{2}$ results in the most efficient suppression of the side lobes, so that their level becomes even lower than in the triangular pulse.

□

Overall, the trapezoidal waveform demonstrates the following useful properties: (1) its main lobe is a bit narrower and side lobes trace much lower than in the rectangular pulse and (2) its main lobe is a bit wider and the nearest side lobes are still better suppressed than in the triangular pulse.

2.5.3 Asymmetric Triangular Pulse

Besides its own interesting properties and practical usefulness, an asymmetric triangular pulse demonstrates an ability to transfer to several other important waveforms.

Consider Fig. 2.34a. Like other pulses with linear sides, this one $y(t)$ may be differentiated that leads to the complex rectangular waveform $x(t)$ (Fig. 2.34a). Similarly to (2.107), we write the spectral density of $x(t)$, invoke the integration property of the transform, make the transformations, and arrive at the spectral density of $y(t)$,

$$\begin{aligned} Y(j\omega) &= \frac{A}{j\omega}\left[\frac{\sin(\omega\tau/2)}{\omega\tau/2}e^{j\frac{\omega\tau}{2}} - \frac{\sin(\omega\tau_s/2)}{\omega\tau_s/2}e^{-j\frac{\omega\tau_s}{2}}\right] \\ &= \frac{A}{2}\left[\tau\frac{\sin^2(\omega\tau/2)}{(\omega\tau/2)^2} + \tau_s\frac{\sin^2(\omega\tau_s/2)}{(\omega\tau_s/2)^2}\right] \\ &\quad -j\frac{A}{\omega}\left[\frac{\sin(\omega\tau)}{\omega\tau} - \frac{\sin(\omega\tau_s)}{\omega\tau_s}\right]. \end{aligned} \tag{2.111}$$

To determine $Y(j\omega)$ at $\omega = 0$, one needs to solve an uncertainty 0/0 in the imaginary part of (2.111). Then use an extension $\sin x = x - \frac{x^3}{3!} + ...$, by which this part becomes zero at $\omega = 0$ and hence

$$Y(j0) = A\frac{\tau + \tau_s}{2}.$$

Now recall that the phase spectral density of a nonperiodic signal is zero at $\omega = 0$. Therefore, the imaginary part in (2.111) must be zero at $\omega = 0$, by definition. Several useful transitions to other waveforms may now be observed.

2.5.3.1 Transition to Symmetric Triangular Pulse

By $\tau_s = \tau$, the triangular pulse becomes symmetric and (2.111) readily transforms to

$$Y(j\omega) = A\tau\frac{\sin^2(\omega\tau/2)}{(\omega\tau/2)^2}$$

that is consistent with (2.90) earlier derived for the twice shorter duration.

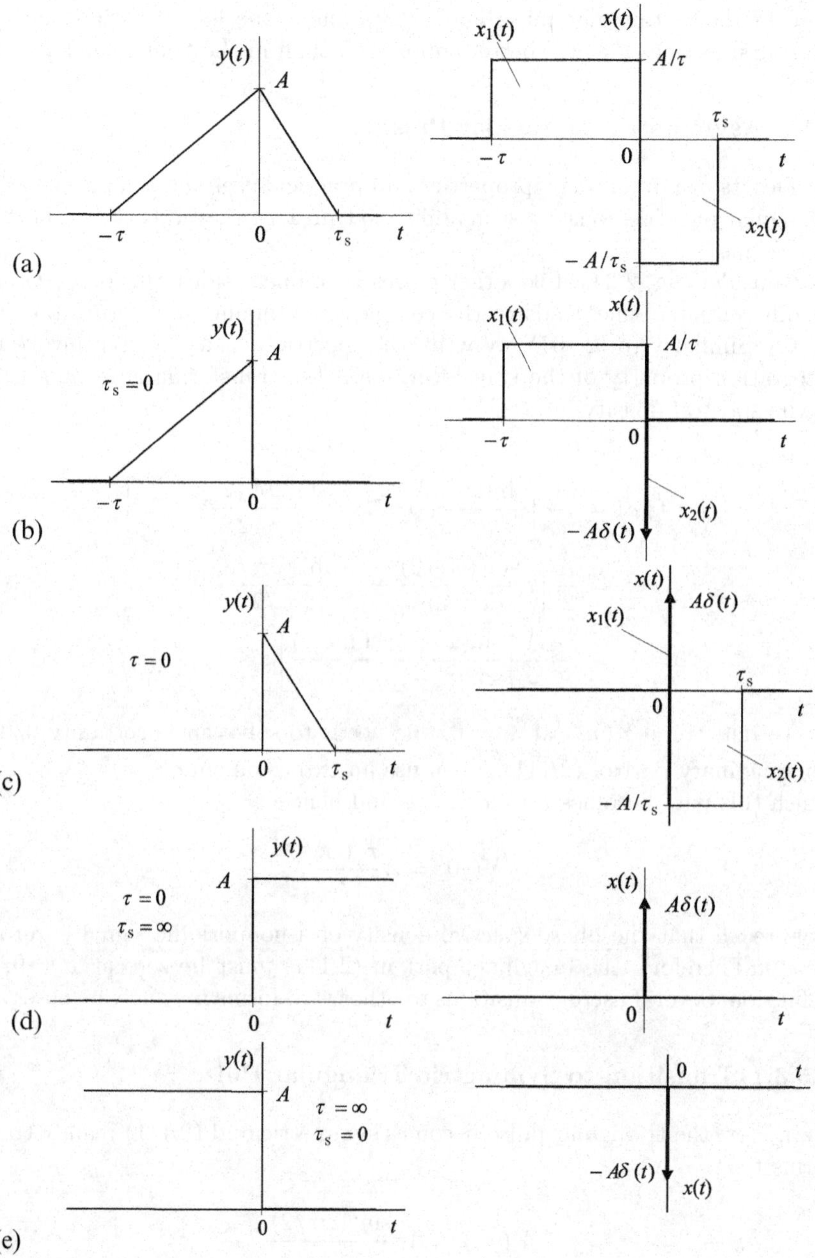

Fig. 2.34 Waveforms (left) and time derivatives (right) of an asymmetric triangular pulse: (**a**) $\tau \neq \tau_s$, (**b**) $\tau_s = 0$, (**c**) $\tau = 0$, (**d**) $\tau = 0$ and $\tau_s = \infty$, and (**e**) $\tau = \infty$ and $\tau_s = 0$.

2.5.3.2 Transition to Truncated Ramp Pulse (Positive Slope)

By $\tau_s \to 0$, the triangular pulse becomes truncated ramp with a positive slope (Fig. 2.34b) and its time derivative is composed with the rectangular and delta pulses (Fig. 2.34b). The spectral density (2.111) thus transforms to

$$\begin{aligned} Y(j\omega) &= \frac{A}{j\omega}\left[\frac{\sin(\omega\tau/2)}{\omega\tau/2}e^{j\frac{\omega\tau}{2}} - 1\right] \\ &= \frac{A\tau}{2}\frac{\sin^2(\omega\tau/2)}{(\omega\tau/2)^2} - j\frac{A}{\omega}\left[\frac{\sin(\omega\tau)}{\omega\tau} - 1\right] \end{aligned} \quad (2.112)$$

and the magnitude spectral density calculates

$$|Y(j\omega)| = \frac{A\tau}{2}\sqrt{\frac{\sin^4(\omega\tau/2)}{(\omega\tau/2)^4} + \frac{4}{\omega^2\tau^2}\left[\frac{\sin(\omega\tau)}{\omega\tau} - 1\right]^2}\,. \quad (2.113)$$

2.5.3.3 Transition to Truncated Ramp Pulse (Negative Slope)

If $\tau \to 0$, we have a truncated ramp pulse with a negative slope (Fig. 2.34c) having the time derivative as in (Fig. 2.34c). Here, the spectral density (2.111) degenerates to

$$Y(j\omega) = \frac{A}{j\omega}\left[1 - \frac{\sin(\omega\tau_s/2)}{\omega\tau_s/2}e^{-j\frac{\omega\tau_s}{2}}\right]. \quad (2.114)$$

2.5.3.4 Transition to the Unit-step Pulse $u(t)$

When we let $\tau \to 0$ and $\tau_s \to \infty$, the pulse goes to the unit step (Fig. 2.34d), which time derivative is shown in Fig. 2.34d. Accordingly, (2.111) becomes (2.72),

$$Y(j\omega) = \frac{A}{j\omega} + A\pi\delta(\omega)\,.$$

2.5.3.5 Transition to Reverse Unit-step Pulse $u(-t)$

The case of $\tau_s \to 0$ and $\tau \to \infty$ produces the reverse unit-step function (Fig. 2.34e) and its time derivative (Fig. 2.34e). Hereby, (2.111) transforms to

$$Y(j\omega) = -\frac{A}{j\omega} + A\pi\delta(\omega)\,.$$

Figure 2.35 shows an evolution of the magnitude spectral density associated with (2.111) if to fix τ and vary τ_s/τ from 0 to 1. It is neatly seen that asymmetry ($\tau \neq \tau_s$) smoothes the side lobes but increases their level up to

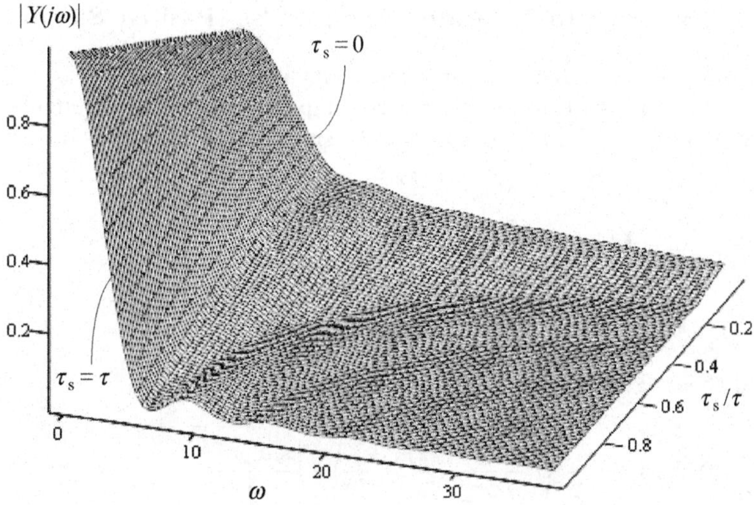

Fig. 2.35 Evolution of the magnitude spectral density associated with (2.111) for $A = 1$, $\tau = 1$, and $\tau_s = \text{var}$.

the maximum value with $\tau_s = 0$. Moreover, the peak value of the main lobe reduces. We then conclude that, on the whole, asymmetry does not produce a positive effect in the spectral performance of a triangular pulse. One exception, however, may be pointed out (Example 2.17). By $\tau_s = \tau/\sqrt{2}$, the spectral density diminishes almost monotonously and with almost the same level of the side lobes as in the symmetric pulse.

Example 2.17. Given an asymmetric triangular pulse with $\tau_s = \tau/\sqrt{2}$ (Fig. 2.34a). By (2.111), its spectral density becomes

$$Y(j\omega) = \frac{A\tau}{2}\left[\frac{\sin^2(\omega\tau/2)}{(\omega\tau/2)^2} + \frac{1}{\sqrt{2}}\frac{\sin^2(\omega\tau/2\sqrt{2})}{(\omega\tau/2\sqrt{2})^2}\right] - j\frac{A}{\omega}\left[\frac{\sin(\omega\tau)}{\omega\tau} - \frac{\sin(\omega\tau/\sqrt{2})}{\omega\tau/\sqrt{2}}\right]. \quad (2.115)$$

Figure 2.36 illustrates (2.115) along with the magnitude densities corresponding to $\tau_s = 0$ (truncated ramp function) and $\tau_s = \tau$ (symmetric triangular pulse). It follows that (2.115) contributes for with a smoothed shape, minimum level of side lobes, and intermediate peak value at zero.

□

2.6 Spectrums of Periodic Pulses

In line with single pulses, *periodic signals* $x(t) = x(t \pm nT)$ of various waveforms are used widely in electronic, radio, and wireless systems. The *periodic*

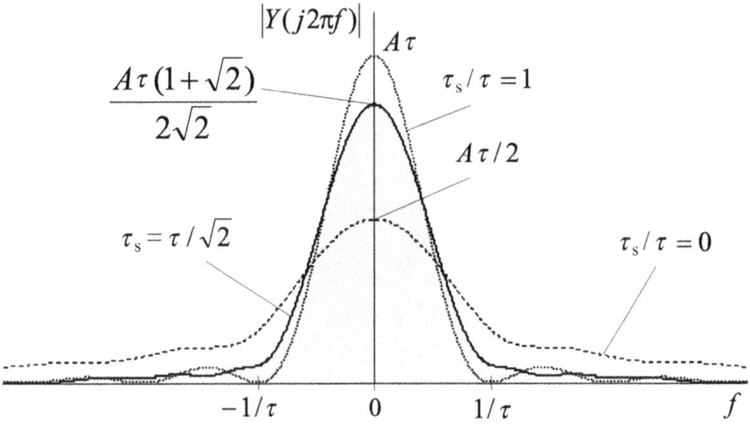

Fig. 2.36 Magnitude spectral densities of an asymmetric triangular pulse for $\tau_s = \tau/\sqrt{2}$, $\tau_s/\tau = 0$, and $\tau_s/\tau = 1$.

pulse signal is also called the *periodic pulse* or *periodic pulse-train.* It may be *causal* (finite or real physical) or *noncausal* (infinite or mathematical). Spectral analysis of periodic signals is based on the Fourier series (2.14) involving the spectral density of its waveform. Analysis and synthesis problems are solved here, respectively, by

$$C_k = \frac{1}{T}X(j2\pi kF), \tag{2.116}$$

$$x(t) = \frac{1}{T}\sum_{k=-\infty}^{\infty} X(j2\pi kF)e^{j2\pi kFt}, \tag{2.117}$$

where the continuous angular frequency ω is taken at discrete points $\omega_k = 2\pi kF = 2\pi k/T$. The coefficients C_k characterizes a complex amplitude of the kth spectral line as a function of kF, meaning that the spectral lines of the *magnitude* and *phase* spectra appear at the frequencies

$$f_k = \pm\frac{k}{T} = \pm kF, \quad k = 0, 1, \ldots \tag{2.118}$$

Here, $F = 1/T$ is called the *fundamental* (or *principle*) *frequency* that is a minimum distance between the spectral lines.

The coefficient C_k may be rewritten as (2.16),

$$C_k = |C_k|\, e^{-j\Psi_k}, \tag{2.119}$$

where $|C_k|$ and Ψ_k are the magnitude and phase spectrums, respectively. Then the synthesis problem (2.117),

$$x(t) = \sum_{k=-\infty}^{\infty} C_k e^{j2\pi kFt}, \tag{2.120}$$

for the one-sided spectrum is solved by

$$x(t) = C_0 + 2\sum_{k=1}^{\infty} |C_k| \cos(2\pi kFt - \Psi_k)\,, \tag{2.121}$$

where C_0 is the magnitude with $k = 0$.

2.6.1 Periodic Rectangular Pulse

A periodic rectangular pulse-train is a classical test signal that is characterized by the pulse duration τ, amplitude A, and period of repetition T (Fig. 2.37).

For the known spectral density (2.80), an analysis problem is solved here by

$$C_k = A\frac{\tau}{T}\frac{\sin(k\pi F\tau)}{k\pi F\tau} = \frac{A}{q}\frac{\sin(k\pi/q)}{k\pi/q}\,, \tag{2.122}$$

where $q = T/\tau$ is a *period-to-pulse duration ratio* and a reciprocal of q is called a *duty cycle*. Accordingly, the *magnitude spectrum* $|C_k|$ and *phase spectrum* Ψ_k are defined by

$$|C_k| = \frac{A}{q}\left|\frac{\sin(k\pi/q)}{k\pi/q}\right|\,, \tag{2.123}$$

$$\Psi_k = \begin{cases} \pm 2\pi n, & \frac{\sin(k\pi/q)}{k\pi/q} \geqslant 0 \\ \pm\pi \pm 2\pi n, & \frac{\sin(k\pi/q)}{k\pi/q} < 0 \end{cases}, \quad n = 0, 1, \ldots \tag{2.124}$$

The synthesis problem is solved in the form of

$$x(t) = \frac{A}{q} + \sum_{k=1}^{\infty} \frac{2A}{q}\left|\frac{\sin(k\pi/q)}{k\pi/q}\right| \cos(2\pi kFt - \Psi_k)\,. \tag{2.125}$$

One may now observe that the spectrum envelope is shaped by the scaled spectral density of a single rectangular pulse. Since the spectrum reaches zero

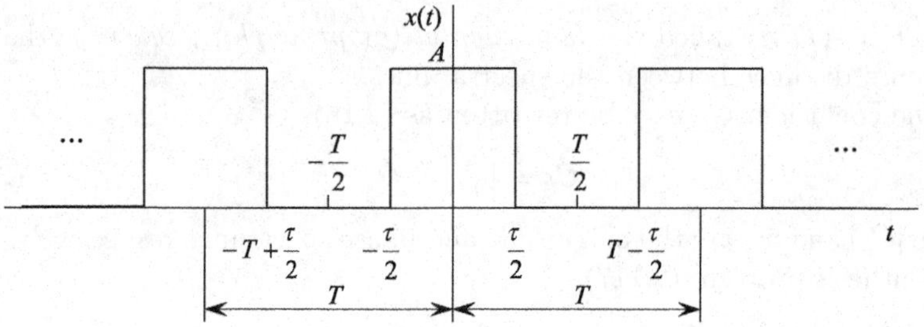

Fig. 2.37 Rectangular pulse-train.

when $\sin k\pi = 0$, then the relevant "zero" frequencies f_n are obtained by the relation

$$\frac{k\pi}{q} = k\pi\frac{\tau}{T} = \pi(kF)\tau = \pi f_n\tau = \pm n\pi\,, \quad n = 1, 2, ... \tag{2.126}$$

to yield

$$f_n = \pm\frac{n}{\tau}\,. \tag{2.127}$$

Bounded by the envelope, the spectral lines appear at the frequencies (2.118).

The spectrum of a rectangular pulse-train may now be shown in the form given in Fig. 2.5 that we do in the following examples.

Example 2.18. Given a rectangular pulse-train (Fig. 2.37) with $q = 3$. By (2.123) and (2.124) with $n = 0$, its magnitude and phase spectrums, $|C_k|$ and Ψ_k, become as in Fig. 2.38a and b, respectively.

□

Example 2.19. Given a rectangular pulse-train (Fig. 2.37) with $q = 7.4$ that is shifted on $\tau/2$. Its magnitude spectrum is calculated by (2.123). Its phase spectrum, by applying the time-shifting property of the Fourier series

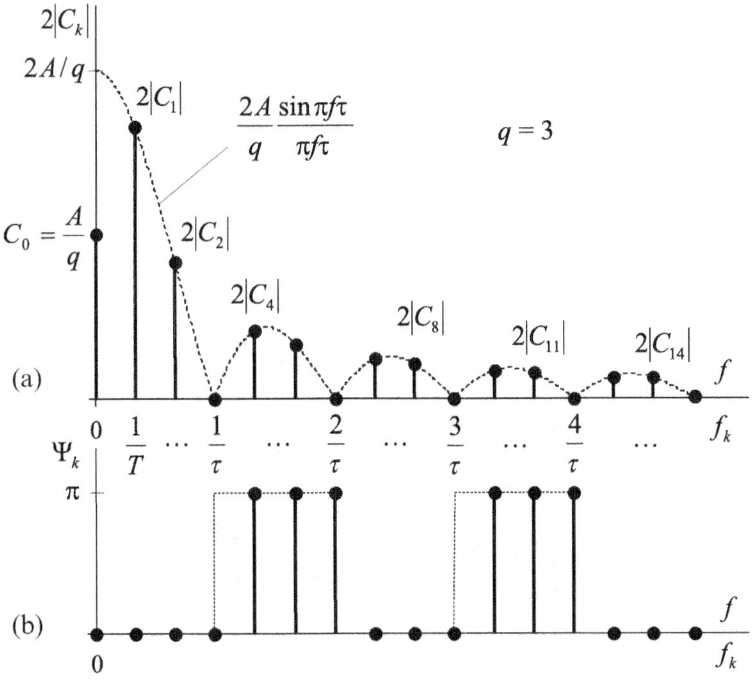

Fig. 2.38 One-sided spectrum of a rectangular pulse-train for $q = 3$: **(a)** magnitude and **(b)** phase.

to (2.124), becomes, by $n = 0$,

$$\Psi_k = \begin{cases} k\pi/q, & \frac{\sin(k\pi/q)}{k\pi/q} \geqslant 0 \\ k\pi/q + \pi, & \frac{\sin(k\pi/q)}{k\pi/q} < 0 \end{cases} . \tag{2.128}$$

Figure 2.39 illustrates the spectrum of this signal. It follows that the total phase ranges within infinite bounds having a constant slope coursed by the time shift $\tau/2$.

□

2.6.2 Triangular Pulse-train

A symmetric triangular pulse-train is shown in Fig. 2.40. By the spectral density (2.89), an analysis problem is solved to yield

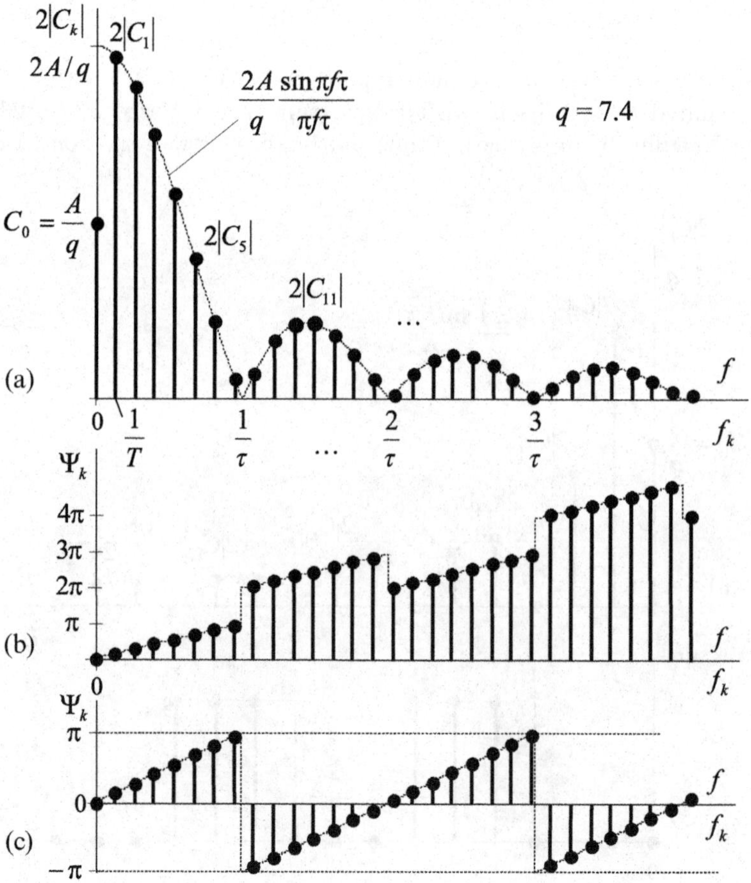

Fig. 2.39 One-sided spectrum of a rectangular pulse-train for $q = 7.4$ and time shift $\tau/2$: **(a)** magnitude, **(b)** total phase, and **(c)** modulo 2π phase.

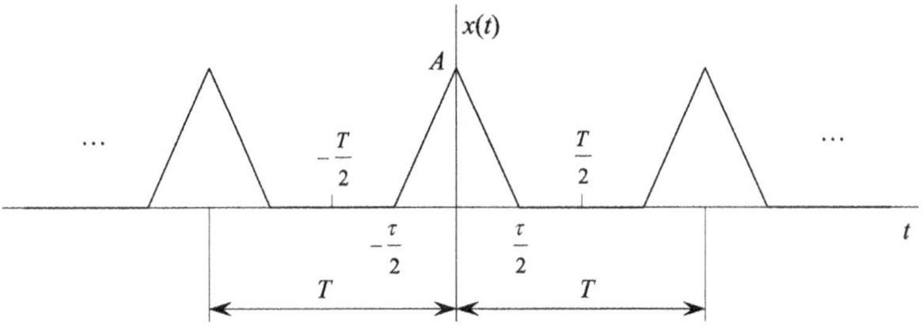

Fig. 2.40 Triangular pulse-train.

$$C_k = \frac{A}{2q} \frac{\sin^2(k\pi/2q)}{(k\pi/2q)^2}, \quad (2.129)$$

$$C_0 = \frac{A}{2q}. \quad (2.130)$$

We notice that the phase spectrum is zero here, $\Psi_k = 0$, for all k.

Accordingly, the synthesis problem is solved by

$$x(t) = \frac{A}{2q} + \frac{A}{q} \sum_{k=1}^{\infty} \frac{\sin^2(k\pi/2q)}{(k\pi/2q)^2} \cos 2\pi kFt. \quad (2.131)$$

It may easily be shown that zero points in the spectrum envelope are determined by $\pi kF\tau/2 = \pm n\pi, n = 1, 2, \ldots$ and thus $f_n = 2n/\tau$. The spectral lines are still placed at the same frequencies $f_k = kF = k/T$ that is stated by (2.118).

Example 2.20. Given a periodic triangular pulse with amplitude A and $q = 3$. Its magnitude spectrum is shown in Fig. 2.41.

□

2.6.3 Periodic Gaussian Pulse

A periodic Gaussian pulse-train is combined with an infinite number of shifted Gaussian waveforms, which durations are theoretically infinite. Practically, however, the latter fact may be neglected and we see the train as in Fig. 2.42.

By (2.96), the coefficient C_k is defined here to be

$$C_k = \frac{A\sqrt{\pi}}{\alpha T} e^{-\frac{(\pi kF)^2}{\alpha^2}}, \quad (2.132)$$

$$C_0 = \frac{A\sqrt{\pi}}{\alpha T}. \quad (2.133)$$

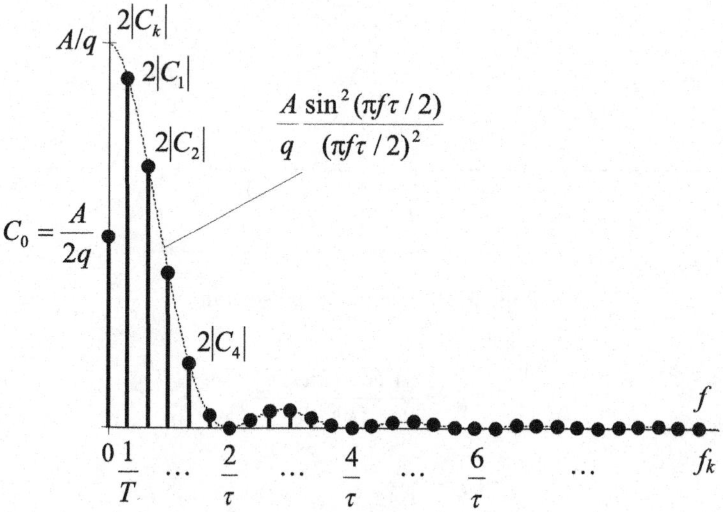

Fig. 2.41 Magnitude spectrum of a triangular pulse-train with $q = 3$

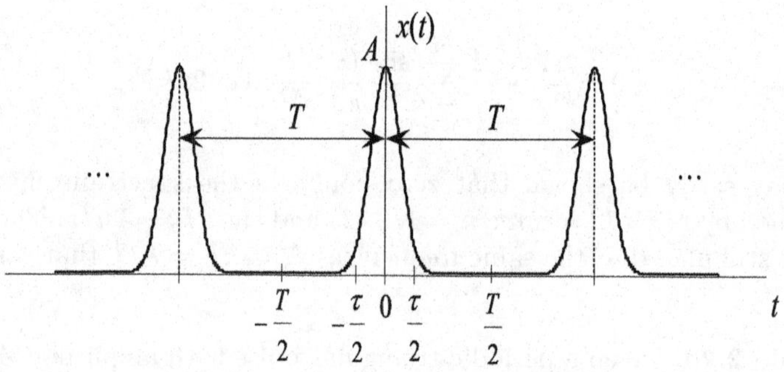

Fig. 2.42 Periodic Gaussian pulse-train.

Like in the triangular-train case, the phase spectrum is zero here, $\Psi_k = 0$. Therefore, a synthesis restores a signal with

$$x(t) = \frac{A\sqrt{\pi}}{\alpha T} + \sum_{k=1}^{\infty} \frac{2A\sqrt{\pi}}{\alpha T} e^{-\frac{(\pi k F)^2}{\alpha^2}} \cos 2\pi k F t, \tag{2.134}$$

To make possible for the spectrum to be analyzed, it is in order to suppose that the pulse duration τ is fixed for some level A/β, $\beta > 1$. This level was earlier introduced for a single Gaussian waveform and we just recall that β produces $\tau = 2\sqrt{\ln \beta}/\alpha$ and $\alpha = 2\sqrt{\ln \beta}/\tau$.

Example 2.21. Given a periodic Gaussian pulse with $\alpha = 2\sqrt{\ln \beta}/\tau, \beta = 10, \tau = 1, A = 1$, and $q = 4$. Figure 2.43 shows the magnitude spectrum of this train calculated by (2.132) and (2.133).

□

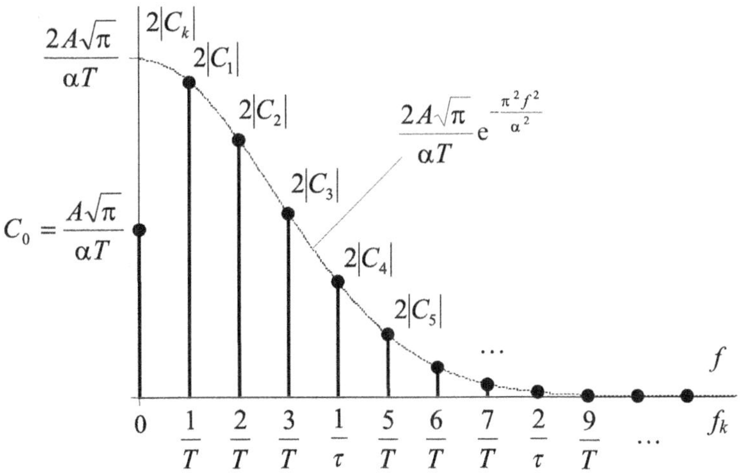

Fig. 2.43 Magnitude spectrum of a periodic Gaussian pulse-train with $\alpha = 2\sqrt{\ln\beta}/\tau$, $\beta = 10$, $\tau = 1$, $A = 1$, and $q = 4$.

Example 2.22. A Gaussian pulse-train is performed with $A = 4\,\text{V}, \tau = 10^{-6}\,\text{s}$, and $T = 6 \times 10^{-6}\,\text{s}$. Its spectrum is given by (2.132) and (2.133). By $\beta = 10$, the attenuation coefficient is calculated to be $\alpha = 2\sqrt{\ln\beta}/\tau \cong 3 \times 10^6$ rad/s. The spectral line at zero has the magnitude $C_0 = A\sqrt{\pi}/\alpha T \cong 0.4\,\text{V}$. The spectrum envelope at zero frequency calculates $2A\sqrt{\pi}/\alpha T \cong 0.8\,\text{V}$. The fundamental frequency is defined to be $F = 1/T = 166 \times 10^3\,\text{Hz}$. Finally, the spectrum width ΔF is derived at the level of 0.1 by the relation $e^{-\frac{\pi^2 \Delta F^2}{\alpha^2}} = 0.1$ that produces $\Delta F = \alpha/\pi\sqrt{\ln 10} = 1.466 \times 10^6\,\text{Hz}$.

□

2.6.4 Periodic Sinc-shaped Pulse

A sinc-shaped pulse-train is the other representative of periodic signals shaped with the waveform of infinite length, namely with $x(t) = A\sin\alpha t/\alpha t$. Frequently, duration τ is evaluated for zero points of the main lobe by $\tau = 2\pi/\alpha$. Truncating, however, causes errors in the frequency domain, since the side lobes attenuate slowly and the spectrum stretches. Figure 2.44 gives an example of this train.

We have already learned that the spectral density of a sinc waveform differs from all others by its uniform shape within the frequency range $\pm 1/\tau$ (Fig. 2.24). Such a property saves in the periodic pulse, which magnitude spectrum is calculated by

$$|C_k| = \begin{cases} \frac{A}{2q}, & k \leqslant q \\ 0, & k > q \end{cases}, \qquad C_0 = \frac{A}{2q}, \tag{2.135}$$

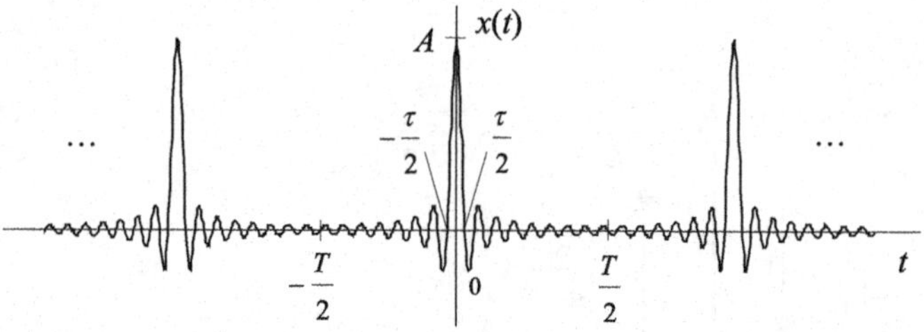

Fig. 2.44 A periodic sinc-pulse.

and which phase spectrum consists of zeros, $\Psi_k = 0$, over all frequencies. The train is synthesized with

$$x(t) = \frac{A}{2q} + \sum_{k=1}^{\infty} \frac{A}{q} \cos 2\pi kFt\,. \tag{2.136}$$

What is important is that (2.135) and (2.136) are rigorous only if period T tends toward infinity. Otherwise, the Gibbs phenomenon occurs in the spectrum envelope. Two other practically important observations may be made:

- If q is integer plus 0.5 the spectrum is strongly uniform, but the Gibbs effect in the envelope is large. That is because the spectral lines appear at the mean line of the oscillating envelope.
- When q is integer, the spectrum is not absolutely uniform, but the Gibbs effect is small. Oscillations of the envelope are efficiently attenuated here. However, the spectral lines are not placed at its mean line.

Example 2.23. Given a sinc pulse-train with $q = T/\tau = 15.5$ and $q = T/\tau = 15$ (Fig. 2.45a and b, respectively). The magnitude spectra of these signals are shown in Fig. 2.45c and d, respectively. It is seen that the case (Fig. 2.45a) is preferable whenever a uniform spectrum is needed.

□

Example 2.24. A sinc pulse-train is specified with $A = 5\,\text{V}, \tau = 2 \times 10^{-6}\,\text{s}$, and $T = 8 \times 10^{-6}\,\text{s}$. Its spectral line at zero frequency has the magnitude $A\tau/2T = 0.625\,\text{V}$. The value of the spectrum envelope at zero frequency is $A\tau/T = 1.25\,\text{V}$. The cut-off frequency calculates $f_e = 1/\tau = 5 \times 10^5\,\text{Hz}$ and the spectral lines are placed at the frequencies multiple to $F = 1/T = 1.25 \times 10^5\,\text{Hz}$.

□

2.7 Spectral Densities of Pulse-Bursts

To increase the energy of signals, impulse wireless electronic systems often employ finite sets of pulses called the *pulse-bursts* or *nonperiodic trains*.

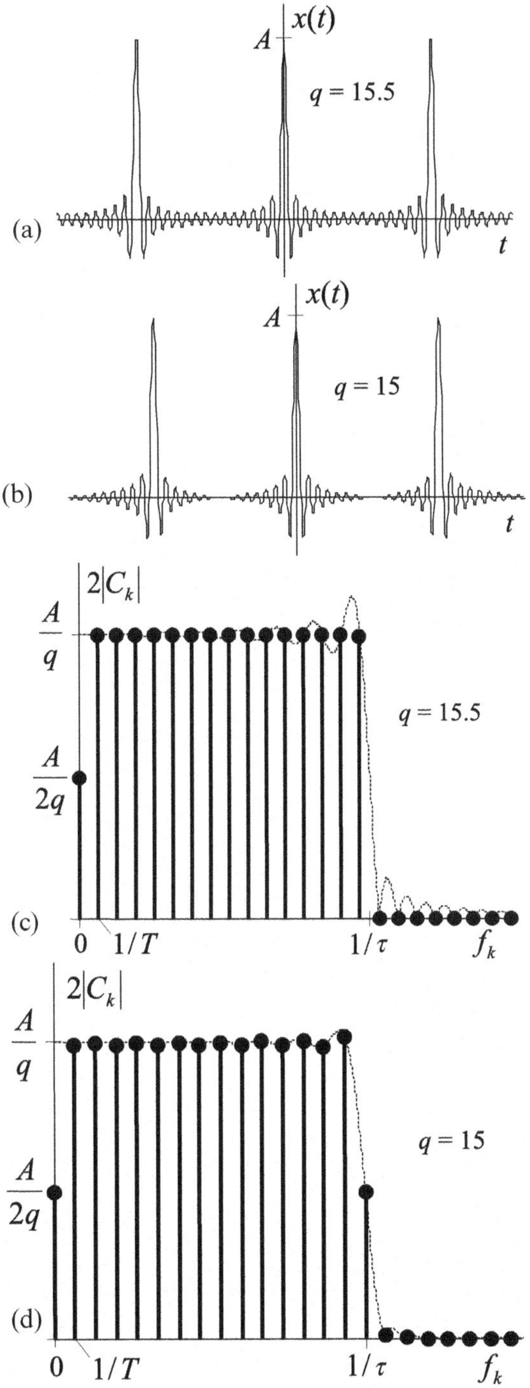

Fig. 2.45 Sinc pulse-train: **(a)** with $q = 15.5$, **(b)** with $q = 15.0$, **(c)** magnitude spectrum for $q = 15.5$, and **(d)** magnitude spectrum for $q = 15.0$.

In a single pulse-burst with period of pulses T, energy is concentrated about the frequencies multiple to $1/T$. Depending on the number N of pulses in the burst, we have two limiting cases. If N tends toward infinity, the burst evolves to the pulse-train and, by $N = 1$, it becomes a single pulse. Thus, the properties of a pulse-burst are intermediate between those featured to a single pulse and pulse-train.

2.7.1 General Relations

Commonly, it may be supposed that the pulse-burst $y(t)$ is composed by $N > 1$ pulses $x_1(t)$, $x_2(t)$, ..., $x_N(t)$ of the same waveform with a constant period T, as it is sketched in Fig. 2.46. We will also suppose that the energy of this burst is finite and hence the Fourier transform exists.

To solve an analysis problem, it is in order to exploit the superposition principle of the transform and first write

$$y(t) = x_1(t) + x_2(t) + \ldots + x_N(t)\,.$$

Since every pulse in the burst is merely a shifted version of the first one $x(t) \equiv x_1(t)$, the relation becomes

$$y(t) = x(t) + x(t-T) + \ldots + x[t-(N-1)T]\,. \tag{2.137}$$

By the time-shifting property, the Fourier transform of (2.137) may be performed as

$$\begin{aligned} Y(j\omega) &= X(j\omega) + X(j\omega)\mathrm{e}^{-j\omega T} + \ldots + X(j\omega)\mathrm{e}^{-j\omega(N-1)T} \\ &= X(j\omega)\left[1 + \mathrm{e}^{-j\omega T} + \ldots + \mathrm{e}^{-j\omega(N-1)T}\right] = X(j\omega)\Phi(j\omega) \end{aligned} \tag{2.138}$$

where an auxiliary function $\Phi(j\omega)$ is calculated by the geometrical progression with the first term $a_1 = 1$, last term $a_N = a_1 g^{N-1} = g^{N-1}$, and $g = \mathrm{e}^{-j\omega T}$. The function then calculates

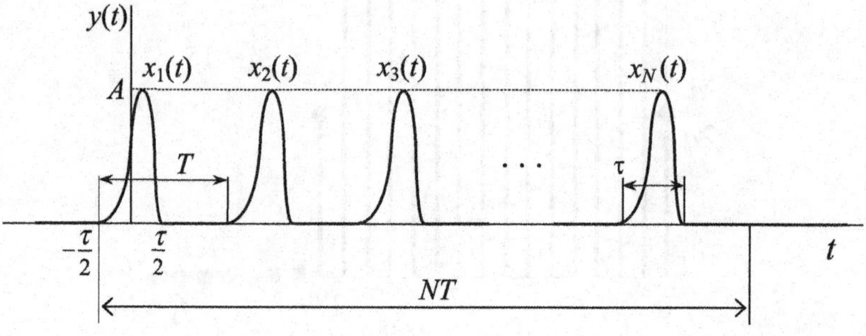

Fig. 2.46 Pulse-burst.

$$\Phi(j\omega) = \sum_{n=0}^{N-1} \mathrm{e}^{-jn\omega T} = \frac{a_1 - a_N g}{1-g}$$
$$= \frac{1-\mathrm{e}^{-j\omega(N-1)T}\mathrm{e}^{-j\omega T}}{1-\mathrm{e}^{-j\omega T}} = \frac{1-\mathrm{e}^{-j\omega NT}}{1-\mathrm{e}^{-j\omega T}} .$$

Searching for the most appropriate form, we arrive at

$$\Phi(j\omega) = \frac{\mathrm{e}^{-j\omega N\frac{T}{2}}\left(\mathrm{e}^{j\omega N\frac{T}{2}} - \mathrm{e}^{-j\omega N\frac{T}{2}}\right)}{\mathrm{e}^{-j\omega\frac{T}{2}}\left(\mathrm{e}^{j\omega\frac{T}{2}} - \mathrm{e}^{-j\omega\frac{T}{2}}\right)}$$
$$= \frac{\sin(\omega NT/2)}{\sin(\omega T/2)}\mathrm{e}^{-j\omega(N-1)\frac{T}{2}} . \tag{2.139}$$

The generalized form of the spectral density (2.138) then becomes

$$Y(j\omega) = X(j\omega)\frac{\sin(\omega NT/2)}{\sin(\omega T/2)}\mathrm{e}^{-j\omega(N-1)\frac{T}{2}} , \tag{2.140}$$

where $X(j\omega)$ is the transform of a pulse waveform.

An observation shows that, under certain conditions, both sine functions in (2.140) may reach zero and we thus will be in need to resolve an uncertainty

$$\frac{\sin(\omega NT/2)}{\sin(\omega T/2)} = \frac{\sin(\pi NfT)}{\sin(\pi fT)} = \frac{0}{0} .$$

This may be done by the L'Hôupital[6] rule,

$$\lim_{z\to a}\frac{f_1(z)}{f_2(z)} = \lim_{z\to a}\frac{\partial f_1(z)/\partial z}{\partial f_2(z)/\partial z} ,$$

and we go to

$$\frac{\sin(\pi NfT)}{\sin(\pi fT)} = \lim_{z\to f}\frac{\sin(\pi NzT)}{\sin(\pi zT)} = \lim_{z\to f}\frac{\partial\sin(\pi NzT)/\partial z}{\partial\sin(\pi zT)/\partial z}$$
$$= N\frac{\cos(\pi NfT)}{\cos(\pi fT)} . \tag{2.141}$$

The problem with an uncertainty 0/0 may appear in (2.140) when $\pi NfT = m\pi$, $m = 1, 2, \ldots$, and $\pi fT = p\pi$, $p = 1, 2, \ldots$ This gives two special cases:

[6] Guillaume François Antoine Marquis de L'Hôupital, French mathematician, 1661–2 February 1704.

- When $m = pN$, then, by (2.141), we have

$$N\frac{\cos(\pi fNT)}{\cos(\pi fT)} = N\,. \tag{2.142}$$

Thus, the spectral density increases by the factor of N at

$$f_p = \frac{p}{T}, \quad p = 1, 2, \ldots \tag{2.143}$$

- When $m \neq pN$, the spectral density becomes zero at

$$f_m = \frac{m}{NT}, \quad m = 1, 2, \ldots, \quad m \neq pN\,. \tag{2.144}$$

It is seen from (2.140) that the phase spectral density has a nonzero slope that is owing to the time shift of the burst (Fig. 2.46). One already knows that if to place any signal to be symmetric about zero, then its phase performance will get a zero slope. We will use this opportunity to simplify the presentation form further.

2.7.2 Rectangular Pulse-burst

As was mentioned earlier, the rectangular pulse is the most widely used test signal. Furthermore, it is one of the most successful mathematical idealizations or real impulse signals. Therefore, properties of the rectangular pulse-burst are of high interest. We proceed with Fig. 2.47, assuming that the burst consists of N pulses and is placed symmetrically about zero.

By (2.80) and (2.140), its transform reads

$$Y(j\omega) = |Y(j\omega)|\, e^{-j\varphi_y(\omega)} = A\tau\frac{\sin(\omega\tau/2)}{\omega\tau/2}\frac{\sin(\omega NT/2)}{\sin(\omega T/2)} \tag{2.145}$$

and the magnitude and phase spectral densities become, respectively,

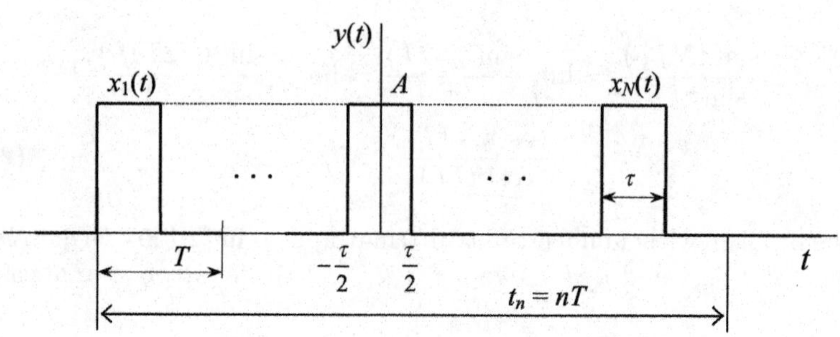

Fig. 2.47 Rectangular pulse-burst.

$$|Y(j2\pi f)| = A\tau \left| \frac{\sin(\pi f \tau)}{\pi f \tau} \frac{\sin(\pi N f T)}{\sin(\pi f T)} \right| \tag{2.146}$$

$$\varphi_y(f) = \begin{cases} \pm 2\pi n, & \frac{\sin(\pi f\tau)}{\pi f\tau}\frac{\sin(\pi N fT)}{\sin(\pi fT)} \geqslant 0 \\ \pm\pi \pm 2\pi n, & \frac{\sin(\pi f\tau)}{\pi f\tau}\frac{\sin(\pi N fT)}{\sin(\pi fT)} < 0 \end{cases}, \quad n = 0, 1, \ldots. \tag{2.147}$$

It follows that the nulls in (2.146) exist, in general, at the frequencies (2.144). Yet, the spectral envelope reaches zero at the frequencies multiple to $1/\tau$,

$$f_l = \frac{l}{\tau}, \quad l = 1, 2, \ldots. \tag{2.148}$$

We also notice that, by $m = pN$, the spectral density is gained by the factor of N that occurs in the main lobes at $f_p = p/T$, $p = 1, 2, \ldots$ (2.143). The side lobes lying in the mid range between the main ones are restricted approximately by the spectral density of a single waveform. The latter observation, however, is less obvious and not actually strong, especially when N is small.

Example 2.25. Given a rectangular pulse-burst with $N = 4$ and $q = 2$. The maximum value of the envelope of the main lobes of its spectral density is calculated by $4A\tau$ at $f = 0$. The envelope of the side lobes has a peak value $A\tau$ at $f = 0$. Spectral nulls appear at frequencies multiple to $1/4T$ and $1/\tau$. Maximums of the main lobes correspond to the frequencies multiple to $1/T$. Figure 2.48 sketches the magnitude spectral density of this signal.

□

Example 2.26. Given a rectangular pulse-burst with $N = 2, q = 1.5, \tau = 1 \times 10^{-6}$ s, and $A = 1$ V. Zero points of the envelope of the spectral density correspond to the frequencies $l/\tau = l \times 10^6$ Hz. Maximums of the main lobes

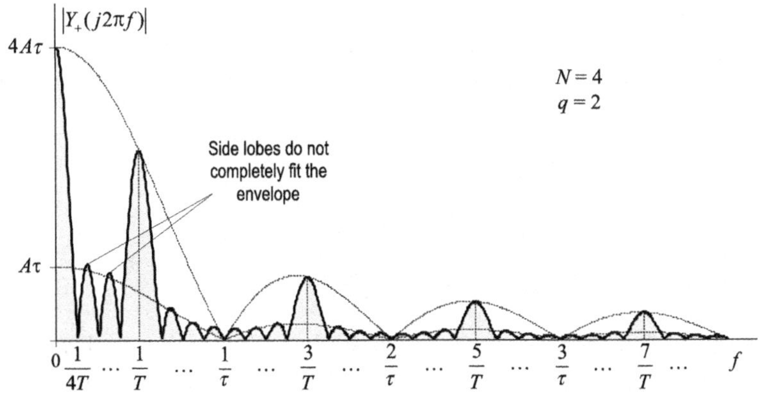

Fig. 2.48 Magnitude spectral density of a rectangular pulse-burst for $N = 4$ and $q = 2$.

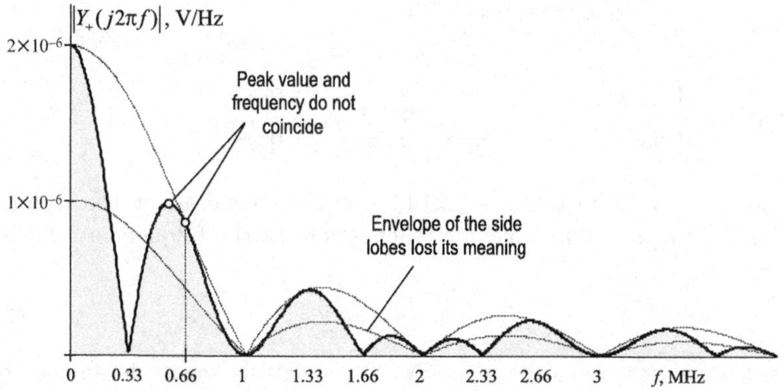

Fig. 2.49 Magnitude spectral density of the rectangular pulse-burst with $N = 4$ and $q = 1.5$.

appear at $p/T = p/1.5\tau \cong 0.666p \times 10^6$ Hz. A peak value of the envelope of the main lobes calculates $2A\tau = 2 \times 10^{-6}$ V and that for the side lobes is $A\tau = 1 \times 10^{-6}$ V. The frequency step between zero points is $m/NT = m/Nq\tau \cong 0.333m \times 10^6$ Hz. Figure 2.49 illustrates the relevant magnitude spectral density. It is neatly seen that the envelope of the side lobes has lost its meaning.

□

2.7.3 Triangular Pulse-burst

Effortlessly, by (2.140) and (2.89), one may now write the spectral density of a symmetric triangular pulse-burst, i.e.,

$$Y(j\omega) = \frac{A\tau}{2} \frac{\sin^2(\omega\tau/4)}{(\omega\tau/4)^2} \frac{\sin(\omega NT/2)}{\sin(\omega T/2)} . \qquad (2.149)$$

The relevant magnitude and phase densities, respectively, are

$$|Y(j2\pi f)| = \frac{A\tau}{2} \frac{\sin^2(\pi f\tau/2)}{(\pi f\tau/2)^2} \left| \frac{\sin(\pi N fT)}{\sin(\pi fT)} \right| , \qquad (2.150)$$

$$\varphi_y(f) = \begin{cases} \pm 2\pi n, & \frac{\sin(\pi N fT)}{\sin(\pi fT)} \geqslant 0 \\ \pm\pi \pm 2\pi n, & \frac{\sin(\pi N fT)}{\sin(\pi fT)} < 0 \end{cases} , \quad n = 0, 1, \ldots . \qquad (2.151)$$

Example 2.27. Given a triangular pulse-burst with $N = 4, q = 3.25$, $\tau = 1 \times 10^{-6}$ s, and $A = 1$ V. Its magnitude spectral density is calculated by (2.150). Zero points of the envelope correspond to the frequencies $2l/\tau = 2l \times 10^6$ Hz. Maximums of the main lobes appear at $p/T = p/3.25\tau \cong 0.3p \times 10^6$ Hz. The maximum value of the envelope of the main lobes calculates

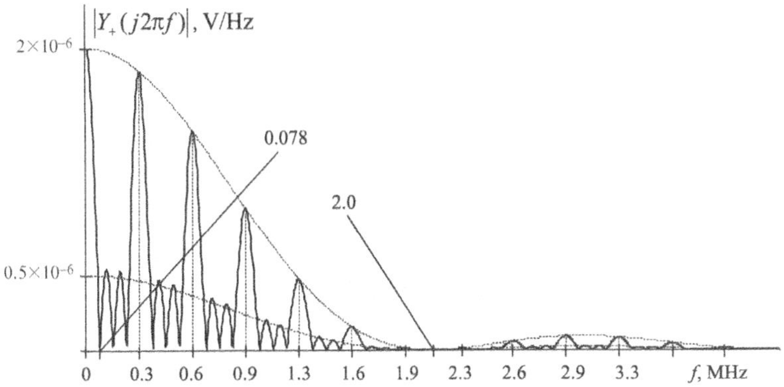

Fig. 2.50 Magnitude spectral density of a triangular pulse-burst for $N = 4$ and $q = 3.25$.

$4A\tau/2 = 2 \times 10^{-6}$ V and that for the side lobes is $A\tau/2 = 0.5 \times 10^{-6}$ V. The span between zero points is given by $m/NT = m/Nq\tau \cong 0.078m \times 10^6$ Hz. Figure 2.50 illustrates this density.

□

2.7.4 Sinc Pulse-burst

In the sinc pulse-burst, we meet the same effect as in the sinc pulse-train. If the pulses are snugly placed, the oscillations are either summed (Fig. 2.45a) or subtracted (Fig. 2.45b) in the intermediate range. This influences the transform of the burst and, if the integration range is restricted (finite), the Gibbs phenomenon is present in its envelope.

In an ideal case, by (2.93) and (2.140), the transform of a symmetric sinc pulse-burst is uniform within the range of $|\omega| \leqslant \alpha$ and 0 otherwise,

$$Y(j\omega) = \begin{cases} \frac{A\pi}{\alpha}\frac{\sin(\omega NT/2)}{\sin(\omega T/2)}, & |\omega| \leqslant \alpha \\ 0, & |\omega| > \alpha \end{cases}. \tag{2.152}$$

Accordingly, the magnitude and phase spectral densities may be written, respectively, as

$$|Y(j2\pi f)| = \begin{cases} \frac{A\pi}{\alpha}\left|\frac{\sin(\pi NfT)}{\sin(\pi fT)}\right|, & |f| \leqslant \frac{\alpha}{2\pi} \\ 0, & |f| > \frac{\alpha}{2\pi} \end{cases}, \tag{2.153}$$

$$\varphi_y(f) = \begin{cases} \pm 2\pi n, & \frac{\sin(\pi NfT)}{\sin(\pi fT)} \geqslant 0 \\ \pm\pi \pm 2\pi n, & \frac{\sin(\pi NfT)}{\sin(\pi fT)} < 0 \end{cases}, \quad n = 0, 1, \ldots. \tag{2.154}$$

Example 2.28. Given a sinc pulse-burst $y(t)$ with $N = 3, q = 3.5, \tau = 1 \times 10^{-6}$ s, and $A = 1$ V (Fig. 2.51). Its magnitude spectral density is calculated

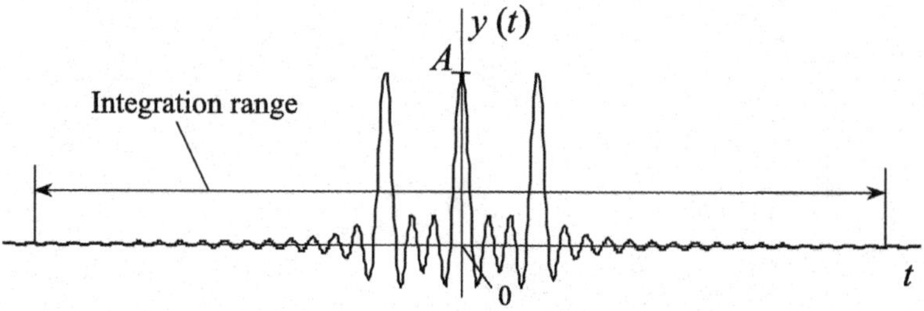

Fig. 2.51 A sinc pulse-burst by $N = 3$ and $q = 3.5$.

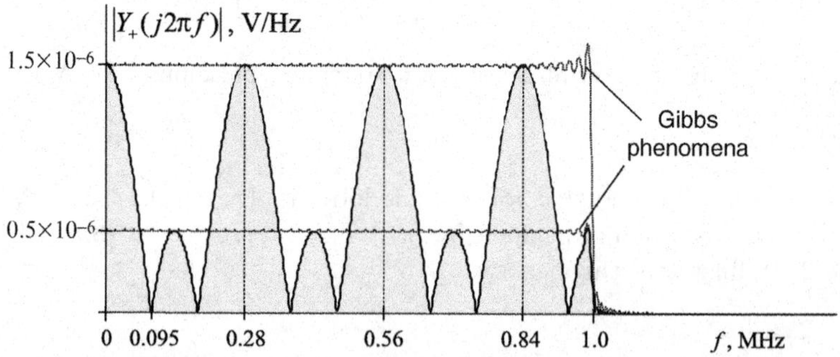

Fig. 2.52 Magnitude spectral density of a sinc pulse-burst with $N = 3$ and $q = 3.5$.

by (2.153). For $\alpha = 2\pi/\tau$, the cut-off frequency becomes $1/\tau = 1 \times 10^6$ Hz. The maximums of the main lobes appear at $p/T = p/3.5\tau \cong 0.28p \times 10^6$ Hz and α is calculated to be $\alpha = 2\pi/\tau = 2\pi \times 10^6$ Hz. The envelope value of the main lobes is given by $3A\pi/\alpha = 1.5 \times 10^{-6}$ V/Hz and that for the side lobes is $A\pi/\alpha = 0.5 \times 10^{-6}$ V/Hz. The frequency span between zero points is $m/NT = m/Nq\tau \cong 0.095m \times 10^6$ Hz. Figure 2.52 illustrates this density. Intentionally, we restricted an integration area by $\pm 30T$ to show that it causes the Gibbs phenomenon to occur in the envelope.

□

2.7.5 Pulse-to-burst-to-train Spectral Transition

It is now a proper place to demonstrate a transition of the magnitude spectral density of a single pulse ($N = 1$) to that of a pulse-burst ($N > 1$) and to that of a pulse-train ($N \to \infty$), by changing N.

Figure 2.53 shows what happens with the transform of the rectangular pulse-burst if $q = 2$ and N changes from 1 to 20. The plot goes over all three above-mentioned signal structures. The case of $N = 1$ corresponds to a single pulse. With $N > 1$, we watch for the density associated with a pulse-burst. Finally, by $N = 20$, the picture becomes close to that featured to the pulse-train ($N \to \infty$) when the spectral density degenerates to the discrete spectrum.

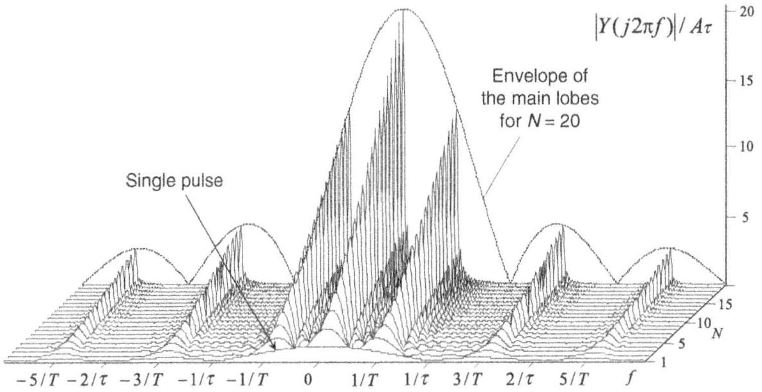

Fig. 2.53 Magnitude spectral densities of a rectangular pulse-burst with $q = 2$ and N changed from 1 to 20.

2.8 Spectrums of Periodic Pulse-bursts

The other representative of signals is the *periodic pulse-burst* that is also called the *pulse-burst-train* or *burst-train*. As the name suggests, a signal has a discrete spectrum described by the Fourier series and, as in any other periodic case, its spectrum envelope is shaped by the spectral density of a single pulse-burst. We already know the rules in picturing the discrete spectrums and may thus start applying this knowledge to the periodic pulse-bursts.

To solve an analysis problem (2.116), one needs to recall spectrums of periodic signals. The methodology implies that the envelope of the spectrum is shaped by the spectral density of a single burst and that the spectral lines are placed at the frequencies multiple to $1/T_1$, where T_1 is the period of repetition of the bursts in the train. Figure 2.54 gives a generalized picture of such a signal.

For the symmetric pulse-burst-train, the transform (2.116) becomes, by (2.140),

$$C_k = \frac{1}{T_1} X(j2\pi f_k) \frac{\sin(\pi N f_k T)}{\sin(\pi f_k T)}, \tag{2.155}$$

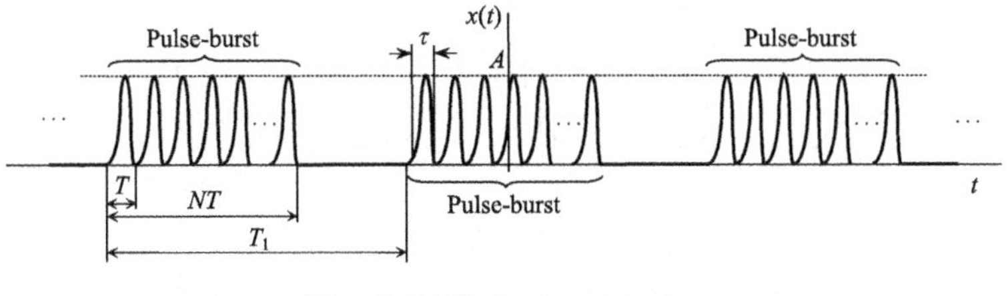

Fig. 2.54 Pulse-burst-train.

where $f_k = kF_1 = k/T_1$ is the frequency of the kth spectral line and $X(j2\pi f_k)$ is a discrete value of the spectral density of a single waveform. In turn, the magnitude and phase spectral densities are given, respectively, by

$$|C_k| = \frac{1}{T_1}\left|X(j2\pi f_k)\frac{\sin(\pi N f_k T)}{\sin(\pi f_k T)}\right|, \tag{2.156}$$

$$\Psi_k = \begin{cases} \pm 2\pi n, & X(j2\pi f_k)\frac{\sin(\pi N f_k T)}{\sin(\pi f_k T)} \geqslant 0 \\ \pm\pi \pm 2\pi n, & X(j2\pi f_k)\frac{\sin(\pi N f_k T)}{\sin(\pi f_k T)} < 0 \end{cases}, \quad n = 0, 1, \ldots \tag{2.157}$$

If to take into consideration that the value of (2.156) with $k = 0$ is $C_0 = (N/T_1)X(0)$, then a synthesis problem (2.117) will be solved by

$$x(t) = \frac{N}{T_1}X(0) + \frac{2}{T_1}\sum_{k=1}^{\infty}\left|X(j2\pi f_k)\frac{\sin(\pi N f_k T)}{\sin(\pi f_k T)}\right| \cos(2\pi f_k t - \Psi_k), \tag{2.158}$$

where $X(0)$ and $X(j2\pi f_k)$ are predefined by the spectral density of a pulse waveform.

Below we apply (2.155)–(2.158) to several widely used waveforms.

2.8.1 Rectangular Pulse-burst-train

The rectangular pulse-burst-train is shown in (Fig. 2.55). By (2.80), the coefficients of the Fourier series (2.155) calculate

$$C_k = \frac{A\tau}{T_1}\frac{\sin(\pi f_k \tau)}{\pi f_k \tau}\frac{\sin(\pi N f_k T)}{\sin(\pi f_k T)} \tag{2.159}$$

and, respectively, the magnitude and phase spectrums become

$$|C_k| = \frac{A\tau}{T_1}\left|\frac{\sin(\pi f_k \tau)}{\pi f_k \tau}\frac{\sin(\pi N f_k T)}{\sin(\pi f_k T)}\right|, \tag{2.160}$$

$$\Psi_k = \begin{cases} \pm 2\pi n, & \frac{\sin(\pi f_k \tau)}{\pi f_k \tau}\frac{\sin(\pi N f_k T)}{\sin(\pi f_k T)} \geqslant 0 \\ \pm\pi \pm 2\pi n, & \frac{\sin(\pi f_k \tau)}{\pi f_k \tau}\frac{\sin(\pi N f_k T)}{\sin(\pi f_k T)} < 0 \end{cases}, \quad n = 0, 1, \ldots \tag{2.161}$$

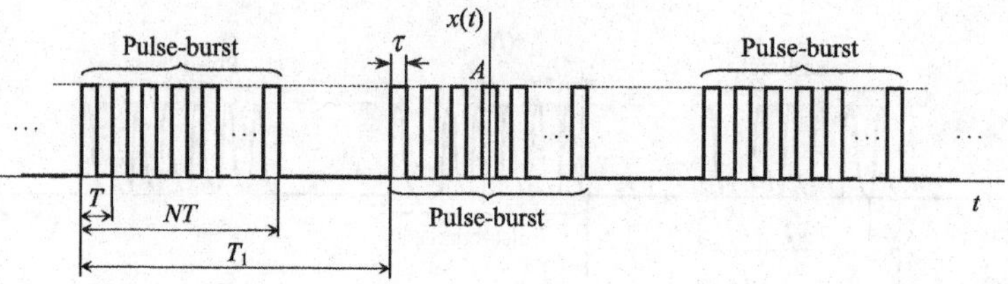

Fig. 2.55 Periodic rectangular pulse-burst.

The value of (2.160), by $k = 0$, is specified to be $C_0 = A\tau N/T_1$ and then a synthesis problem is solved with

$$x(t) = \frac{AN\tau}{T_1} + \frac{2A\tau}{T_1}\sum_{k=1}^{\infty}\left|\frac{\sin(\pi f_k \tau)}{\pi f_k \tau}\frac{\sin(\pi N f_k T)}{\sin(\pi f_k T)}\right| \cos\left(2\pi f_k t - \Psi_k\right). \quad (2.162)$$

Finally, by two auxiliary ratios $q_1 = T_1/\tau$ and $q_2 = T_1/NT$, (2.162) may be rewritten as

$$x(t) = \frac{AN}{q_1} + \frac{2A}{q_1}\sum_{k=1}^{\infty}\left|\frac{\sin(\pi k/q_1)}{\pi k/q_1}\frac{\sin(\pi k/q_2)}{\sin(\pi k/Nq_2)}\right| \cos\left(2\pi f_k t - \Psi_k\right). \quad (2.163)$$

Example 2.29. Given a rectangular pulse-burst-train performed with $T = 1 \times 10^{-6}$ s, $N = 4, q_1 = 16$, and $q_2 = 2$. From $q_2/q_1 = \tau/NT$, we determine $q = T/\tau = 2$ and then go to $\tau = 0.5 \times 10^{-6}$ s. A ratio $q_1 = T_1/\tau = 16$ produces $T_1 = 8 \times 10^{-6}$ s. Figure 2.56 shows the magnitude spectrum of this signal.

□

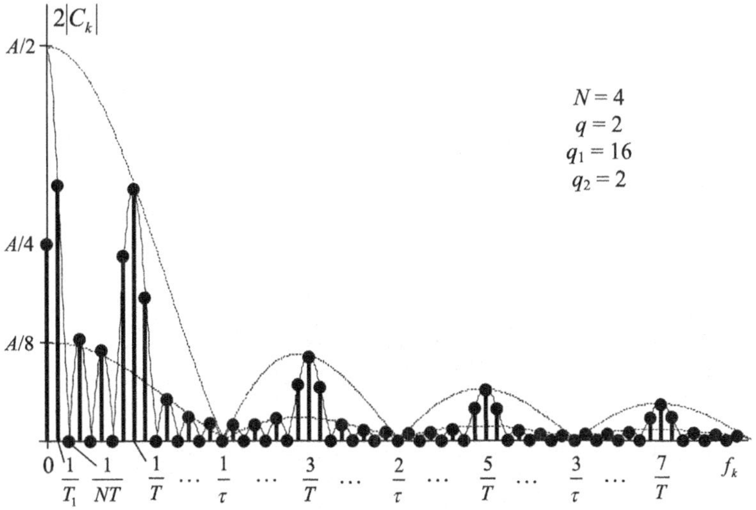

Fig. 2.56 Magnitude spectrum of a rectangular pulse-burst-train (Example 2.29).

2.8.2 Triangular Pulse-burst-train

By (2.155) and (2.90), an analysis problem for the triangular pulse-burst-train is solved with

$$C_k = \frac{A\tau}{2T_1}\frac{\sin^2(\pi f_k \tau/2)}{(\pi f_k \tau/2)^2}\frac{\sin(\pi N f_k T)}{\sin(\pi f_k T)} \tag{2.164}$$

and thus the magnitude and phase spectrums are determined, respectively, by

$$|C_k| = \frac{A\tau}{2T_1}\frac{\sin^2(\pi f_k \tau/2)}{(\pi f_k \tau/2)^2}\left|\frac{\sin(\pi N f_k T)}{\sin(\pi f_k T)}\right| , \tag{2.165}$$

$$\Psi_k = \begin{cases} \pm 2\pi n, & \frac{\sin(\pi N f_k T)}{\sin(\pi f_k T)} \geqslant 0 \\ \pm\pi \pm 2\pi n, & \frac{\sin(\pi N f_k T)}{\sin(\pi f_k T)} < 0 \end{cases}, \quad n = 0, 1, \ldots . \tag{2.166}$$

Since $C_0 = AN\tau/2T_1$, then a synthesis may be provided in two equal forms of

$$x(t) = \frac{AN\tau}{2T_1} + \frac{A\tau}{T_1}\sum_{k=1}^{\infty}\frac{\sin^2(\pi f_k \tau/2)}{(\pi f_k \tau/2)^2}\left|\frac{\sin(\pi N f_k T)}{\sin(\pi f_k T)}\right| \cos(2\pi f_k t - \Psi_k) \tag{2.167}$$

$$= \frac{AN}{2q_1} + \frac{A}{q_1}\sum_{k=1}^{\infty}\frac{\sin^2(\pi k/2q_1)}{(\pi k/2q_1)^2}\left|\frac{\sin(\pi k/q_2)}{\sin(\pi k/Nq_2)}\right| \cos(2\pi f_k t - \Psi_k), \tag{2.168}$$

where q_1 and q_2 are as in (2.163).

Example 2.30. Given a periodic triangular pulse-burst performed with $T = 1 \times 10^{-6}$ s, $N = 3, q_1 = 12$, and $q_2 = 2$. From $q_2/q_1 = \tau/NT$, we determine $q = T/\tau = 2$ and then calculate $\tau = 0.5 \times 10^{-6}$ s. Figure 2.57 illustrate the magnitude spectrum of this signal.

□

2.8.3 Periodic Sinc Pulse-burst

An analysis problem for the sincpulse-burst-train is solved by

$$C_k = \begin{cases} \frac{A\pi}{\alpha T_1}\frac{\sin(\pi N f_k T)}{\sin(\pi f_k T)}, & |f_k| \leqslant \alpha/2\pi \\ 0, & |f_k| > \alpha/2\pi \end{cases} \tag{2.169}$$

through the magnitude and phase spectrums, respectively,

$$|C_k| = \begin{cases} \frac{A\pi}{\alpha T_1}\left|\frac{\sin(\pi N f_k T)}{\sin(\pi f_k T)}\right|, & |f_k| \leqslant \alpha/2\pi \\ 0, & |f_k| > \alpha/2\pi \end{cases}, \tag{2.170}$$

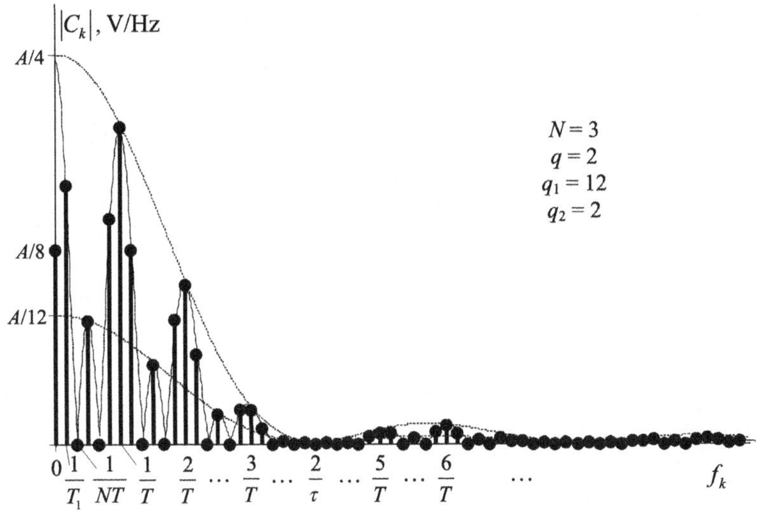

Fig. 2.57 Magnitude spectrum of the triangular pulse-burst-train (Example 2.30).

$$\Psi_k = \begin{cases} \pm 2\pi n, & \frac{\sin(\pi N f_k T)}{\sin(\pi f_k T)} \geqslant 0 \\ \pm\pi \pm 2\pi n, & \frac{\sin(\pi N f_k T)}{\sin(\pi f_k T)} < 0 \end{cases}, \quad n = 0, 1, \ldots . \tag{2.171}$$

For the infinite integration range, a synthesized signal becomes

$$x(t) = \begin{cases} \frac{AN\pi}{\alpha T_1} + \frac{2AN\pi}{\alpha T_1} \sum_{k=1}^{\infty} \left| \frac{\sin(\pi N f_k T)}{\sin(\pi f_k T)} \right| \cos(2\pi f_k t - \Psi_k), & f_k \leqslant \alpha/2\pi \\ 0, & f_k > \alpha/2\pi \end{cases} . \tag{2.172}$$

Example 2.31. Given a sinc pulse-burst-train with $T = 1 \times 10^{-6}$ s, $N = 3, q = 3.5, q_1 = 21$, and $q_2 = 2$. The magnitude spectrum calculated by (2.170) is shown in Fig. 2.58. Assuming $\alpha = 2\pi/\tau$, the synthesis problem is solved in two equal forms of

$$\begin{aligned} x(t) &= \begin{cases} \frac{AN\tau}{2T_1} + \frac{AN\tau}{T_1} \sum_{k=1}^{\infty} \left| \frac{\sin(\pi N f_k T)}{\sin(\pi f_k T)} \right| \cos(2\pi f_k t - \Psi_k), & f_k \leqslant 1/\tau \\ 0, & f_k > 1/\tau \end{cases} \\ &= \begin{cases} \frac{AN}{2q_1} + \frac{AN}{q_1} \sum_{k=1}^{\infty} \left| \frac{\sin(\pi k/q_2)}{\sin(\pi k/N q_2)} \right| \cos(2\pi f_k t - \Psi_k), & f_k \geqslant 1/\tau \\ 0, & f_k > 1/\tau \end{cases} . \end{aligned} \tag{2.173}$$

To show an appearance of the Gibbs phenomenon, the integration range (Fig. 2.51) is restricted here by $\pm 30T$.

□

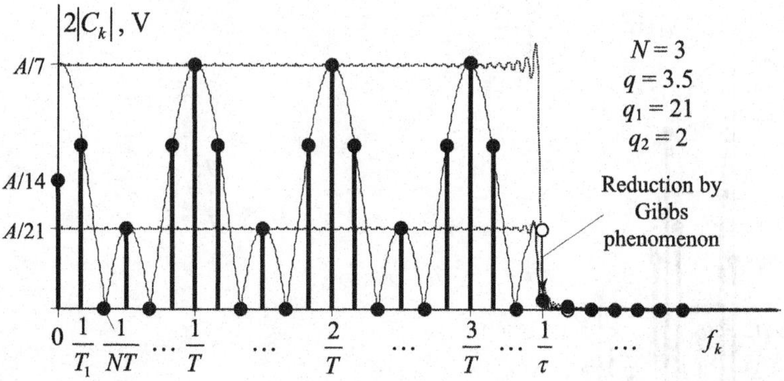

Fig. 2.58 Magnitude spectrum of a periodic sinc pulse-burst (Example 2.31).

2.9 Signal Widths

Every signal has some width (finite or infinite) in the time domain and distributes its spectrum over some frequency range. Since the waveforms differ and because the contributions of their different parts may be of different interest, there is no universal measure of the signal width. One of the convenient comparative measures, for certain shapes of function, is the *equivalent width* that is the area of the function divided by its central ordinate. The equivalent width may be calculated through the signal function (then it is also called the *effective duration*) or its Fourier transform (then it is also called the *effective bandwidth*) owing to the splendid property: the wider a function is the narrower is its spectrum. Such a measure, however, does not fit all needs and the other one, the *mean square width,* is also used. Ignoring these measures, the concept of the *halfpower bandwidth* is also widely used in applications as defined at the half power level.

2.9.1 Equivalent Width

The *equivalent width* of a signal is the width of the rectangle whose height is equal to the central ordinate and whose area is the same as that of the function of a signal. Assuming that the signal is placed to have the center at zero, its equivalent width is specified by

$$W_x = \frac{1}{x(0)} \int_{-\infty}^{\infty} x(t)\mathrm{d}t\,. \qquad (2.174)$$

It then follows that, if $x(0) = 0$, the equivalent width does not exist. For all other cases where $x(0) \neq 0$ and $\int_{-\infty}^{\infty} x(t)\mathrm{d}t$ exist, there exists an "equivalent width." Frequently, but not obligatory, the signal is placed to have its maximum at zero. This case is illustrated in Fig. 2.59.

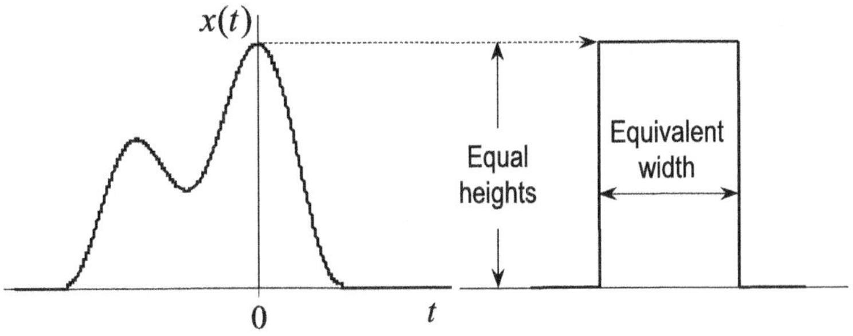

Fig. 2.59 The equivalent rectangle of a signal.

We now notice that $\int_{-\infty}^{\infty} x(t)\mathrm{d}t$ is nothing more than $X(j0)$ and that $x(0)$ is nothing less than $\frac{1}{2\pi}\int_{-\infty}^{\infty} X(j\omega)\mathrm{d}\omega$. Thus the equivalent width in the frequency domain is calculated (in Hertz) by

$$W_X = \frac{1}{2\pi X(0)} \int_{-\infty}^{\infty} X(j\omega)\mathrm{d}\omega \tag{2.175}$$

and we arrive at the useful relation

$$W_x = \frac{1}{W_X} \tag{2.176}$$

that follows from the similarity theorem, according to which the compression and expansion of a signal and its spectral density behave reciprocally. The relation (2.176) results in the *uncertainty principle*

$$W_x W_X \geqslant 1$$

that states that the widths W_x and W_X cannot be independently specified.

The other measure of the effective duration of a signal $x(t)$, distributed about $t = 0$, is specified by

$$W_{|x|} = \frac{1}{x(0)} \int_{-\infty}^{\infty} |x(t)|\,\mathrm{d}t\,. \tag{2.177}$$

Similar to (2.174), the effective bandwidth is then given with

$$W_{|X|} = \frac{1}{2\pi X(0)} \int_{-\infty}^{\infty} |X(j\omega)|\,\mathrm{d}\omega\,. \tag{2.178}$$

By simple manipulations, it may be shown that the uncertainty principle relates also to the latter pair of functions, i.e., we may write that

$$W_{|x|} W_{|X|} \geqslant 1\,.$$

Example 2.32. Given a rectangular pulse of duration τ and amplitude A. Its width in the time domain is trivial, that is $W_x = \tau$. In the frequency domain, exploiting (2.80) and using an identity $\int\limits_0^{\infty} \frac{\sin bx}{x} dx = \frac{\pi}{2}\mathrm{sgn} b$, we obtain

$$W_X = \frac{1}{2\pi} \int\limits_{-\infty}^{\infty} \frac{\sin(\omega\tau/2)}{\omega\tau/2} d\omega = \frac{2}{\pi\tau} \int\limits_{0}^{\infty} \frac{\sin(\omega\tau/2)}{\omega} d\omega = \frac{1}{\tau}$$

that fits (2.176). We then deduce that the equivalent width $W_X = 1/\tau$ is a half of the bandwidth of the main lobe of the spectral density of a rectangular pulse that is illustrated in Fig. 2.60.

□

Example 2.33. Given a triangular pulse of duration τ and amplitude A. The pulse width in the time domain is calculated by $W_x = \tau/2$. In the frequency domain, by (2.90) and an identity $\int_0^{\infty} (\sin^2 bx/x^2) dx = \pi b/2$, the width calculates

$$W_X = \frac{1}{2\pi} \int\limits_{-\infty}^{\infty} \frac{\sin^2(\omega\tau/4)}{(\omega\tau/4)^2} d\omega = \frac{16}{\pi\tau^2} \int\limits_{0}^{\infty} \frac{\sin^2(\omega\tau/4)}{\omega^2} d\omega = \frac{2}{\tau} .$$

We notice that the equivalent width of the triangular pulse (Fig. 2.61) is twice smaller than that of the rectangular pulse. Therefore, its equivalent spectral width is twice wider by (2.176).

□

2.9.2 Central Frequency and Mean Square Widths

The equivalent width is not always the best measure of width. There is the other measure of wide applications called the mean square width. To introduce, let us be interested of the signal width in the frequency domain. We may then calculate the mean square value $\langle \omega^2 \rangle = \int_{-\infty}^{\infty} \omega^2 X(j\omega) d\omega$, normalize it, and go to the mean square width (in Hertz)

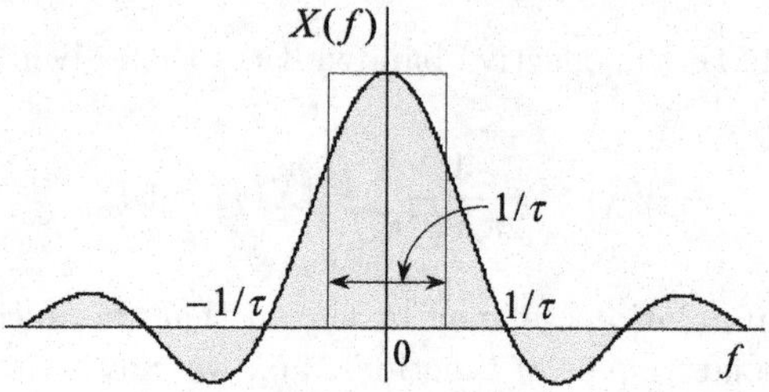

Fig. 2.60 Equivalent width of a rectangular pulse in the frequency domain.

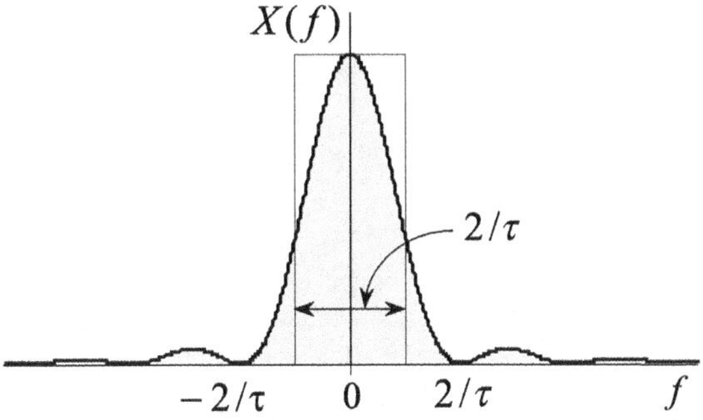

Fig. 2.61 Equivalent width of a triangular pulse in the frequency domain.

$$W_{XMS1}^2 = \frac{\int\limits_{-\infty}^{\infty} \omega^2 X(j\omega)\mathrm{d}\omega}{2\pi \int\limits_{-\infty}^{\infty} X(j\omega)\mathrm{d}\omega} \tag{2.179}$$

that is appropriate for centered signals.

Example 2.34. Given a Gaussian pulse $x(t) = A\mathrm{e}^{-(\alpha t)^2}$, whose spectral density is determined by (2.96) as $X(j\omega) = A\sqrt{\pi}/\alpha \mathrm{e}^{-\omega^2/4\alpha^2}$. Using identities $\int_{-\infty}^{\infty} \mathrm{e}^{-px^2}\mathrm{d}x = \sqrt{\frac{\pi}{p}}$ and $\int_{-\infty}^{\infty} x^2\mathrm{e}^{-px^2}\mathrm{d}x = \sqrt{\pi}p^{-3/2}$, the width (2.179) is calculated to be

$$W_{XMS1}^2 = \frac{\int\limits_{-\infty}^{\infty} \omega^2 \mathrm{e}^{-\frac{\omega^2}{4\alpha^2}}\mathrm{d}\omega}{2\pi \int\limits_{-\infty}^{\infty} \mathrm{e}^{-\frac{\omega^2}{4\alpha^2}}\mathrm{d}\omega} = \frac{\alpha^2}{\pi}$$

and thus $W_{XMS1} = \alpha/\sqrt{\pi}$. It may easily be shown that for the Gaussian pulse, the latter measure is equal to the equivalent width, produced by (2.175),

$$W_X = \frac{1}{2\pi}\int\limits_{-\infty}^{\infty} \mathrm{e}^{-\frac{\omega^2}{4\alpha^2}}\mathrm{d}\omega = \frac{\alpha}{\sqrt{\pi}}\,.$$

We go to the same result by the reciprocal of the signal width in the time domain, namely (2.174) produces $W_x = \int_{-\infty}^{\infty} \mathrm{e}^{-(\alpha t)^2}dt = \frac{\sqrt{\pi}}{\alpha}$. The width of this signal is illustrated in Fig. 2.62.

□

If a signal in not centered and its spectrum is distributed about some "center of weight" ω_0, the other measure must be used. In applications, it may also be in need first to determine this central frequency $\omega_0 = 2\pi f_0$ of the

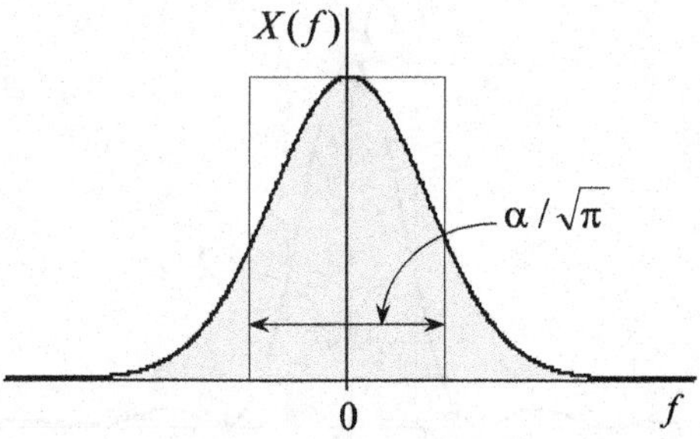

Fig. 2.62 Width (2.178) of a Gaussian pulse in the frequency domain.

measured spectrum. To obtain, the following quantity is usually minimized:

$$\int_0^\infty (\omega - \omega_0)^2 X(j\omega)\, d\omega . \tag{2.180}$$

To derive ω_0, we take the first derivative with respect to ω_0, set it to zero, and go to

$$\omega_0 = \frac{\int_0^\infty \omega X(j\omega)\, d\omega}{\int_0^\infty X(j\omega)\, d\omega} . \tag{2.181}$$

The mean square width for the noncentered signal may now be introduced as follows:

$$W_{XMS2}^2 = \frac{\int_0^\infty (\omega - \omega_0)^2 X(j\omega) d\omega}{2\pi \int_0^\infty X(j\omega) d\omega} = \frac{\langle \omega^2 \rangle - 2\omega_0 \langle \omega \rangle + \omega_0^2 \int_0^\infty X(j\omega) d\omega}{2\pi \int_0^\infty X(j\omega) d\omega} .$$

By (2.181), this formula becomes

$$W_{XMS2}^2 = W_{XMS1}^2 - 2\pi f_0^2 . \tag{2.182}$$

2.9.3 Signal Bandwidth

Setting aside the above given definitions, they often follow the concept of the *bandwidth* that is denoted as BW. The special case is the ***absolute bandwidth B*** Hz that is when a signal possesses a spectrum only over a contiguous frequency range of BW Hz, and is zero elsewhere. In reality, signals having an absolute bandwidth cannot be met at least thanks to noise. Therefore, an appropriate level to measure the BW should be introduced. Such a level conventionally corresponds to the 1/2 power (3 dB) frequencies ω_L and ω_H. The 3 dB definition comes from the fact that $10\log 0.5 = -3\text{dB}$. Accordingly, the BW is specified to be BW $= \omega_H - \omega_L$ for the narrowband signals and BW $= \omega_H$ for the baseband signals (Fig. 2.63a).

The edge frequencies correspond to the 1/2 power frequencies defined for the baseband signal (Fig. 2.63a) by

$$\frac{|X(j\omega_H)|^2}{|X(0)|^2} = \frac{1}{2} \tag{2.183}$$

and for the narrowband signal (Fig. 2.63b) by

$$\frac{|X(j\omega_H)|^2}{|X(j\omega_0)|^2} = \frac{1}{2}. \tag{2.184}$$

and

$$\frac{|X(j\omega_L)|^2}{|X(j\omega_0)|^2} = \frac{1}{2}. \tag{2.185}$$

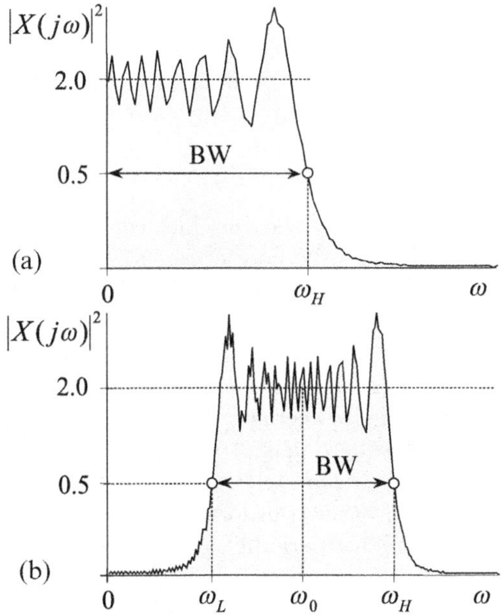

Fig. 2.63 The bandwidth of a signal: **(a)** baseband and **(b)** narrowband.

It follows that ω_L and ω_H may also be derived through the relevant amplitude ratios given by

$$\frac{|X(j\omega_H)|}{|X(0)|} = \frac{1}{\sqrt{2}}, \quad \frac{|X(j\omega_H)|}{|X(j\omega_0)|} = \frac{1}{\sqrt{2}}, \quad \text{and} \quad \frac{|X(j\omega_L)|}{|X(j\omega_0)|} = \frac{1}{\sqrt{2}}.$$

Measured spectra cannot obligatory be uniform about its center. Furthermore, the edges may oscillate at the 3 dB level. This means that the measurement of the BW must be provided with some care. For example, in Fig. 2.63a, it may be appropriate to use an average value of the spectrum as a value at zero.

Example 2.35. Given a baseband signal with the spectral density $X(j\omega) = 1/(1 + j\omega)$. By (2.183), we go to $|1 + j\omega_H| = \sqrt{2}$ and derive $\omega_H = 1$.

□

Example 2.36. Given a Gaussian signal with the spectral density $X(j\omega) = A\sqrt{\pi}/\alpha e^{-\omega^2/4\alpha^2}$. By (2.183), its edge frequency is calculated to be $\omega_H = 2\alpha\sqrt{\ln\sqrt{2}}$ and thus $BW = 2\alpha\sqrt{\ln\sqrt{2}}$. It follows that this value differs by the factor of $\pi W_X/BW = \frac{1}{2}\sqrt{\frac{\pi}{\ln\sqrt{2}}} \cong 1.5$ from that associated with the effective width.

□

2.10 Summary

Signal presentation in the frequency domain is intended to solve an analysis problem that implies a determination of the spectral content of a signal. The following rules are obeyed in translating a signal to the frequency domain and synthesizing it back to the time domain:

- Any periodic signal satisfying the Dirichlet conditions may be presented in the frequency domain by the Fourier series.
- Spectrum of a periodic signal is performed with the magnitude and phase spectrums.
- If the synthesis problem is solved with the finite Fourier series, the Gibbs phenomenon accompanies the restored signal.
- By the Parceval theorem, an average power of a periodic signal is equal to the infinite sum of the square magnitudes in the frequency domain.
- Analysis and synthesis problems of the nonperiodic signals are solved with the direct and inverse Fourier transforms, respectively.
- The spectral density of a nonperiodic signal is presented by the magnitude and phase spectral densities.
- By the Rayleigh theorem (or Parceval's relation), a total energy of a signal is given by the integral of its square modulus or square magnitude of its spectral density.

- The Gaussian pulse has unique equal shapes in the time and frequency domains.
- Complex signals allow for forming spectrums of the given shape.
- A pulse-burst has a spectral density that is intermediate between those featured to a single pulse and pulse-train.
- In the time domain, signals are compared for the effective duration and, in the frequency domain, for the effective bandwidth.
- The compression and expansion of a signal and its spectral density behave reciprocally, i.e., the narrower is a signal in the time domain, the wider is its spectral performance in the frequency domain.

2.11 Problems

2.1. Explain what the orthogonal functions are and why they can be used to extend a periodic signal to the series?

2.2. Interpret in simple words the meaning of the Dirichlet conditions.

2.3. Give a graphical proof that the Gibbs phenomenon exists in the synthesized signal even if the length of the series is infinite.

2.4 (Fourier series). Prove mathematically the following properties of the Fourier series: (a) conjugation, (b) time reversal, and (c) modulation.

2.5. Bring a derivation of the Fourier transform as a limiting case of the Fourier series.

2.6. Using the properties of the Fourier series (Table A.1) and Fourier series of basic signals (Table B.1), define the coefficients C_k of the following periodic signals with period T:

1. $x(t) = \mathrm{e}^{-j\Omega t} - 2\mathrm{e}^{j\Omega t}$
2. $x(t) = 2\cos 2\Omega t - \sin \Omega t$
3. $x(t) = \frac{1}{2t}\sin(-2t) + \cos 2t$
4. $x(t) = \left(2\mathrm{e}^{-2t} - t\mathrm{e}^{-3t}\right) u(t)$
5. $x(t) = t\left(2t\mathrm{e}^{-2t} + \mathrm{e}^{-3t}\right) u(t)$

2.7 (Fourier transform). Prove mathematically the following properties of the Fourier transforms: (a) time integration, (b) linearity, and (c) correlation.

2.8. Using the integral identities (Appendix D) and properties of the Fourier transform (Table A.2), solve an analysis problem for the following signals:

1. $x(t) = t\mathrm{e}^{j10t}u(t)$

2. $x(t) = \int\limits_{-\infty}^{\infty} \frac{\sin 2\theta}{\theta[4+(t-\theta)^2]} d\theta$

3. $x(t) = t(2-3t)u(t)$

4. $x(t) = \frac{e^{-2|t|}}{4+t^2}$

5. $x(t) = 2e^{-2|t|-4t^2}$

2.9. Using the Fourier transform, derive the spectral densities of the ramp pulses, (2.112) and (2.114).

2.10. Consider a complex rectangular pulse (Fig. 2.27). Show that two pulses generated by this complex pulse with $A = B$ and $A = -B$ are orthogonal. Give proofs in the time and frequency domains.

2.11. Given the spectral density (2.114) of a ramp pulse. Derive and investigate its magnitude and phase spectral densities.

2.12. Using the Fourier transforms of signals given in Table B.1, solve a synthesis problem for the given spectral densities:

1. $X(j\omega) = \frac{1+\omega^2+2(2+j\omega)}{(2+j\omega)(1+\omega^2)}$

2. $X(j\omega) = \frac{5+2(j\omega)+(j\omega)^2}{2+(j\omega)+2(j\omega)^2-(j\omega)^3}$

3. $X(j\omega) = \frac{2(1-j\omega)}{8+2(j\omega)-(j\omega)^2}$

4. $X(j\omega) = \frac{2}{\omega}(\sin\omega + \sin 2\omega)$

5. $X(j\omega) = \frac{2}{\omega}\sin\frac{3\omega}{2}\left(1 + \frac{2}{3\omega}\sin\frac{3\omega}{2}\right)$

2.13. Solve an analysis problem for the waveform given in Fig. 2.64 under your personal number M (e.g., your number in the list). Use the integral identities given in Appendix D. If the pulse is not given analytically, write its function using the plot given. Calculate the spectral density numerically using the applied software (Matlab or Mathcad). Sketch the magnitude spectral density derived analytically and calculated numerically. Make conclusions.

2.14 (Periodic signal). Given a periodic pulse of the waveform (Fig. 2.64). Write its Fourier series for arbitrary T.

2.15. Calculate special points of the spectrum of a periodic signal (Problem 2.14) for $T = (M+1)\tau$, where τ is the pulse duration. If τ is not given, take it approximately at the level of 0.1. Sketch the magnitude spectrum of the train and compare with that obtained numerically.

2.16. A pulse-train of the waveform (Fig. 2.64) is delayed on τ/M. Write its total phase and the modulo 2π phase. Sketch both phase spectra.

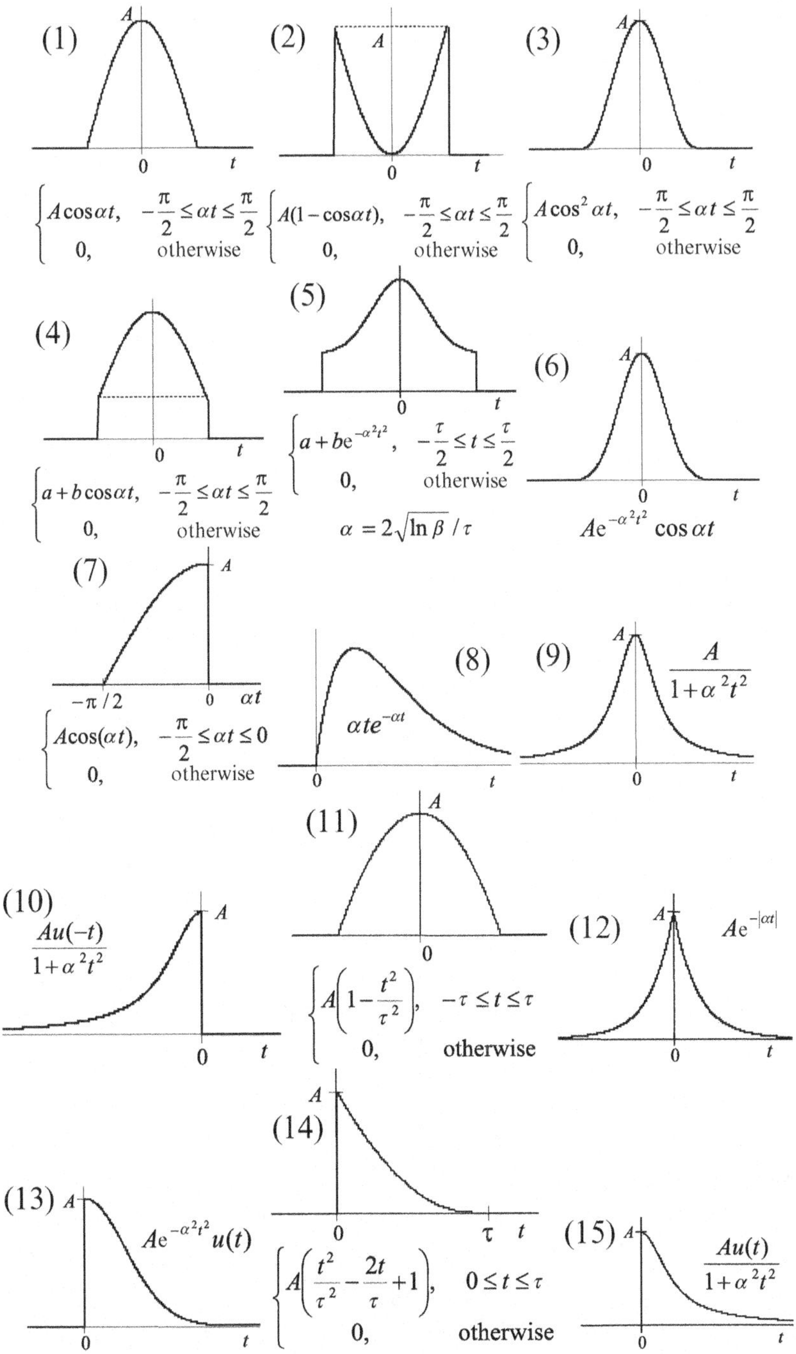

Fig. 2.64 Single waveforms.

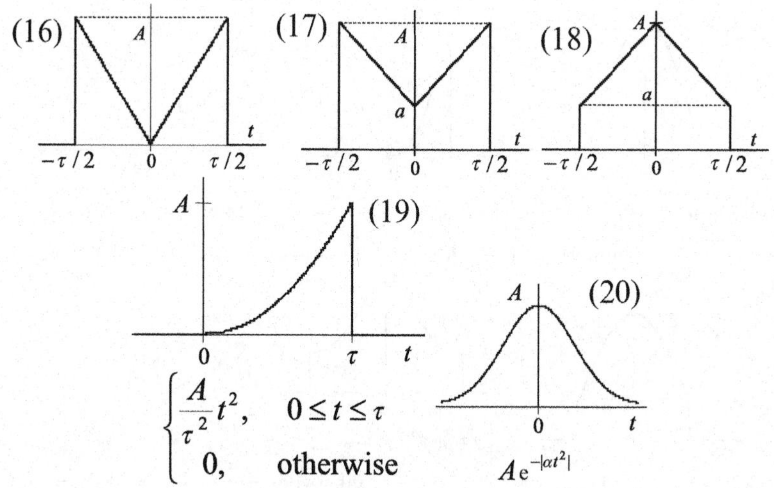

Fig. 2.64 *Continued*

2.17 (Pulse-burst). Given a pulse-burst of the waveform (Fig. 2.64) with the number of pulses $N = M$. Write the Fourier transform of this signal for arbitrary T.

2.18. The pulse-burst of the waveform (Fig. 2.64) is performed with $T = (M+1)\tau$ and $N = M$. Sketch its Fourier transform and compare with that obtained numerically.

2.19 (Gibbs phenomenon). Given a sinc pulse-burst with $N = 4$ and $q = 2.3$, 3.0, 3.5, and 4.1. Study numerically the Gibbs phenomenon for the integration range $\pm 200T$, $\pm 100T$, $\pm 30T$, and $\pm 10T$. Make a conclusion about the errors induced.

2.20. Given a waveform (Fig. 2.64). Using Matlab, calculate numerically and exhibit the transition of the spectral densities similarly to Fig. 2.53.

2.21 (Periodic pulse-burst). Given a pulse-burst-train of the waveform (Fig. 2.64). The pulse-burst if formed with the number of pulses $N = M$. Write the Fourier series for this signal.

2.22. The pulse-burst-train of the waveform (Fig. 2.64) has $T = 3\tau$ and period of repetition of the pulse-bursts is $T_1 = 3MT$. Write its magnitude spectrum, sketch it, and compare with that calculated numerically.

2.23. Given a sinc pulse-burst-train with $N = 4$ and $q = 2.3$, 3.0, 3.5, and 4.1. The period of repetition of the pulse-bursts is $T_1 = 3MT$. Calculate the spectrum numerically and explain errors induced by the Gibbs phenomenon to the magnitude spectrum for the integration range $\pm 200T$, $\pm 100T$, $\pm 30T$, and $\pm 10T$.

2.24 (Width). Given a waveform (Fig. 2.64). Calculate its equivalent duration in the time domain and the equivalent width in the frequency domain.

2.25. Determine the central frequency and mean square width of the Fourier transform of a given waveform (Fig. 2.64).

2.26. Define the bandwidth of the Fourier transform of a given waveform (Fig. 2.64).

3

Signals Modulation

3.1 Introduction

Useful signals are usually baseband signals thus existing at low frequencies. Therefore, with any necessity of transmitting such signals on a distance, we instantly face several problems:

- The signal rapidly attenuates and thus the transmission distance becomes unpleasantly short.
- All transmitted signals occupy almost the same frequency range and thus the problem with signals recognition and separation arises.
- Perturbation and distortion of signals by medium (noise and interference) is large.

It is well known from electromagnetic theory that, for efficient radiation of electrical energy from an antenna, the latter must be at least in the order of magnitude of a waveform in size, i.e., $\lambda = c/f$, where c is the velocity of light, f is the signal frequency, and λ is the wavelength. For example, if $f = 1\,\text{kHz}$, then $\lambda = 300$ km, and it becomes clear that an antenna of this size is not practical for efficient transmission.

To pass over these problems, it will be useful to replace the signal spectrum to higher frequencies where the wavelength is shorter, attenuation is significantly lower, and hence the transmission distance is larger. Furthermore, if we make this replacement, we may place signals to lie in different frequency ranges avoiding problems with their interplay. Finally, it is known that the effect of medium at high frequencies is much lesser than that at low ones. So, we arrive at the necessity to replace somehow the useful signal to high frequencies. Such an incredibly useful and efficient technique is called *modulation*.

Historically, the first practical implementation of modulation was fixed in wire telephony in 1878 that was perfected by Bell.[1] Soon after, in 1896,

[1] Alexander Graham Bell, Scottish-born inventor, 3 March 1847–2 August 1922.

wireless telegraphy employing radio waves was demonstrated by Popov.[2] Since then, modulation has passed over several stages of its improvement, and nowadays modern systems transmit and receive information exploiting very efficient and perfect modulation methods. Notwithstanding this fact, in bottom, modulation may be accomplished over either the amplitude or the angular measure of a carrier signal. Therefore, the roots of modulation are entirely in elementary signals even though we may expect for some new efficient methods of its implementation.

3.2 Types of Modulation

In electronic and optical systems, modulation is intended to transfer useful signals on a distance and this is independent on the system type, operation principle, and solved problems. This universality of modulation is explained by the fact that useful signals occupy usually a low-frequency band. Therefore, they cannot be transmitted straightforwardly over a long distance and the radio or optical channels are exploited.

3.2.1 Fundamentals

Irrespective of particular applications, the following definitions for modulation may be given:

***Modulation*:**

- The transmission of a signal by using it to vary a carrier wave; changing the carrier's amplitude or frequency or phase.
- The process by which some characteristic of a carrier signal is varied in accordance with a modulating signal.

□

In other words, if one signal is used to change a parameter (or parameters) of some other signal, such a process is said to be modulation. Thus, any radio electronic system may fulfill its functional aim if modulation is implemented in its principle of operation.

It follows from the above-given definition that modulating and modulated signals may not obligatory be useful. In particular, signal modulation by noise produces a problem of signal immunity against nearby channels.

By employing modulation, plenty of problems may efficiently be solved as associated with information quantization, signal spectrum conversion, signals jamming immunity, etc. In a modern view, modulation techniques have wide facilities to solve problems coupled with transfer of useful signals. And this is in spite of the fact that there are only two ways to accomplish modulation: through the amplitude and angle of a carrier signal. Respecting these ways, the following types of modulation are distinguished.

---

[2] Alexander Stepanovich Popov, Russian scientist, 16 March 1859–13 January 1906.

3.2.1.1 Amplitude Modulation

Amplitude modulation (AM) is said to be the process by which the amplitude of a carrier signal is varied in accordance with the modulating signal.

Most commonly, the AM signal may be performed by

$$y(t) = A(t)\cos(\omega_0 t + \psi_0)\,, \tag{3.1}$$

where $\omega_0 = 2\pi f_0$ is a natural (angular) constant carrier frequency, f_0 is a carrier frequency, and ψ_0 is a constant phase. Here $A(t)$ is a real-valued amplitude function (frequently called the envelope) , which bears the transmitted message in communications or the necessary properties of the pulse in other wireless electronic systems.

3.2.1.2 Angular Modulation

Angular modulation is the process by which the angle of a carrier signal is varied in accordance with the modulating signal. With angular modulation, the general form of a signal is as follows:

$$y(t) = A_0 \cos \Psi(t)\,, \tag{3.2}$$

where the amplitude A_0 is supposed to be constant and $\Psi(t)$ is the instantaneous angle of modulation given in radians. The instantaneous frequency of a signal (3.2) may be defined as

$$\omega(t) = \lim_{\Delta t \to 0} \frac{\Psi(t+\Delta t) - \Psi(t)}{\Delta t} = \frac{\mathrm{d}\Psi(t)}{\mathrm{d}t}\,, \tag{3.3}$$

implying that, inversely,

$$\Psi(t) = \int_0^t \omega(t)\mathrm{d}t + \psi_0\,, \tag{3.4}$$

where ψ_0 is some constant phase.

There are two common ways to vary $\Psi(t)$: *frequency modulation* (FM) and *phase modulation* (PM). We may then say that FM and PM are two processes by which the frequency and phase, respectively, of a carrier signal are varied in accordance with the modulating signal.

3.2.1.3 Frequency Modulation

Frequency modulation (FM) is the process by which the frequency $\omega(t) = \omega_0 + \Delta\omega(t)$, where ω_0 is some constant frequency usually called the carrier and $\Delta\omega(t)$ is a variable add bearing a message of a carrier signal, is varied in accordance with the modulating signal. The phase (3.4) then becomes

$$\Psi(t) = \omega_0 t + \int_0^t \Delta\omega(t)\mathrm{d}t + \psi_0 \,. \tag{3.5}$$

3.2.1.4 Phase Modulation

Phase modulation (PM) is the process by which the variable phase constituent $\Delta\psi(t)$ bears the message. In accordance with (3.5), this term is the integral measure of $\Delta\omega(t)$, namely $\Delta\psi(t) = \int_0^t \Delta\omega(t)\mathrm{d}t$, and therefore the function (3.4) is written as

$$\Psi(t) = \omega_0 t + \Delta\psi(t) + \psi_0 \,. \tag{3.6}$$

It is clear that the PM effect is nothing more than the integral measure of the FM effect or the FM effect is the differential measure of the PM effect. This means that two types of angular modulation may be considered from the common position and thus its separation to FM and PM is conditional in a sense. All three types of modulation AM, PM, and FM may be caused by the continuous (analog or impulse) and digital modulating signals.

In analog communications, analog waveforms are transmitted and received. Herewith, AM, PM, FM, quadrature amplitude modulation (QAM), and pulse amplitude modulation (PAM) may be exploited. In digital communications, digitized versions of analog waveforms are transmitted and received. The sampled and quantized versions of the analog examples of modulation are called amplitude shift keying (ASK), phase shift keying (PSK), frequency shift keying (FSK), quadrature amplitude shift keying (QASK), digital PAM. Other types of modulation are also used, such as phase difference shift keying (PDSK), pulse time modulation (PTM), etc.

In continuous-wave radars, FM is exploited to provide the necessary mark to allow range information to be calculated. In pulse radars, it is a common practice to use a pulsed wave for modulating the amplitude of a carrier signal. In this form of modulation, the continuous wave is broken up into very short pulse-bursts separated by long periods of silence during which no wave is transmitted. Also, linear frequency modulation (LFM) and FSK are employed.

In navigation systems such as the GPS and global navigation system (GLONASS), the carrier phase is modulated.

3.3 Amplitude Modulation

If is the defined as follows:

> ***Amplitude modulation:*** *AM* is the process by which the amplitude of a carrier signal is varied in accordance with the modulating signal. □

AM may be realized in different forms (analog and digital), even though its basic principles remain unchanged. In accordance with (3.1), the carrier signal with AM may generally be written as

$$y(t) = A(t)\cos(\omega_0 t + \psi_0)\,. \tag{3.7}$$

Since AM results in variations in the amplitude $A(t)$, the latter may be performed by

$$A(t) = A_0 + a(t) = A_0[1 + k_\mathrm{a} m(t)]\,, \tag{3.8}$$

where $a(t)$ is a variable part of the amplitude caused by AM,

$$m(t) = \frac{a(t)}{k_\mathrm{a} A_0}$$

is the baseband *message signal*, and k_a is the *amplitude sensitivity.*

3.3.1 Simplest Harmonic AM

We start with the simplest AM, assuming that the amplitude $A(t)$ of a RF carrier harmonic signal is modulated by some other harmonic signal of lower frequency. In this case, it is supposed that the message signal is

$$m(t) = m_0 \cos(\Omega t + \vartheta_0)\,, \tag{3.9}$$

where m_0 is the *peak modulating amplitude* or the *amplitude deviation*, $\Omega = 2\pi F$ is a natural (angular) modulation frequency, F is a modulation frequency, and ϑ_0 is a constant phase. If such a message is supposed, then $\mu = k_\mathrm{a} m_0$ is said to be the *modulation factor.*

By (3.8) and (3.9), the AM signal (3.7) acquires the form of

$$y(t) = A_0[1 + \mu\cos(\Omega t + \vartheta_0)]\cos(\omega_0 t + \psi_0) \tag{3.10}$$

that allows transferring to the signal spectrum, by the simple transformations,

$$\begin{aligned} y(t) &= A_0\cos(\omega_0 t + \psi_0) + A_0\mu\cos(\Omega t + \vartheta_0)\cos(\omega_0 t + \psi_0) \\ &= A_0\cos(\omega_0 t + \psi_0) + \frac{A_0\mu}{2}\cos[(\omega_0 - \Omega)t + \psi_0 - \vartheta_0] + \\ &\quad + \frac{A_0\mu}{2}\cos[(\omega_0 + \Omega)t + \psi_0 + \vartheta_0]\,. \end{aligned} \tag{3.11}$$

It follows, by (3.11), that the spectral line at the carrier frequency ω_0 is accompanied with two side spectral lines at the frequencies $\omega_0 \pm \Omega$. These lines are characterized by equal amplitudes $A_0\mu/2$. It is also seen that the spectrum (3.11) is symmetric. The carrier phase is saved at ω_0, whereas the phases of the side spectral lines possess shifts $\pm\vartheta_0$ at $\omega_0 \pm \Omega$, respectively, induced by

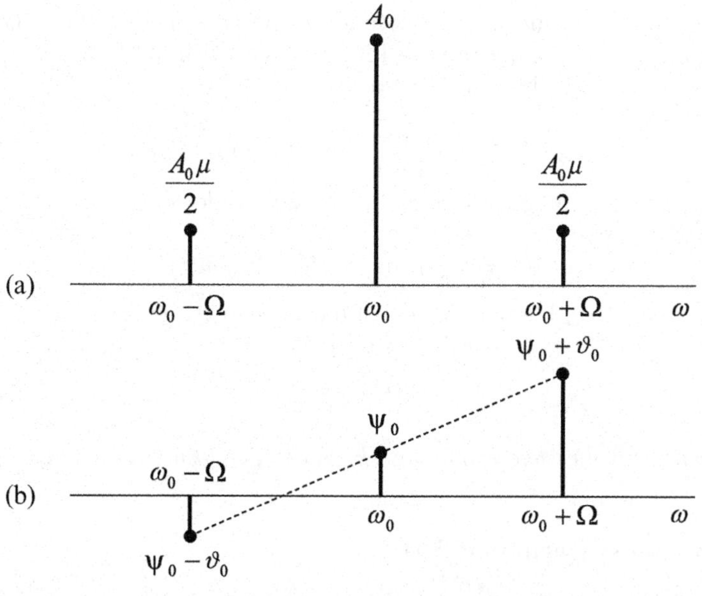

Fig. 3.1 Spectrum of AM signal with simplest modulation: **(a)** magnitude and **(b)** phase.

the modulating signal. One may also conclude that the case of $\psi_0 = \vartheta_0 = 0$ makes the phase spectrum to be identically zero. Figure 3.1 illustrates the magnitude and phase spectra associated with (3.10).

To provide undistorted modulation, the modulation factor is usually set to be less than unity, $0 < \mu < 1$. The case of a unit modulation factor, $\mu = 1$, is termed the *total AM*. AM may also be realized with overmodulation, if $\mu > 1$. Figure 3.2 sketches all three typical situations with AM.

An AM signal may alternatively be performed by the vector diagram combined with three rotated vectors as shown in Fig. 3.3. Here, the vector of a carrier signal with amplitude A_0 is rotated on an angle $\omega_0 t + \psi_0$ with respect to the basic coordinate vector. The vector with the amplitude $A_0\mu/2$ corresponds to the left-hand side spectral line at the frequency $\omega_0 - \Omega$. This vector has a phase $-\Omega t - \vartheta_0$ and is rotated clockwise with respect to the carrier vector. The vector of the right-hand side spectral line with the amplitude $A_0\mu/2$ and frequency $\omega_0 + \Omega$ has a phase $\Omega t + \vartheta_0$. It is rotated counterclockwise regarding the carrier vector.

An important peculiarity of AM is that the sideline vectors are rotated symmetrically about the carrier signal vector and therefore no angular modulation occurs. The maximum amplitude is attained when two side vectors coincide in direction with the carrier signal vector, i.e., when $\pm\Omega t \pm \vartheta_0 = 2\pi n$ for arbitrary integer n. When the side line vectors are forwarded opposite to the carrier signal vector, $\pm\Omega t \pm \vartheta_0 = \pi n$, the AM signal amplitude attains its minimum value.

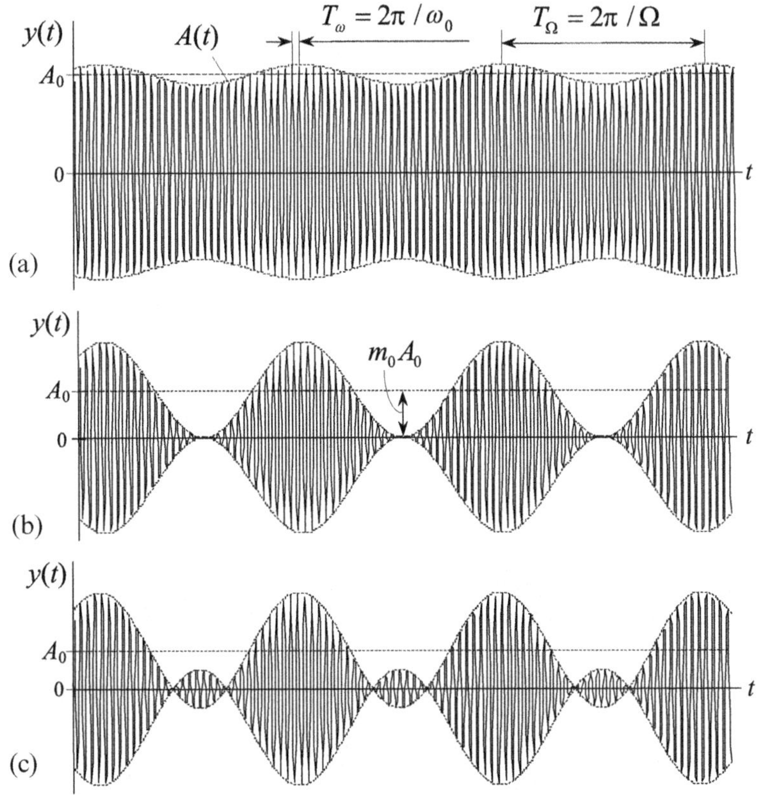

Fig. 3.2 AM of a carrier signal with different modulation factors: **(a)** $\mu < 1$, **(b)** $\mu = 1$, and **(c)** $\mu > 1$.

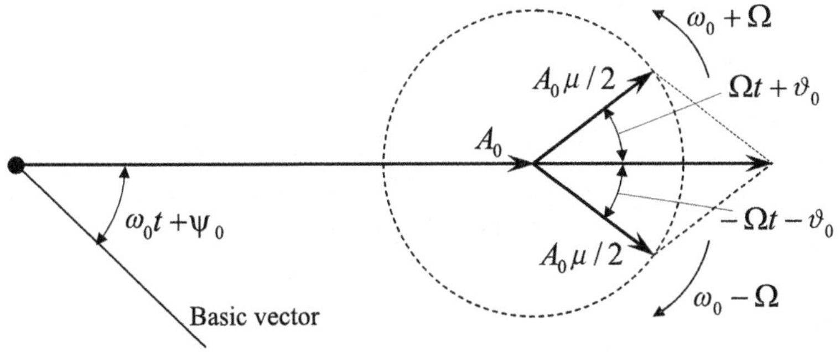

Fig. 3.3 Vector diagram of a signal with simplest harmonic AM for $\mu < 1$.

3.3.2 General Relations with AM

Simplest AM is not the one that is widely used. In applications, modulation is perfected by baseband signals of complex waveforms. Therefore, spectrum of an AM signal also becomes complex and its analysis often entails difficulties. Nevertheless, there are some common features of an AM signal allowing us to speak about its generalized spectrum.

The carrier signal is periodic by nature (causal or noncausal), whereas the modulating signal may be either periodical (e.g., harmonic) or not (single pulse). Hence, a signal with periodic AM will be characterized by the Fourier series and that with nonperiodic AM by the Fourier transform.

A generalized AM signal (3.7) may be assumed to be a product of the multiplication of two functions,

$$y(t) = x(t)z(t)\,, \tag{3.12}$$

where $x(t)$ is a modulating signal with known transform $X(j\omega)$ and $z(t)$ is a complex carrier signal with a unit amplitude

$$z(t) = e^{j(\omega_0 t + \psi_0)}\,. \tag{3.13}$$

For convenience and without loss of generality, the phase angle ψ_0 in (3.13) may be omitted and incorporated, if necessary, by using the time-shifting property of the Fourier transforms. If so, then (3.12) may be written as $y(t) = x(t)e^{j\omega_0 t}$.

By the multiplication property of the transform (2.59), the spectral density of (3.12) becomes

$$Y(j\omega) = \frac{1}{2\pi}X(j\omega) * Z(j\omega)\,. \tag{3.14}$$

Since the spectral density of a carrier signal $z(t)$ (3.13) is performed with the delta function,

$$Z(j\omega) = 2\pi\delta(\omega - \omega_0)\,, \tag{3.15}$$

then (3.14) transfers to

$$Y(j\omega) = X(j\omega - j\omega_0)\,. \tag{3.16}$$

We then conclude that the spectral density of a complex AM signal is one-sided and nothing more than the spectral density of a modulating signal removed to the carrier frequency.

The other generalization gives a real signal (3.7), which transform is

$$\begin{aligned} Y(j\omega) &= \int_{-\infty}^{\infty} x(t)\cos(\omega_0 t + \psi_0)\, e^{-j\omega t} dt \\ &= \frac{1}{2}\left[X(j\omega - j\omega_0)e^{j\psi_0} + X(j\omega + j\omega_0)e^{-j\psi_0}\right] = Y_-(j\omega) + Y_+(j\omega)\,. \end{aligned} \tag{3.17}$$

Observing (3.17), we conclude that the spectral density of a real AM signal is double-sided having the LSBs and USBs, respectively,

$$Y_-(j\omega) = \frac{1}{2}X(j\omega - j\omega_0)e^{j\psi_0}\,, \tag{3.18}$$

$$Y_+(j\omega) = \frac{1}{2}X(j\omega + j\omega_0)e^{-j\psi_0}\,. \tag{3.19}$$

It is important to notice that each $Y_-(j\omega)$ and $Y_+(j\omega)$ bears a half energy of the modulating signal.

Figure 3.4 sketches the generalized spectral densities of the modulating **(a)** and modulated **(b)** signals with AM. Intentionally, we restricted the width of the modulating signal with some maximum value $\Omega_{\max}$. As it may be observed, the spectral density of a modulated signal **(b)** consists of two shifted versions of the spectral density of a modulating signal **(a)** with a half amplitude. The sidebands locate around the negative and positive carrier frequencies, $-\omega_0$ and ω_0, occupying the ranges from $-\omega_0 - \Omega_{\max}$ to $-\omega_0 + \Omega_{\max}$ and $\omega_0 - \Omega_{\max}$ to $\omega_0 + \Omega_{\max}$, respectively.

Frequently, spectrum of a modulating signal is restricted with some minimum and maximum frequencies, $\Omega_{\min}$ and $\Omega_{\max}$, respectively. An example is a human voice that occupies a range from 300 to 3000 Hz. In this case, the LSB and USB of the AM spectrum is also performed with two sidebands as shown in Fig. 3.5.

Hence, an AM signal occupies the frequency range of

$$\Delta\omega_{\mathrm{AM}} = (\omega_0 + \Omega_{\max}) - (\omega_0 - \Omega_{\max}) = 2\Omega_{\max}\,.$$

Furthermore, the Fourier transform of an AM signal is symmetric about ω_0 and thus its two sidebands bear the same information. This splendid property of AM is exploited in the single side band (SSB) communications, by which only one of the two sidebands is transmitted and received saving the frequency resources of a channel.

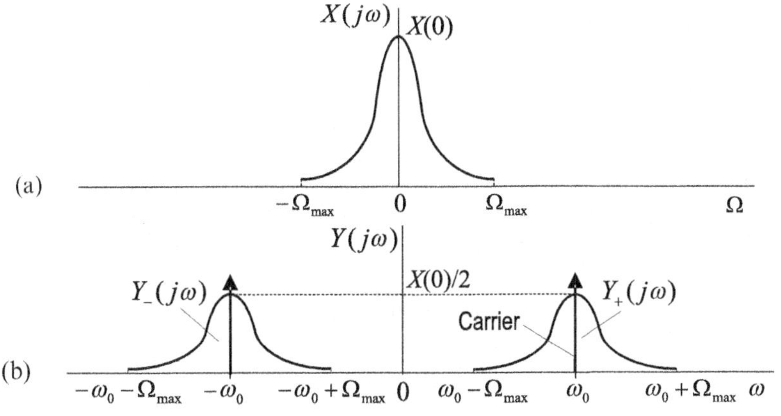

Fig. 3.4 Generalized spectral densities with AM: **(a)** modulating signal, and **(b)** AM signal.

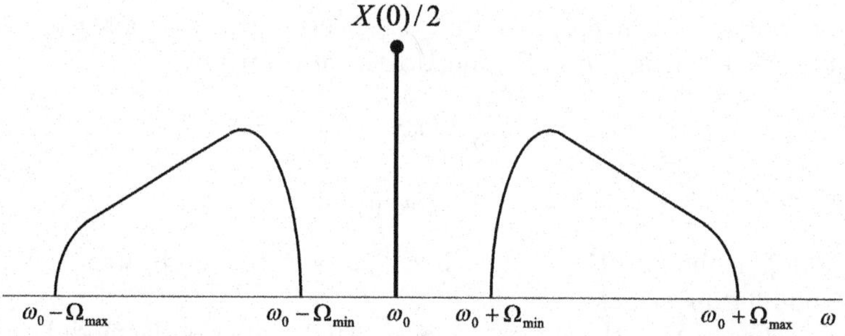

Fig. 3.5 The upper sideband of an AM signal with the band-pass modulating signal.

3.3.3 Carrier and Sideband Power of AM Signal

To optimize channels with AM, it is of importance to know the power of a carrier signal and fraction of the total power contained in the sidebands. This is because

- Carrier bears no information so it is just wasted power.
- AM inevitably varies the amplitude of a carrier signal and therefore some care should be done to match a channel with such a variable signal to avoid distortions.

An instantaneous power of the electric current i_{AM} modulated with a simplest AM may be assumed to be generated on some real load R, i.e.,

$$P(t) = i_{\mathrm{AM}}^2 R = \frac{v_{\mathrm{AM}}^2}{R} = \frac{A_0^2}{R}(1 + \mu\cos\Omega t)^2 \cos^2\omega_0 t\,, \tag{3.20}$$

where v_{AM} is the electric voltage induced by i_{AM} on R.

An average power calculated over the period $T_\omega = 2\pi/\omega_0$ of a carrier signal is given by

$$\langle P\rangle_\omega = \frac{1}{T_\omega}\int_{-T_\omega/2}^{T_\omega/2} P(t)\mathrm{d}t = \frac{A_0^2}{RT_\omega}\int_{-T_\omega/2}^{T_\omega/2}(1 + \mu\cos\Omega t)^2\cos^2\omega_0 t\,\mathrm{d}t$$

$$\cong \frac{A_0^2}{RT_\omega}(1 + \mu\cos\Omega t)^2\int_{-T_\omega/2}^{T_\omega/2}\cos^2\omega_0 t\,\mathrm{d}t = P_0(1 + \mu\cos\Omega t)^2\,, \tag{3.21}$$

where the power of a pure carrier signal is

$$P_0 = \frac{A_0^2}{2R}\,. \tag{3.22}$$

It follows from (3.21) that the maximum and minimum average power is, respectively,

$$\langle P\rangle_{\omega\,\max} = P_0(1+\mu)^2\,, \tag{3.23}$$

$$\langle P\rangle_{\omega\,\min} = P_0(1-\mu)^2\,. \tag{3.24}$$

Thus, by $\mu = 1$, we rely on $\langle P\rangle_{\omega\,\max} = 4P_0$ and $\langle P\rangle_{\omega\,\min} = 0$.

The signal power averaged over the period $T_\Omega = 2\pi/\Omega$ is calculated by

$$\langle P\rangle_\Omega = \frac{1}{T_\Omega}\int\limits_{-T_\Omega/2}^{T_\Omega/2} P_0(1+\mu\cos\Omega t)^2 \mathrm{d}t$$

$$= \frac{P_0}{T_\Omega}\int\limits_{-T_\Omega/2}^{T_\Omega/2} \left(1+2\mu\cos\Omega t+\mu^2\cos^2\Omega t\right)\mathrm{d}t \tag{3.25}$$

that, by an identity $\int\cos^2 x\mathrm{d}x = 1/2(\sin x\cos x) + 1/2(x)$, becomes

$$\langle P\rangle_\Omega = P_0\left(1+\frac{\mu^2}{2}\right) = P_0 + \frac{\mu^2 P_0}{2}\,. \tag{3.26}$$

By $\mu = 1$ and $\mu = 0$, (3.26) becomes $\langle P\rangle_\Omega = 3P_0/2$ and $\langle P\rangle_\Omega = P_0$, respectively.

It is seen that the total signal power (3.26) consists of the carrier power P_0 and the sidebands power $P_0\mu^2/2$, and hence even with the 100% AM, $\mu = 1$, the sidebands power is $P_0/2$, meaning that about 67% of the total signal power is just wasted with the carrier.

3.4 Amplitude Modulation by Impulse Signals

In digital and impulse systems, AM is achieved with impulse waveforms and an analysis problem is solved with (3.15) and (3.16). So, the spectral density of a modulated signal is learned by the spectral density of a modulating impulse signal. We notice that practical implementations of impulse AM may differ resulting in different system structures.

3.4.1 AM by a Rectangular Pulse

A typical and even classical RF pulse has a constant carrier frequency and a rectangular envelope (Fig. 3.6). Most commonly, it may be described by

$$y(t) = x(t)z(t)\,, \tag{3.27}$$

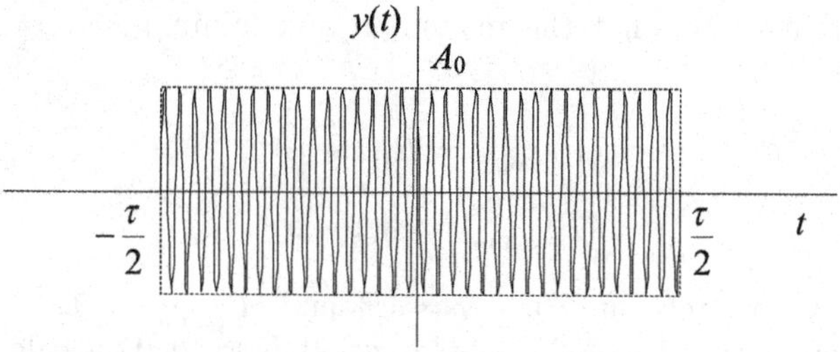

Fig. 3.6 RF impulse signal with a rectangular envelope.

where the modulating rectangular impulse signal $x(t)$ and the real carrier signal $z(t)$ are, respectively,

$$x(t) = A_0 \left[u\left(t + \frac{\tau}{2}\right) - u\left(t - \frac{\tau}{2}\right) \right] , \tag{3.28}$$

$$z(t) = \cos(\omega_0 t + \psi_0) . \tag{3.29}$$

The Fourier transforms of (3.27), by (3.17), is

$$Y(j\omega) = \frac{1}{2} e^{-j\psi_0} X(j\omega + j\omega_0) + \frac{1}{2} e^{j\psi_0} X(j\omega - j\omega_0) , \tag{3.30}$$

where the shifted spectral densities of the rectangular envelope are, by (2.80),

$$X(j\omega \pm j\omega_0) = A_0 \tau \frac{\sin[(\omega \pm \omega_0)\tau/2]}{(\omega \pm \omega_0)\tau/2} . \tag{3.31}$$

For the sake of simplicity, we set below $\psi_0 = 0$, remembering that any phase shift may easily be incorporated by using the time-shifting property of the Fourier transforms. Yet, we will analyze only the USB in (3.31), since the LSB is just its reflection. So, we write the USBs of the magnitude and phase spectral densities associated with (3.30) and (3.31), respectively, as

$$|Y_+(j\omega)| = \frac{A_0 \tau}{2} \left| \frac{\sin[(\omega - \omega_0)\tau/2]}{(\omega - \omega_0)\tau/2} \right| , \tag{3.32}$$

$$\varphi_y(\omega) = \begin{cases} 0 \pm 2\pi n, & \frac{\sin[(\omega - \omega_0)\tau/2]}{(\omega - \omega_0)\tau/2} \geqslant 0 \\ \pm\pi \pm 2\pi n, & \frac{\sin[(\omega - \omega_0)\tau/2]}{(\omega - \omega_0)\tau/2} < 0 \end{cases} , \quad n = 0, 1, \ldots . \tag{3.33}$$

Figure 3.7 illustrates (3.32) and (3.33). Conventionally, we show the phase density has taken the values of 0 and $-\pi$ to the right of the carrier frequency and 0 and π to the left. An important finding follows instantly: to concentrate the signal energy in the narrow band around the carrier frequency f_0, as it very often is desirable, the duration τ must be large. But the pulse usually is required to be short in time. This contradiction is usually solved in applications in an optimum way.

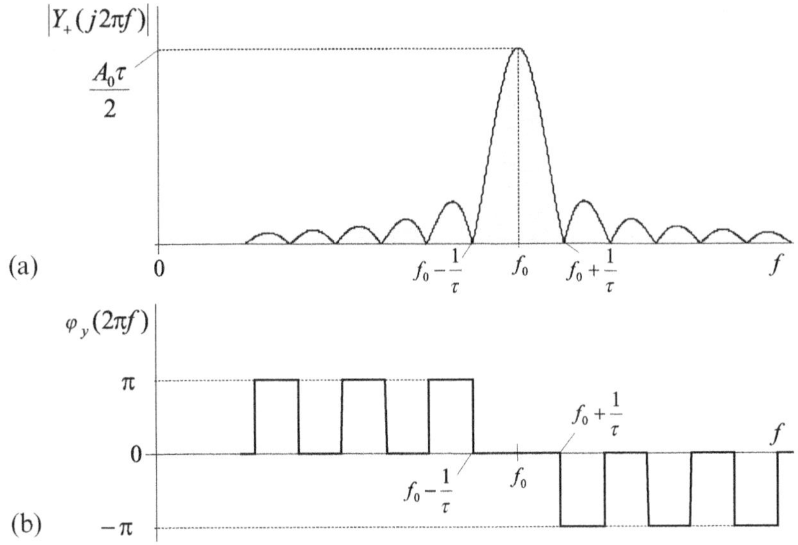

Fig. 3.7 Spectral density of an AM signal with a rectangular envelope: **(a)** magnitude and **(b)** phase.

3.4.2 Gaussian RF Pulse

The Gaussian waveform is of importance, e.g., for radars and remote sensing. A near Gaussian envelope may also appear in the original rectangular RF pulse after passed through the narrowband channel. This notable signal is described by

$$y(t) = A_0 e^{-(\alpha t)^2} \cos(\omega_0 t + \psi_0) , \tag{3.34}$$

where an attenuation coefficient α is often coupled with the pulse duration τ. Figure 3.8 illustrates (3.34).

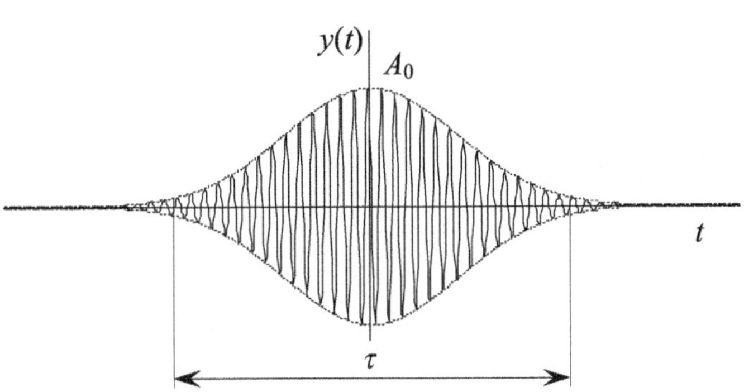

Fig. 3.8 Gaussian RF pulse.

The Fourier transform of (3.34) is performed around ω_0 by

$$Y_+(j\omega) = \frac{1}{2}X(j\omega - j\omega_0)e^{j\psi_0} = \frac{A_0\sqrt{\pi}}{2\alpha}e^{-\frac{(\omega-\omega_0)^2}{4\alpha^2}}e^{j\psi_0}, \tag{3.35}$$

producing the magnitude and phase densities, respectively,

$$|Y_+(j\omega)| = \frac{A_0\sqrt{\pi}}{2\alpha}e^{-\frac{(\omega-\omega_0)^2}{4\alpha^2}}, \tag{3.36}$$

$$\varphi_{y+}(\omega) = \psi_0 . \tag{3.37}$$

Figure 3.9 sketches (3.36) and (3.37). As it is seen, the phase density is performed here only with a constant phase ψ_0 and we recall that any time shift of a pulse on t_0 will result in the additional linear phase drift ωt_0.

3.4.3 AM by Pulse-Bursts

We have already mentioned that the pulse-bursts are used in applications to increase power of signals and thereby increase the signal-to-noise ratio at receiver. The waveform of the RF pulse-burst may be arbitrary as it is shown in Fig. 3.10.

By using (2.140) and relating this relation to the carrier frequency ω_0, the USB of the spectral density of the RF pulse-burst becomes as follows:

$$Y_+(j\omega) = \frac{1}{2}X(j\omega - j\omega_0)\frac{\sin[(\omega-\omega_0)NT/2]}{\sin[(\omega-\omega_0)T/2]}e^{-j\omega(N-1)\frac{T}{2}}, \tag{3.38}$$

producing the magnitude and phase spectral densities, respectively,

$$|Y_+(j2\pi f)| = \frac{1}{2}\left|X(j2\pi f - j2\pi f_0)\frac{\sin[\pi N(f-f_0)T]}{\sin[\pi(f-f_0)T]}\right| \tag{3.39}$$

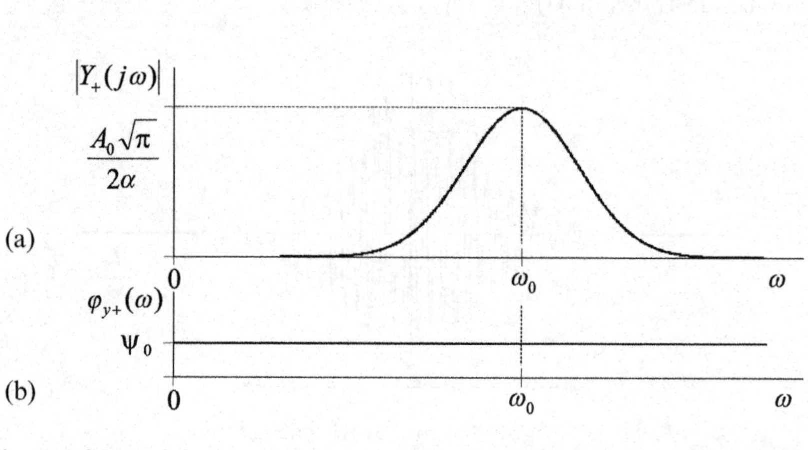

Fig. 3.9 Spectral density of a Gaussian RF pulse: **(a)** magnitude and **(b)** phase.

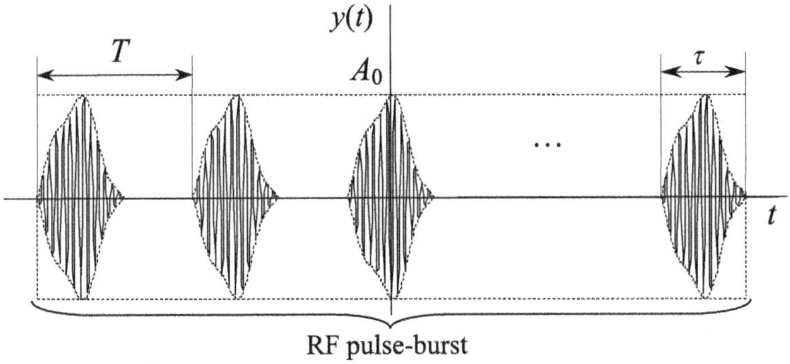

Fig. 3.10 RF pulse-burst.

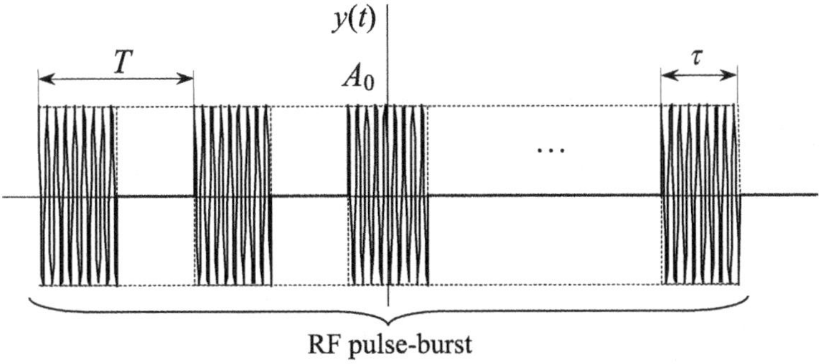

Fig. 3.11 RF rectangular pulse-burst.

$$\varphi_{y+}(f) = \begin{cases} 0 \pm 2\pi n, & X(j2\pi f - j2\pi f_0)\dfrac{\sin[\pi N(f-f_0)T]}{\sin[\pi(f-f_0)T]} \geqslant 0 \\ \\ \pm\pi \pm 2\pi n, & X(j2\pi f - j2\pi f_0)\dfrac{\sin[\pi N(f-f_0)T]}{\sin[\pi(f-f_0)T]} < 0 \end{cases}, \qquad n = 0, 1, \ldots \tag{3.40}$$

Example 3.1. Given a symmetric RF rectangular pulse-burst with $q = 2$ and $N = 4$ (Fig. 3.11), which envelope is analyzed in Example 2.25. An analysis problem is solved by (3.38) and (2.80) in the form of

$$Y_+(j\omega) = \frac{A_0\tau}{2}\frac{\sin[(\omega-\omega_0)\tau/2]}{(\omega-\omega_0)\tau/2}\frac{\sin[(\omega-\omega_0)NT/2]}{\sin[(\omega-\omega_0)T/2]}e^{j\psi_0} \tag{3.41}$$

that leads to magnitude and phase spectral densities, respectively,

$$|Y_+(j2\pi f)| = \frac{A_0\tau}{2}\left|\frac{\sin[\pi(f-f_0)\tau]}{\pi(f-f_0)\tau}\frac{\sin[\pi N(f-f_0)T]}{\sin[\pi(f-f_0)T]}\right|, \tag{3.42}$$

$$\varphi_{y+}(f) = \begin{cases} 0 \pm 2\pi n, & \dfrac{\sin[\pi(f-f_0)\tau]}{\pi(f-f_0)\tau}\dfrac{\sin[\pi N(f-f_0)T]}{\sin[\pi(f-f_0)T]} \geqslant 0 \\ \pm\pi \pm 2\pi n, & \dfrac{\sin[\pi(f-f_0)\tau]}{\pi(f-f_0)\tau}\dfrac{\sin[\pi N(f-f_0)T]}{\sin[\pi(f-f_0)T]} < 0 \end{cases}, \quad n = 0, 1, \ldots \tag{3.43}$$

Figure 3.12 illustrates (3.42) along with the modulating signal.

□

Example 3.2. Given a symmetric RF rectangular pulse-burst with $q = 1.5, N = 2, \tau = 1 \times 10^{-6}$s, $f_0 = 1 \times 10^8 Hz$, and $A_0 = 1V$. Zero points of the envelope of the spectral density follow with the step $k/\tau = k \times 10^6$ Hz. Maximums of the main side lobes correspond to $l/T = l/1.5\tau \cong 0.666n \times 10^6$ Hz. Maximum values of the envelopes are $A_0\tau = 1 \times 10^{-6}$ V/Hz and $A_0\tau/2 = 5 \times 10^{-7}\, V/Hz$ at $f = f_0$. Nulls of the spectral density follow with the span $m/NT = m/Nq\tau = 0.333m \times 10^6\, Hz$.

The relevant magnitude spectral density is shown in Fig. 3.13. It is neatly seen that the maximums of the main lobes are shifted for the frequencies calculated. Yet, the envelope of the side lobes lost its meaning. These effects occur when the burst is combined with a small number of pulses.

□

Example 3.3. Given an RF triangular pulse-burst with $q = 3.25, N = 2, \tau = 1 \times 10^{-6}$ s, $f_0 = 1 \times 10^8$ Hz, and $A_0 = 1$ V. By (2.90) and (3.38), its spectral

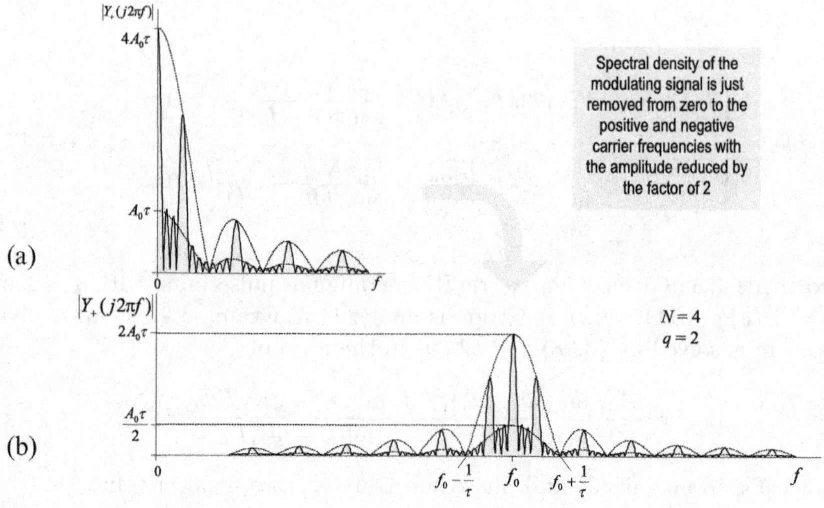

Fig. 3.12 Magnitude spectral density of an RF rectangular pulse-burst with $q = 2$ and $N = 4$: **(a)** modulating signal and **(b)** modulated signal.

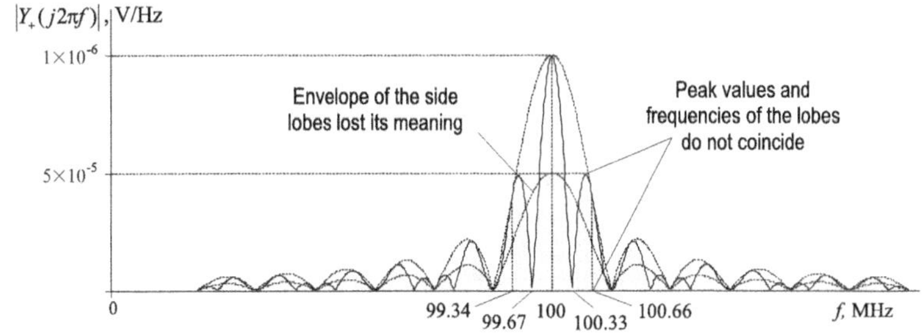

Fig. 3.13 Magnitude spectral density of an RF rectangular pulse-burst with $q = 1.5, N = 2, \tau = 1 \times 10^{-6}$ s, $f_0 = 1 \times 10^8$ Hz, and $A_0 = 1\,\text{V}$.

density becomes

$$Y_+(j\omega) = \frac{A\tau}{4}\left\{\frac{\sin[(\omega-\omega_0)\tau/4]}{(\omega-\omega_0)\tau/4}\right\}^2 \frac{\sin[(\omega-\omega_0)NT/2]}{\sin[(\omega-\omega_0)T/2]}\,. \tag{3.44}$$

and the magnitude and phase densities are given by, respectively,

$$|Y_+(j2\pi f)| = \frac{A\tau}{4}\frac{\sin^2[\pi(f-f_0)\tau/2]}{[\pi(f-f_0)\tau/2]^2}\left|\frac{\sin[\pi(f-f_0)NT]}{\sin[\pi(f-f_00)T]}\right|, \tag{3.45}$$

$$\varphi_{y+}(f) = \begin{cases} 0 \pm 2\pi n, & \dfrac{\sin[\pi(f-f_0)NT]}{\sin[\pi(f-f_0)T]} \geqslant 0 \\ \pm\pi \pm 2n\pi, & \dfrac{\sin[\pi(f-f_0)NT]}{\sin[\pi(f-f_0)T]} < 0 \end{cases}, \quad n = 0, 1, \ldots \tag{3.46}$$

The nulls of (3.45) are calculated at $2k/\tau = 2k \times 10^6$ Hz. The peak values of the main lobes are expected at $l/T = l/3.5\tau \cong 0.3l \times 10^6$ Hz. Maxima of the main and side lobes, $4E\tau/4 = 1 \times 10^{-6}\,V/\text{Hz}$ and $E\tau/4 = 2.5 \times 10^{-7}\,\text{V/Hz}$, respectively, are placed at f_0. The frequency span between the nulls of the spectral density is $m/NT = m/Nq\tau \cong 7.8m \times 10^4$ Hz. Figure 3.14 illustrates this spectral performance.

□

Example 3.4. Given an RF sinc pulse-burst with $q = 3.5, N = 3, \tau = 1 \times 10^{-6}$ s, $f_0 = 1 \times 10^8$ Hz, and $A_0 = 1\,\text{V}$. By $\alpha = 2\pi/\tau$, its spectral density is provided to be

$$Y_+(j\omega) = \begin{cases} \dfrac{A\tau}{4}\dfrac{\sin[(\omega-\omega_0)NT/2]}{\sin[(\omega-\omega_0)T/2]}, & |\omega| \leqslant 2\pi/\tau \\ 0, & |\omega| > 2\pi/\tau \end{cases} \tag{3.47}$$

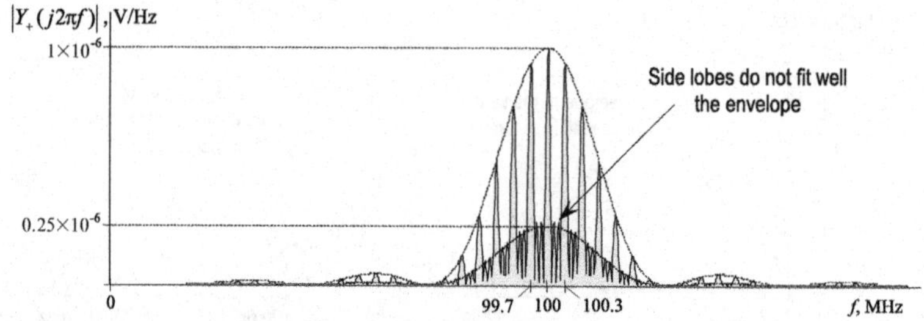

Fig. 3.14 Magnitude spectral density of an RF burst of triangular pulses with $q = 3.25, N = 4, \tau = 1 \times 10^{-6}$ s, $f_0 = 1 \times 10^8$ Hz, and $A_0 = 1$ V.

and defines the the magnitude and phase spectral densities by

$$|Y_+(j2\pi f)| = \begin{cases} \dfrac{A\tau}{4}\left|\dfrac{\sin[\pi(f-f_0)NT]}{\sin[\pi(f-f_0)T]}\right|, & |f| \leqslant \frac{1}{\tau} \\ 0, & |f| > \frac{1}{\tau} \end{cases}, \tag{3.48}$$

$$\varphi_{y+}(f) = \begin{cases} 0 \pm 2\pi n, & \dfrac{\sin[\pi(f-f_0)NT]}{\sin[\pi(f-f_0)T]} \geqslant 0 \\ \pm\pi \pm 2n\pi, & s\dfrac{\sin[\pi(f-f_0)NT]}{\sin[\pi(f-f_0)T]} < 0 \end{cases}, \quad n = 0, 1, \ldots \tag{3.49}$$

The cutoff frequency for (3.47) is calculated by $1/\tau = 1 \times 10^6$ Hz. The frequencies corresponding to maxima of the main lobes are defined by $l/T = l/3.5\tau \cong 2.8l \times 10^5$ Hz. The envelope value of the main lobes is $3A_0\pi/2\alpha = 3A_0\tau/4 = 7.5 \times 10^{-7}$ V/Hz at f_0 and that of the side lobes is $A_0\pi/2\alpha = A_0\tau/4 = 2.5 \times 10^{-7}$ V/Hz. Spectral nulls are watched at $m/NT = m/Nq\tau \cong 9.5m \times 10^4$ Hz. The magnitude density of this RF signal is given in Fig. 3.15. Note that the Gibbs phenomenon appears here owing to the finite integration range.

□

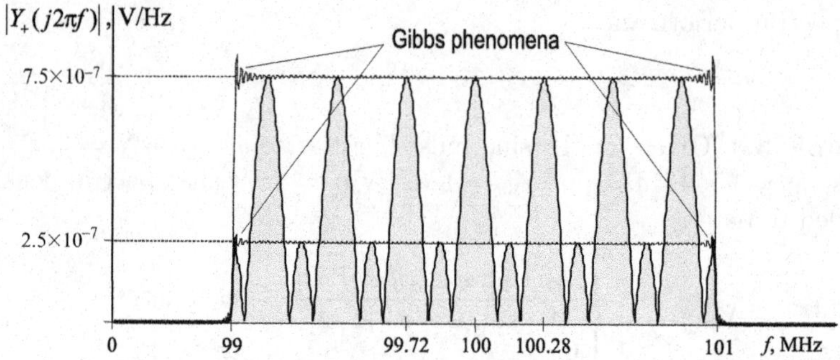

Fig. 3.15 Magnitude spectral density of an RF sinc pulse-burst with $q = 3.25, N = 2, \tau = 1 \times 10^{-6}$ s, $f_0 = 1 \times 10^8$ Hz, and $A_0 = 1$ V.

3.5 Amplitude Modulation by Periodic Pulses

The RF pulse-train and pulse-burst-train are representatives of AM signals with periodic impulse AM. In line with other periodic signals, their spectra can easily be sketched employing the Fourier transform of the modulating signal.

3.5.1 Spectrum of RF Signal with Periodic Impulse AM

If to perform the one-sided spectrum of a periodic modulating signal (2.14) by the series

$$x(t) = C_0 + 2\sum_{k=1}^{\infty} C_k e^{jk\Omega t} = C_0 + 2\sum_{k=1}^{\infty} |C_k|\, e^{j(k\Omega t - \Psi_k)}\,, \tag{3.50}$$

and assign a carrier signal to be $z(t) = e^{j\omega_0 t}$, then a real AM signal (3.12) would be performed by

$$\begin{aligned} y(t) &= C_0 \cos\omega_0 t + 2\sum_{k=1}^{\infty} |C_k| \cos(k\Omega t - \Psi_k)\cos\omega_0 t \\ &= C_0 \cos\omega_0 t + \sum_{k=1}^{\infty} |C_k| \cos[(\omega_0 + k\Omega)t - \Psi_k] \\ &\quad + \sum_{k=1}^{\infty} |C_k| \cos[(\omega_0 - k\Omega)t + \Psi_k] \\ &= C_0 \cos\omega_0 t + \sum_{k=1}^{\infty} C_k \cos(\omega_0 + k\Omega)t + \sum_{k=1}^{\infty} C_k \cos(\omega_0 - k\Omega)t\,, \end{aligned} \tag{3.51}$$

As it may easily be observed, a set of spectral lines (3.51) is placed at the frequencies $\omega_0 \pm k\Omega$, $k = 0, 1, \ldots$, with the complex amplitudes C_k determined by the modulating waveform.

3.5.2 AM by Periodic Rectangular Pulse

The RF signal modulated by a periodic rectangular pulse-train is shown in Fig. 3.16. The magnitude and phase spectra of the modulating signal are given by (2.123) and (2.124), respectively,

$$|C_k| = \frac{A}{q}\left|\frac{\sin(k\pi/q)}{k\pi/q}\right|, \quad C_0 = \frac{A}{q}\,, \tag{3.52}$$

$$\Psi_k = \begin{cases} \pm 2\pi n, & \dfrac{\sin(k\pi/q)}{k\pi/q} \geqslant 0 \\ \pm\pi \pm 2\pi n, & \dfrac{\sin(k\pi/q)}{k\pi/q} < 0 \end{cases}, \quad n = 0, 1, \ldots \tag{3.53}$$

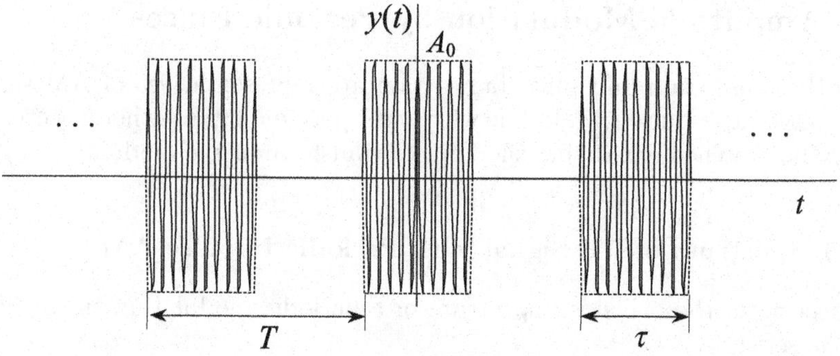

Fig. 3.16 RF rectangular pulse-train.

Using (3.52) and (3.53), spectrum of an AM signal (3.51) may be rewritten as

$$
\begin{aligned}
y(t) &= \frac{A}{q}\cos 2\pi f_0 t \\
&+\frac{A}{q}\sum_{k=1}^{\infty}\left|\frac{\sin(k\pi/q)}{k\pi/q}\right|\cos\left[2\pi\left(f_0+kF\right)t-\Psi_k\right] \\
&+\frac{A}{q}\sum_{k=1}^{\infty}\left|\frac{\sin(k\pi/q)}{k\pi/q}\right|\cos\left[2\pi\left(f_0-kF\right)t+\Psi_k\right],
\end{aligned}
\qquad (3.54)
$$

where $F = 2\pi\Omega$ and $q = T/\tau$.

Evolution of the magnitude spectrum (3.54) by changing q is shown in Fig. 3.17, allowing for several important generalizations:

- By $q = 1$, the RF rectangular pulse-train becomes a pure harmonic unmodulated signal. Its magnitude spectrum is represented by the only spectral line placed at $f = f_0$. Therefore, the spectrum envelope loses its meaning.
- With $q > 1$, the spectral lines appear around f_0 with the span multiple to $1/T$. The envelope of these lines is shaped by the scaled transform of a single pulse.
- By $q \to \infty$, the spectral lines become snugly placed and the discrete spectrum degenerates to the continuous one.

Overall, Fig. 3.17 gives an idea about how a single spectral line associated with a purely harmonic signal may evolve to the discrete spectrum of a periodic RF AM signal and then to the spectral density associated with the non periodic RF AM signal.

3.5.3 AM by Periodic Pulse-Burst

Having no peculiarities in the derivation procedure, the Fourier series of the RF signal with AM by a periodic pulse-burst may be written straightforwardly

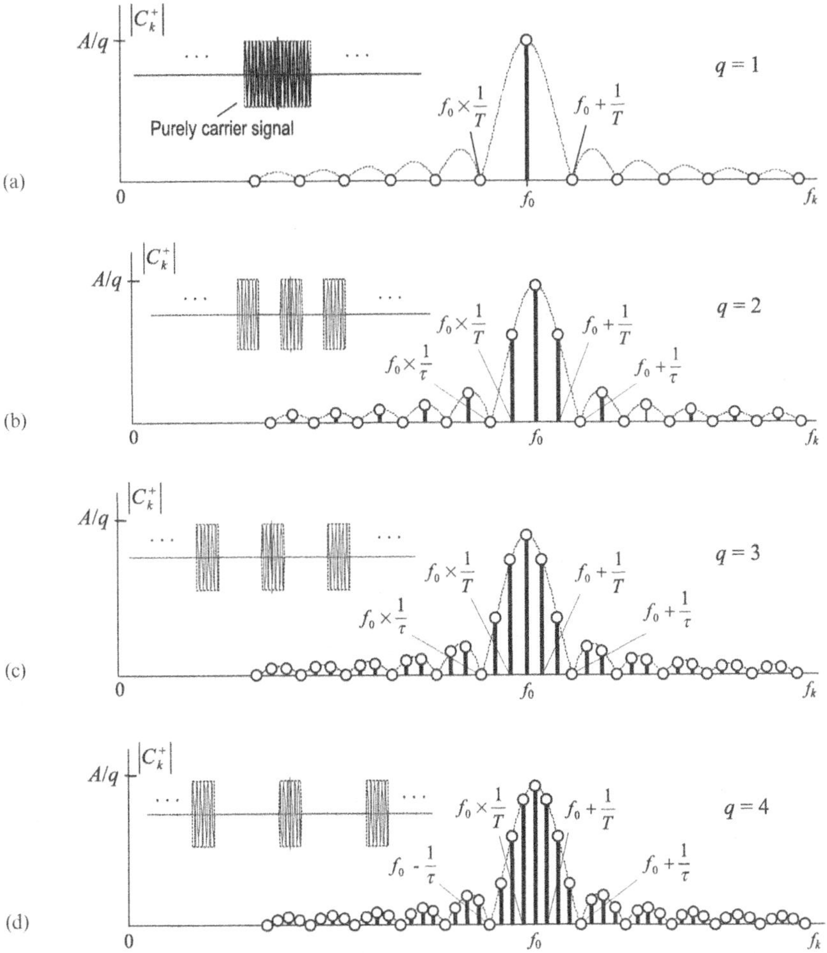

Fig. 3.17 Magnitude spectrum of an RF rectangular pulse-train.

using (3.51). Indeed, if one knows the coefficients C_k of the modulating signal, then (3.51) is a solution of the problem. By the generalized density of the pulse-burst (2.140), an analysis problem of such a signal is solved with

$$
\begin{aligned}
y(t) &= \frac{N}{T_1} X(0) \cos \omega_0 t \\
&+ \frac{1}{T_1} \sum_{k=0}^{\infty} \left| X(j2\pi f_k) \frac{\sin(\pi N f_k T)}{\sin(\pi f_k T)} \right| \cos[(\omega_0 + k\Omega)t - \Psi_k] \\
&+ \frac{1}{T_1} \sum_{k=0}^{\infty} \left| X(j2\pi f_k) \frac{\sin(\pi N f_k T)}{\sin(\pi f_k T)} \right| \cos[(\omega_0 - k\Omega)t + \Psi_k] ,
\end{aligned}
\quad (3.55)
$$

where $X(j2\pi f_k)$ is a discrete value of the spectral density of a single pulse taken at the frequency $f_k = kF_1$ multiple to $F_1 = 1/T_1$, and the phase spectrum is given by

$$\Psi_k = \begin{cases} \pm 2\pi n, & X(j2\pi f_k)\dfrac{\sin(\pi N f_k T)}{\sin(\pi f_k T)} \geqslant 0 \\ \pm\pi \pm 2\pi n, & X(j2\pi f_k)\dfrac{\sin(\pi N f_k T)}{\sin(\pi f_k T)} < 0 \end{cases}, \quad n = 0, 1, \ldots \tag{3.56}$$

Example 3.5. Given an RF signal with AM by a periodic rectangular pulse-burst with $\tau = 1 \times 10^{-6}$ s, $T = 2 \times 10^{-6}$ s, $N = 2$, $f_0 = 1 \times 10^8$ Hz, $T_1 = 2NT$, and $A = 1$ V. By (2.80), the function $X(j2\pi f_k)$ calculates

$$X(j2\pi f_k) = A\tau \frac{\sin(\pi f_k)\tau}{\pi f_k \tau} \tag{3.57}$$

and the series (3.55) becomes

$$\begin{aligned} y(t) = &\frac{AN}{q_1}\cos\omega_0 t \\ &+ \frac{A}{q_1}\sum_{k=1}^{\infty}\left|\frac{\sin(\pi k/q_1)}{\pi k/q_1}\frac{\sin(\pi k/q_2)}{\sin(\pi k/Nq_2)}\right|\cos[2\pi(f_0 + kF)t - \Psi_k] \\ &+ \frac{A}{q_1}\sum_{k=1}^{\infty}\left|\frac{\sin(\pi k/q_1)}{\pi k/q_1}\frac{\sin(\pi k/q_2)}{\sin(\pi k/Nq_2)}\right|\cos[2\pi(f_0 - kF)t + \Psi_k], \end{aligned} \tag{3.58}$$

where $q = T/\tau = 2, q_1 = T_1/\tau = 8$, and $q_2 = T_1/NT = 2$. Nulls of the envelope of the main lobes lie at the frequencies $k/\tau = k \times 10^6$ Hz. Maximums of the main lobes correspond to the frequencies $l/T = 0.5l \times 10^6$ Hz. The magnitude of the spectral line at the carrier frequency is $NA/q_1 = 0.25$ V/Hz. The span between the neighboring nulls of the envelope calculates $m/NT = m/Nq\tau = 0.25m \times 10^6$ HZ. The step between the neighboring spectral lines is $p/T_1 = 0.125 \times 10^6$ Hz. Figure 3.18 illustrates the magnitude spectrum of this signal and we note that, by $N = 2$, there are no side lobes associated with the pulse-burst.

□

3.6 Types of Analog AM

We now know fundamentals of AM and may go on in learning its structural realizations. The most recognized types of analog AM are discussed below along with the modulation and demodulation schemes.

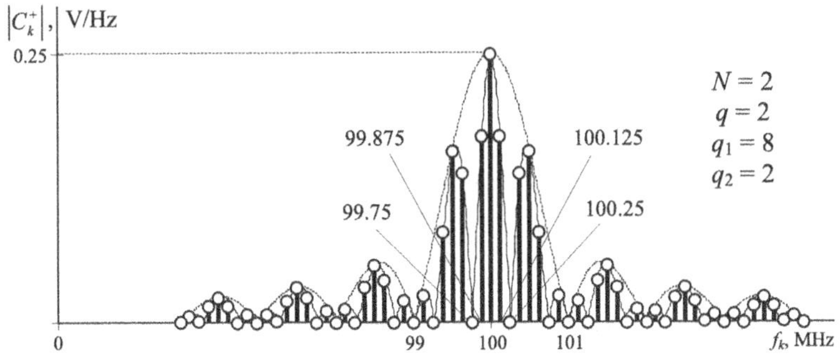

Fig. 3.18 Magnitude spectrum of an RF signal with AM by a periodic rectangular pulse-burst.

3.6.1 Conventional AM with Double Sideband Large Carrier

The most familiar form of AM is called the *conventional AM* or the *double sideband large carrier* (DSB-LC) that we have already learned in Section 3.3. Most generally, it is supposed that the modulating signal is $x(t) = A_0[1 + k_a m(t)]$ and AM signal $y(t) = x(t)z(t)$ appears as a product of the multiplication of $x(t)$ and the carrier signal $z(t) = \cos(\omega_0 t + \psi_0)$. Figure 3.19 illustrates this technique.

The Fourier transform $Y(j\omega)$ of a DSB-LC signal $y(t)$ is given by (3.17) and illustrated in Fig. 3.20. It follows that the spectral content of the modulating signal is removed to the negative and positive carrier frequencies with a scaling factor of $1/2$. Yet, the vector of a carrier signal exists in the modulated signal. When the one-sided spectral density is considered, then the Fourier transform of an AM signal at ω_0 is exactly that of a message signal at zero.

Restoration of a modulating signal $x(t)$ via DSB-LC signal $y(t)$ may be done in two different ways called synchronous and asynchronous demodulation.

3.6.2 Synchronous Demodulation

In the *synchronous demodulation*, a signal $y(t)$ is multiplied with the carrier signal $z(t)$ and then the product of this multiplication passes through the

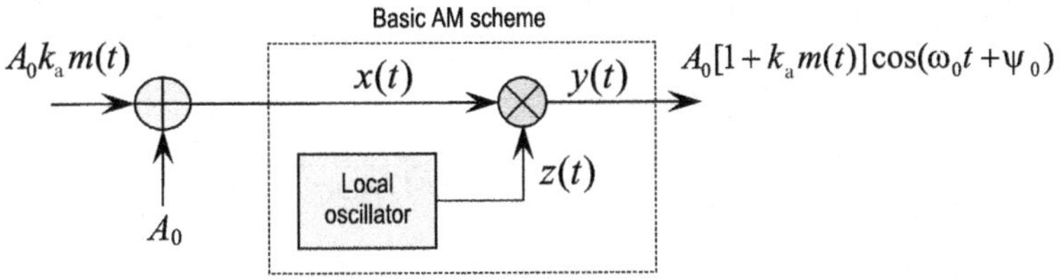

Fig. 3.19 AM scheme for DSB-LC.

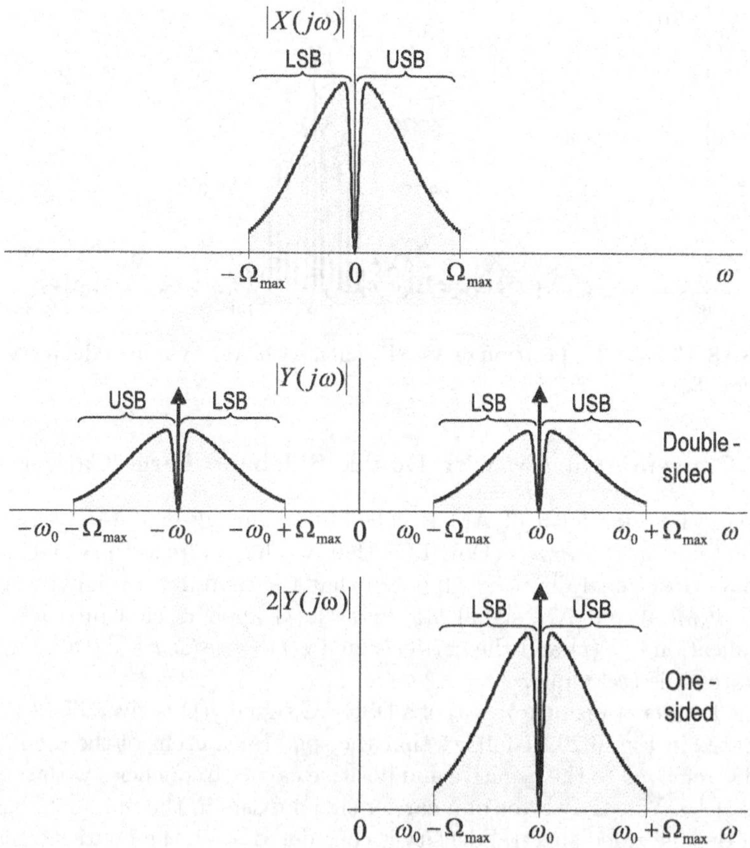

Fig. 3.20 Frequency content of a DSB-LC signal.

low-pass (LP) filter to keep only low-frequency components of the spectrum. The demodulator makes the critical assumption that both the carrier used at the transmitter and the local oscillator used at the receiver have exactly the same frequency and phase. This can only be achieved if two oscillators are synchronized in frequency and phase. Therefore, this type of demodulation is also called *coherent*. In a lack of synchronization, distortion will occur in the demodulated signal.

Figure 3.21 illustrates this scheme. The input of an LP filter is formed to be $y(t)z(t) = x(t)\cos^2(\omega_0 t + \psi_0)$. Extending the squared cosine function into the harmonic series gives $\cos^2\phi = 1/2 + 1/2\cos 2\phi$ and yields

$$y(t)z(t) = \frac{1}{2}x(t) + \frac{1}{2}x(t)\cos(2\omega_0 t + 2\psi_0)\,.$$

If the gain factor of 2 is assumed in the LP filter, the output $\tilde{x}(t)$ appears to be very close to $x(t)$ and the synchronous demodulation is complete.

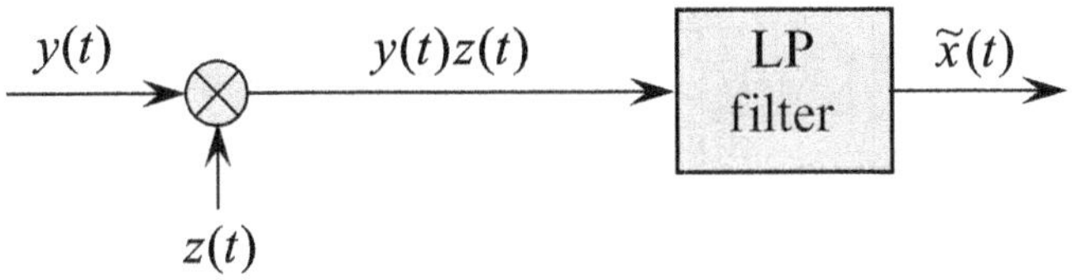

Fig. 3.21 Synchronous demodulation of AM signal.

3.6.3 Asynchronous Demodulation

In the *asynchronous demodulation*, they use a rectifier (Fig. 3.22a) that tracks the envelope of an AM signal as it is shown in Fig. 3.22b. The scheme exploits a simple nonlinear electric-circuit rectifier, being historically the first used in practical applications. This is the simplest demodulator. Its easily seen disadvantage is that the product of a demodulation $\tilde{x}(t)$ is achieved using the time-invariant filter and therefore the output waveform may be far from ideal, just as it is shown in Fig. 3.22b. In contrast, the synchronous approach produces $\tilde{x}(t)$ that usually appears to be much closer to the modulating signal. The effect, however, is achieved using a carrier signal at demodulator that complicates the approach strongly and is its disadvantage.

3.6.4 Square-law Demodulation

The *asynchronous* demodulation of DSB-LC signals may also be provided using what is called *square-law demodulation.* Figure 3.23 shows the relevant scheme.

In this scheme, the AM signal passes through a squaring device to produce

$$\begin{aligned} y^2(t) &= A_0^2[1 + k_a m(t)]^2 \cos^2(\omega_0 t + \psi_0) \\ &= \frac{A_0^2}{2}[1 + k_a m(t)]^2 + \frac{A_0^2}{2}[1 + k_a m(t)]^2 \cos[2(\omega_0 t + \psi_0)] . \end{aligned}$$

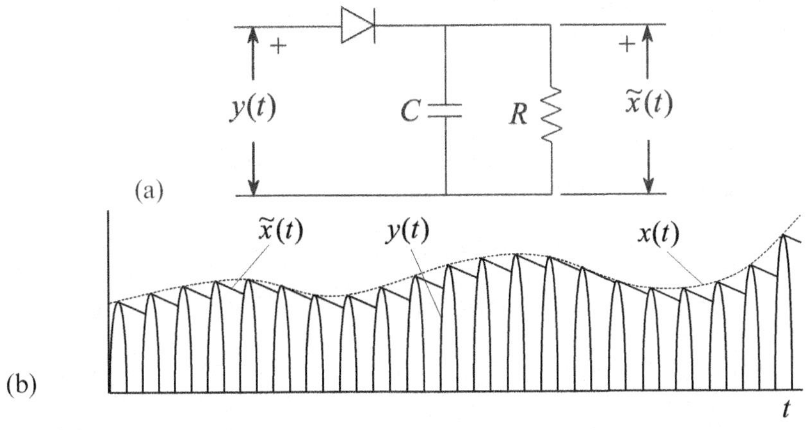

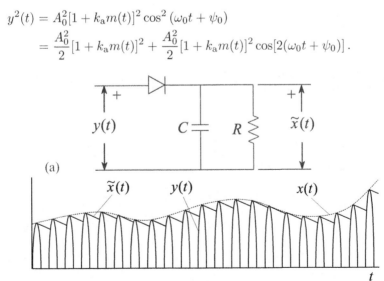

Fig. 3.22 Asynchronous demodulation of AM signal with a rectifier.

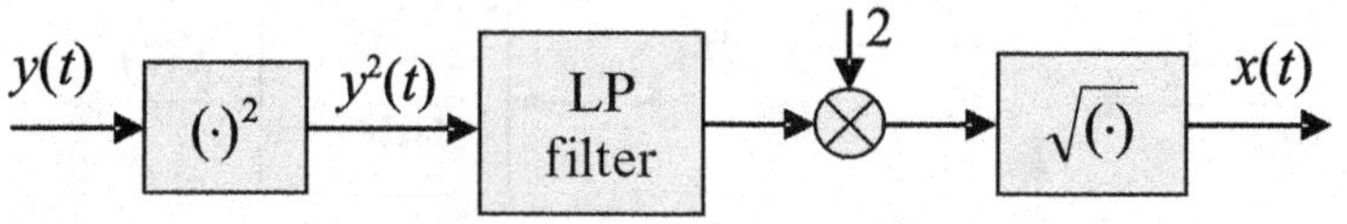

Fig. 3.23 Square-law asynchronous demodulation of AM signals.

After LP filtering with a gain factor of 2 and taking the square root, the demodulated signal becomes

$$x(t) = A_0[1 + k_a m(t)]\,.$$

Let us note that, like the rectifier case, the square-law demodulator does not require a locally generated reference signal.

The following properties of DSB-LC signals may now be outlined:

- Spectral density at ω_0 is a copy of a baseband spectral density of a modulating signal.
- Information in sidebands is redundant.
- Carrier is larger than the rest spectral content.
- At best case, 67% of the signal power is in the carrier.
- Detection is a simple envelope detector.
- Reduces complexity of receivers, therefore, used in commercial broadcast stations.

Overall, the price one pays for wasted power (in the carrier) in this type of AM is a trade-off for simple receiver design. Other types of AM may be even more efficient.

3.6.5 Double Sideband Suppressed Carrier

In AM with the *double sideband suppressed carrier* (DSB-SC), a modulating LP signal is obtained to be $x(t) = A_0 k_a m(t)$ and then the RF modulated signal to be transmitted becomes $y(t) = A_0 k_a m(t)\cos(\omega_0 t + \psi_0)$. The AM scheme for DSB-SC (Fig. 3.24) is then a modification of that shown in Fig. 3.19.

Example 3.6. Given a message signal $x(t) = A_0\mu\cos(\Omega t + \vartheta_0)$. In the DSB-SC scheme, the modulated signal becomes

$$\begin{aligned} y(t) &= A_0\mu\cos(\Omega t + \vartheta_0)\cos(\omega_0 t + \psi_0) \\ &= \frac{A_0\mu}{2}\cos[(\omega_0 - \Omega)t + \psi_0 - \vartheta_0] + \frac{A_0\mu}{2}\cos[(\omega_0 + \Omega)t + \psi_0 + \vartheta_0]\,. \end{aligned}$$

Spectrum of this signal contains two lines placed at the frequencies $\omega_0 \pm \Omega$. It may thus confuse why the spectrum does not consist of a carrier signal, even though it is successfully transmitted at ω_0. To give an explanation of this amusing situation, one needs to consider an interplay of the modulating

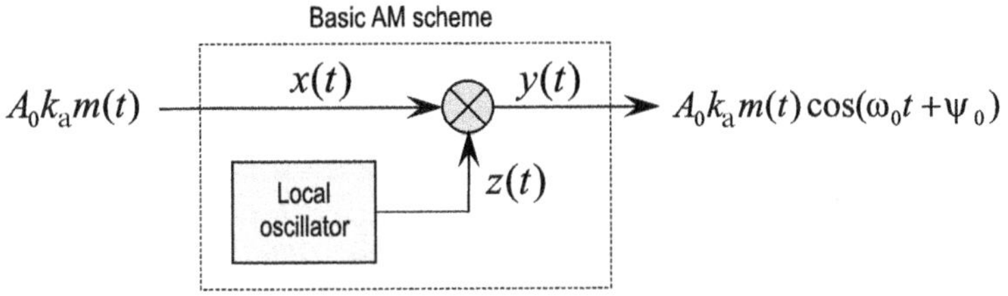

Fig. 3.24 AM scheme for DSB-SC.

and modulated signals in the time domain (Fig. 3.25). It may be seen that the modulated signal phase becomes periodically reversal. Thus, the carrier signal is compensated in average, whereas its instantaneous value is not zero to transmit.

□

The Fourier transform of a DSB-SC signal differs from that shown in Fig. 3.20 by lack of a carrier signal (Fig. 3.26). In order to recover the information signal $x(t)$ from DSB-SC signal, the synchronous demodulation scheme (Fig. 3.21) is used.

Several important features of DSB-SC signal may now be pointed out:

- No wasted power is in the carrier.
- We must have same frequency and phase of a carrier signal at the transmitter and receiver.
- Synchronous detection is the best demodulation scheme here.
- Detection can be done with pilot tone, using the *phase-locked loop* (*PLL*).

3.6.6 Double Sideband Reduced Carrier

In *double sideband reduced carrier* (DSB-RC), the AM signal is formed as DSB-SC with an extra term at the carrier frequency that yields

$$y(t) = A_0 k_a m(t) \cos(\omega_0 t + \psi_0) + A_0 \varepsilon \cos(\omega_0 t + \psi_0)\,.$$

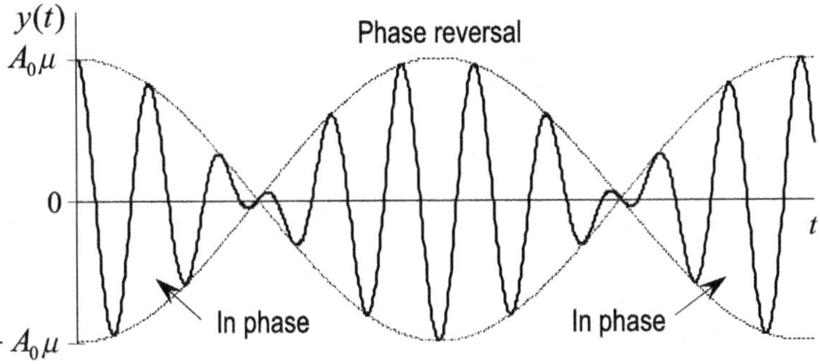

Fig. 3.25 Signals interplay with DSB-SC.

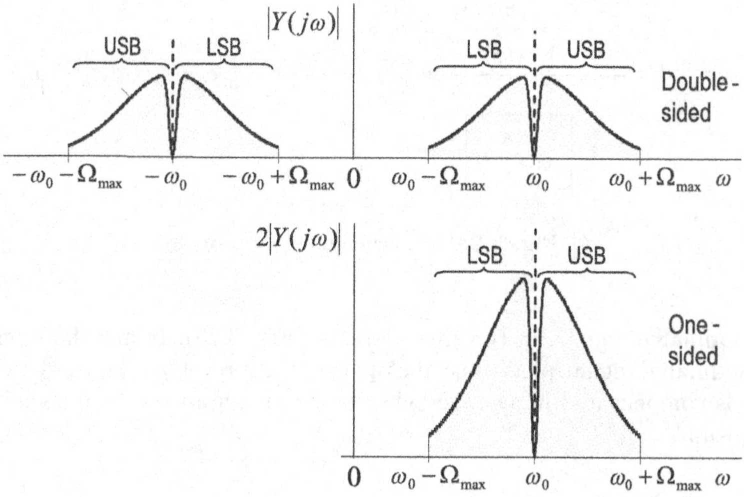

Fig. 3.26 Frequency content of a DSB-SC signal.

In DSB-RC transmission, the carrier is usually transmitted at a level $A_0\varepsilon$ suitable for use as a reference by the receiver, except for the case in which it is reduced to the minimum practical level, i.e., the carrier is suppressed.

3.6.7 Single Sideband

It is seen in Figs. 3.20 and 3.26 that DSB-LC and DSB-SC, respectively, contain same information in LSB and USB. Hence they are not bandwidth efficient. A simple solution is just reducing the bandwidth by a factor of 2 and losing no information if we could somehow transmit only one sideband. That is what is said the *single sideband* (SSB). Here either the LSB or the USB is kept that is usually done in two ways. In the first case, the scheme is complicated with an extremely sharp band-pass or high-pass filter. Alternatively, a Hilbert transformer is used.

3.6.8 SSB Formation by Filtering

If to use a DSB-SC and filter the LSB, then an SSB signal appears as in Fig. 3.27. This technique is most efficient when expensive sharp filter is not required. An example is speech, which has no significant low-frequency content in a range of 0–200 Hz and thus there is no great problem with filtering.

Example 3.7. Given a simplest SSB signal with a suppressed LSB and carrier, i.e.,

$$y(t) = A_0\cos(\omega_0 t + \psi_0) + \frac{A_0\mu}{2}\cos[(\omega_0 + \Omega)t + \psi_0 + \vartheta_0]\,.$$

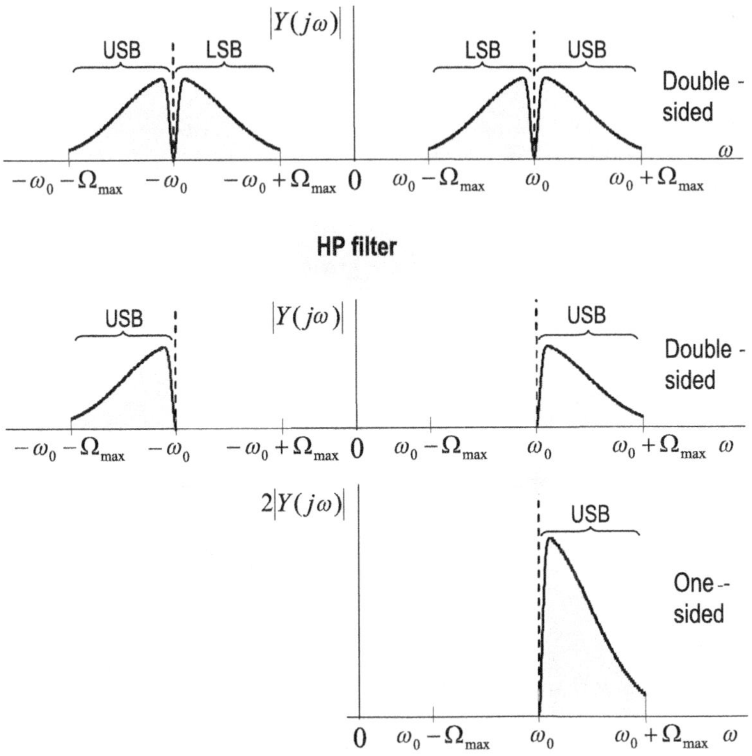

Fig. 3.27 SSB signal with USB saved.

By the transformations, the signal becomes

$$
\begin{aligned}
y(t) &= A_0 \cos(\omega_0 t + \psi_0) + \frac{A_0\mu}{2}\cos(\Omega t + \vartheta_0)\cos(\omega_0 t + \psi_0) \\
&\quad - \frac{A_0\mu}{2}\sin(\Omega t + \vartheta_0)\sin(\omega_0 t + \psi_0) \\
&= A_0\left[1 + \frac{\mu}{2}\cos(\Omega t + \vartheta_0)\right]\cos(\omega_0 t + \psi_0) \\
&\quad - \frac{A_0\mu}{2}\sin(\Omega t + \vartheta_0)\sin(\omega_0 t + \psi_0)\,.
\end{aligned}
$$

The signal envelope is given by

$$
\begin{aligned}
A(t) &= A_0\sqrt{\left[1 + \frac{\mu}{2}\cos(\Omega t + \vartheta_0)\right]^2 + \frac{\mu^2}{4}\sin^2(\Omega t + \vartheta_0)} \\
&= A_0\sqrt{1 + \frac{\mu^2}{4} + \mu\cos(\Omega t + \vartheta_0)}\,.
\end{aligned}
\tag{3.59}
$$

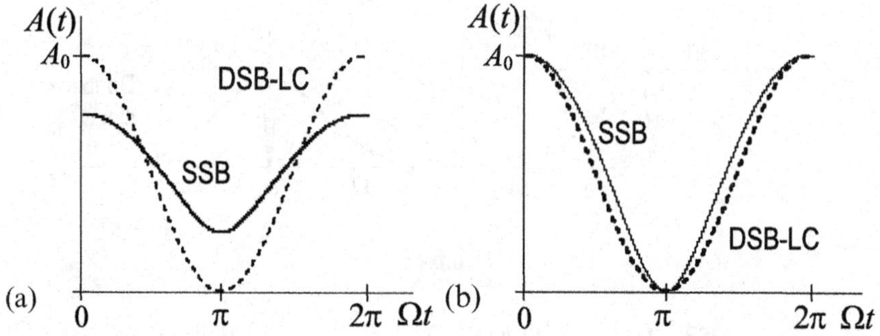

Fig. 3.28 SSB signal with USB saved: **(a)** SSB and DSB-LC and **(b)** SSB and DSB-LC scaled.

Figure 3.28a sketches (3.59) along with the envelope of a DSB-LC signal for $\mu = 1$ and $\theta_0 = 0$. Fig. 3.28b shows (3.59) scaled to the format of a DSB-LC signal envelope. It is seen that an envelope demodulation is accompanied with distortions.

□

3.6.9 SSB Formation by Phase-Shift Method

The second method employs a Hilbert transformer and is sometimes called the *phase-shift method.* We shall now show how an SSB signal can be generated by the scheme in Fig. 3.29. As may be seen, the output signal here is written as

$$\begin{aligned} y(t) &= A_0 k_a m(t)\cos(\omega_0 t + \psi_0) \pm A_0 k_a \hat{m}(t)\sin(\omega_0 t + \psi_0) \\ &= x(t)\cos(\omega_0 t + \psi_0) \pm \hat{x}(t)\sin(\omega_0 t + \psi_0)\,, \end{aligned} \qquad (3.60)$$

where $\hat{m}(t)$ and $\hat{x}(t)$ are the Hilbert transforms of $m(t)$ and $x(t)$, respectively.

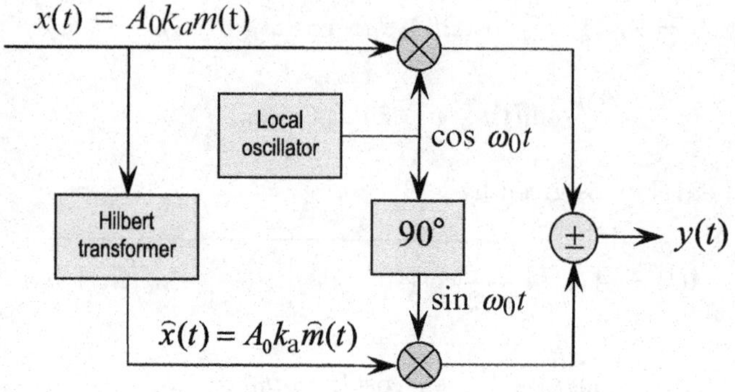

Fig. 3.29 Generation of an SSB signal.

The Fourier transform of (3.60), using (3.17) and setting $\psi_0 = 0$, may be written as

$$Y(j\omega) = \frac{1}{2}\left[X(j\omega - j\omega_0) + X(j\omega + j\omega_0)\right] \pm \frac{1}{2j}\left[\hat{X}(j\omega - j\omega_0) - \hat{X}(j\omega + j\omega_0)\right] . \tag{3.61}$$

If to represent the spectral density of a modulating signal through the lower and USBs as follows:

$$\begin{aligned}
X(j\omega - j\omega_0) &= X_-(j\omega - j\omega_0) + X_+(j\omega - j\omega_0) , \\
X(j\omega + j\omega_0) &= X_+(j\omega + j\omega_0) + X_-(j\omega + j\omega_0) , \\
\hat{X}(j\omega - j\omega_0) &= jX_-(j\omega - j\omega_0) - jX_+(j\omega - j\omega_0) , \\
\hat{X}(j\omega + j\omega_0) &= -jX_+(j\omega + j\omega_0) + jX_-(j\omega + j\omega_0) ,
\end{aligned}$$

then (3.61) will allow for two SSB options, i.e.,

$$Y(j\omega) = X_+(j\omega + j\omega_0) + X_-(j\omega - j\omega_0) , \tag{3.62}$$

$$Y(j\omega) = X_-(j\omega + j\omega_0) + X_+(j\omega - j\omega_0) . \tag{3.63}$$

The first case (3.62) corresponds to an SSB signal with LSB and is shown in Fig. 3.30a. In turn, Fig. 3.30b exhibits how an SSB signal with USB may be formed. As in the DSB-SC case, synchronous detection may be used to recover $x(t)$ from an SSB signal. In a simplified case of $\psi_0 = \theta_0 = 0$, multiplying (3.60) by the carrier waveform yields

$$\begin{aligned}
y(t)\cos\omega_0 t &= x(t)\cos^2(\omega_0 t) \pm \hat{x}(t)\sin(\omega_0 t)\cos\omega_0 t \\
&= \frac{1}{2}x(t) + \frac{1}{2}\left[x(t)\cos 2\omega_0 t \pm \hat{x}(t)\sin 2\omega_0 t\right] .
\end{aligned}$$

The signal $x(t)$ is thus recovered by using an LP filter with a gain factor of 2.

So, an SSB signal utilizes sidebands more efficiently than other AM techniques. We notice that it may use either suppressed carrier, pilot tone, or large-carrier AM.

3.6.10 Quadrature Amplitude Modulation

The QAM is a sort of combined modulations, using which two different signals may be transmitted in the same frequency band without mutual interference. QAM combines phase changes with signal amplitude variations that enable more information to be carried over a limited channel bandwidth. Therefore several varieties of QAM exist.

Assume we have two signals $x_1(t) = A_0 k_a m_1(t)$ and $x_2(t) = A_0 k_a m_2(t)$ to transmit. We may then use one message signal to modulate $\cos\omega_0 t$ and the other one to modulate $\sin\omega_0 t$. The results are added together to produce

$$y(t) = x_1(t)\cos\omega_0 t + x_2(t)\sin\omega_0 t . \tag{3.64}$$

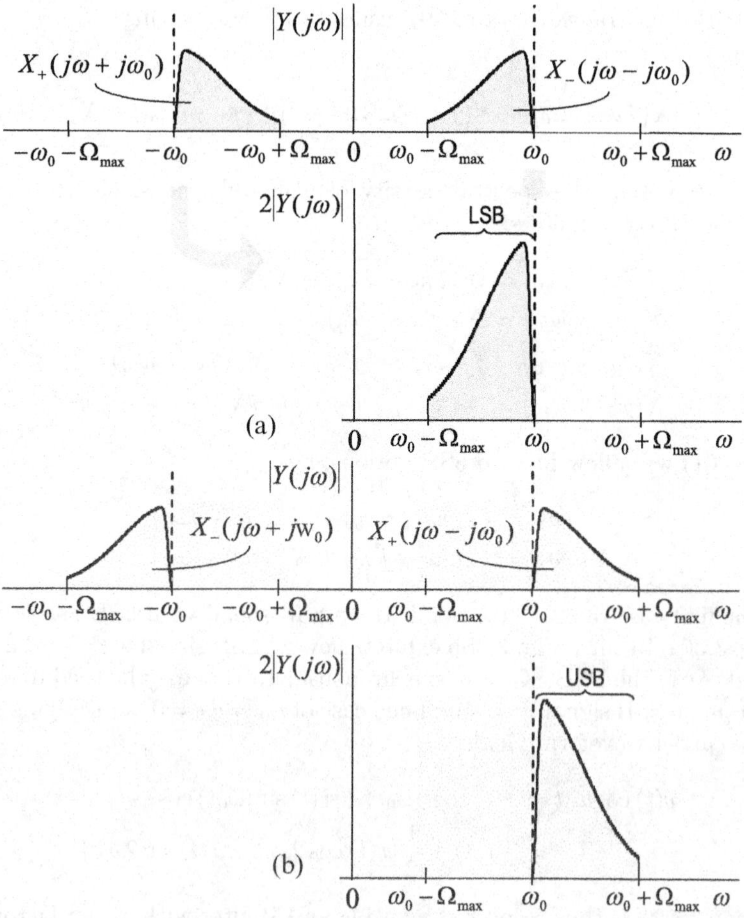

Fig. 3.30 SSB signal: **(a)** with LSB and **(b)** with USB.

The modulated signal $y(t)$ consists now of two modulated carrier signals, which are in phase quadrature with each other. Therefore, they do not interfere and the bandwidth of (3.64) is no greater than that of DSB-SC signal for either message signal alone. The modulation and demodulation scheme for QAM is shown in Fig. 3.31.

At receiver, the demodulator works out to produce

$$
\begin{aligned}
y(t)\cos\omega_0 t &= x_1(t)\cos^2\omega_0 t + x_2(t)\sin\omega_0 t\cos\omega_0 t \\
&= \frac{1}{2}x_1(t) + \frac{1}{2}x_1(t)\cos 2\omega_0 t + \frac{1}{2}x_2(t)\sin 2\omega_0 t\,, \\
y(t)\sin\omega_0 t &= x_1(t)\cos\omega_0 t\sin\omega_0 t + x_2(t)\sin^2\omega_0 t \\
&= \frac{1}{2}x_2(t) - \frac{1}{2}x_2(t)\cos 2\omega_0 t + \frac{1}{2}x_1(t)\sin 2\omega_0 t\,,
\end{aligned}
$$

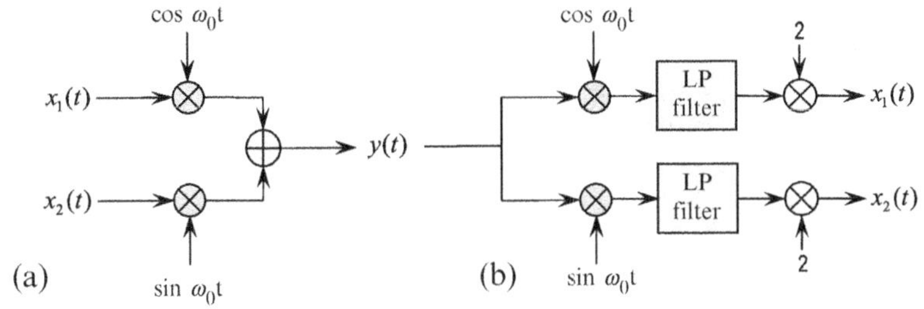

Fig. 3.31 QAM schemes: **(a)** modulation and **(b)** demodulation.

and the LP filters with a gain factor of 2 complete the demodulation procedure. We notice that noncoherent (envelope) detection of QAM is not possible since (3.64) is given as a composition of two message signals.

On the whole, the QAM signal demonstrates efficient utilization of bandwidth and capability for digital M-ary QAM. In the latter case, QAM is very bandwidth-efficient, but susceptible to noise and require linear amplification since it is not power-efficient.

3.6.11 Vestigial Sideband Modulation

Vestigial Sideband (VSB) *modulation* is a type of analog modulation creating a compromise between DSB and SSB. The carrier frequency may or may not be preserved here and one of the sidebands is eliminated through filtering. VSB has frequency bandwidth close to that of SSB without the associated difficulties in building a modulator for SSB. It is also easier to build a demodulator for VSB.

A VSB signal may be generated by filtering a DSB-SC signal (Fig. 3.26) as it is shown in Fig. 3.32. The output is calculated by a convolution

$$y_1(t) = [x(t)\cos\omega_0 t] * h_1(t) \tag{3.65}$$

and its Fourier transform is given by

$$Y_1(j\omega) = \frac{1}{2}[X(j\omega - j\omega_0) + X(j\omega + j\omega_0)]H_1(j\omega)\,. \tag{3.66}$$

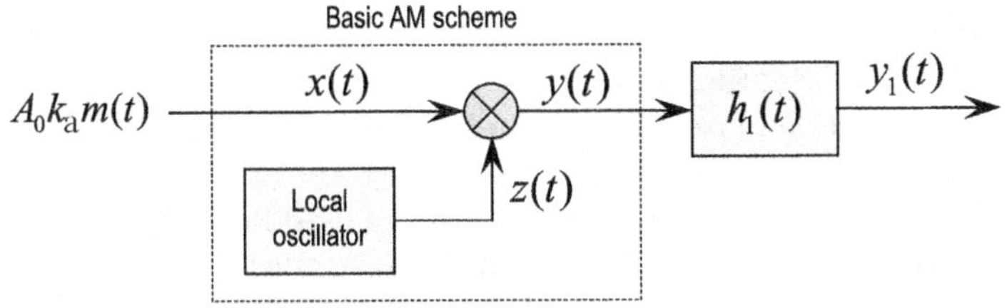

Fig. 3.32 VSB signal scheme.

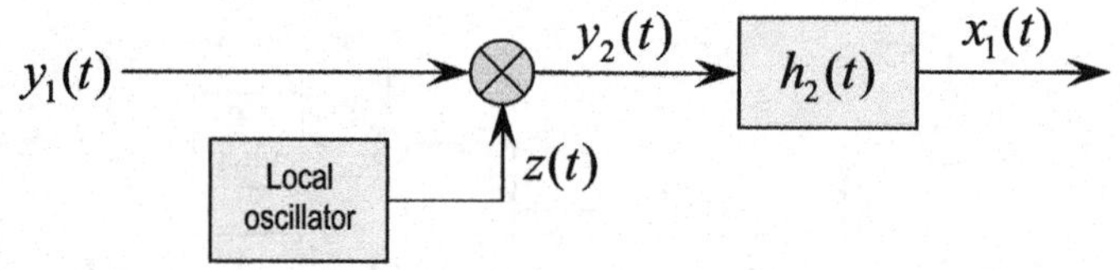

Fig. 3.33 Demodulation of VSB signal.

The received signal (3.65) is demodulated as it is shown in Fig. 3.33 and we have to consider the filter requirements for VSB. In synchronous detection, a signal $y_2(t)$ is formed to be

$$y_2(t) = y_1(t) \cos \omega_0 t \,, \tag{3.67}$$

having the Fourier transform

$$Y_2(j\omega) = \frac{1}{2}[Y_1(j\omega - j\omega_0) + Y_1(j\omega + j\omega_0)] \,. \tag{3.68}$$

Aiming to derive requirements for $h_1(t)$, we substitute (3.66) into (3.68) and write

$$\begin{aligned} Y_2(j\omega) &= \frac{1}{4}[X(j\omega - j2\omega_0) + X(j\omega)]H_1(j\omega - j\omega_0) \\ &+ \frac{1}{4}[X(j\omega) + X(j\omega + j2\omega_0)]H_1(j\omega + j\omega_0) \,. \end{aligned} \tag{3.69}$$

The LP filter $h_2(t)$ rejects the high-frequency terms and passes only the low-frequency components to yield

$$X_1(j\omega) = \frac{1}{4}X(j\omega)\,[H_1(j\omega - j\omega_0) + H_1(j\omega + j\omega_0)] \,.$$

It then follows that for the output $x_1(t)$ to be undistorted version of $x(t)$, the filter transfer function must satisfy the condition

$$H_1(j\omega - j\omega_0) + H_1(j\omega + j\omega_0) = \text{constant} \tag{3.70}$$

that is claimed in the message signal frequency range. It may be shown that (3.70) is satisfied by a filter having an asymmetrical response about the carrier frequency. Additionally, the filter must be designed to get a linear phase response in this range. Figure 3.34 shows an example of a generated and demodulated VSB signal with the filter having a linear slope of a frequency response about ω_0. So, the filter transfer function must be such that the output half of the vestige must fold over and fill any difference between the inner half values.

We notice that the envelope detector may be used for VSB if a carrier term is injected. However, as in SSB, this will be an inefficient use of transmitted power since most of the power must be in the carrier signal.

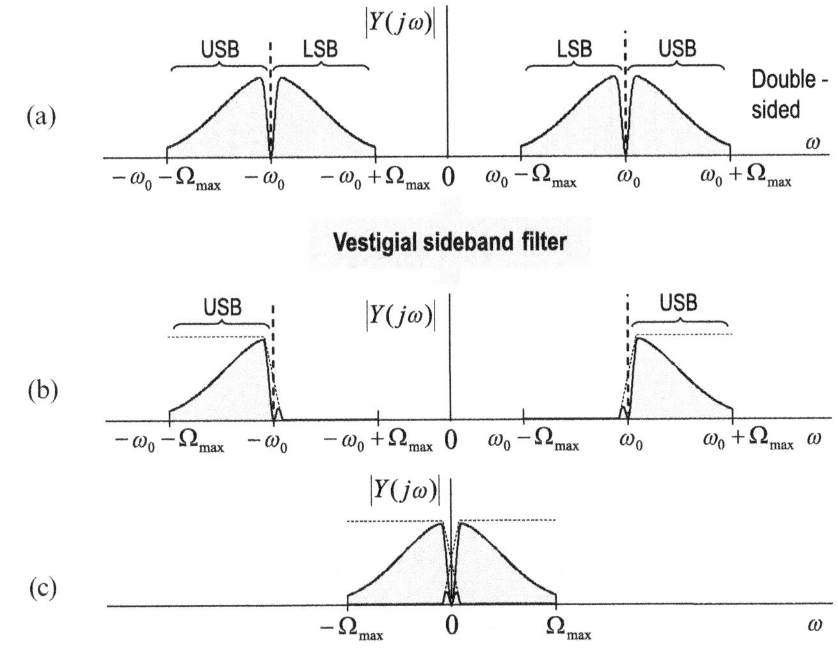

Fig. 3.34 Generation and demodulation of VSB signal: **(a)** DSB-SC signal, **(b)** VSB signal and **(c)** demodulated VSB signal.

3.7 Types of Impulse AM

The other widely used kind of AM employs impulse waveforms. Most frequently, two types of impulse AM are used. A message signal may modulate amplitude of the RF pulse-train in what is called PAM. The amplitude value of an RF pulse may be either "1" or "0" in accordance with a message signal that is called *binary amplitude shift keying* (BASK) or it may be of M levels that is said to be M-*ary amplitude shift keying* (M-ASK).

3.7.1 Pulse Amplitude Modulation

The basic idea of PAM is to use an RF as the carrier signal, which amplitude is modulated by a message signal. The choice of a pulse waveform in the train is made from the standpoint of energy and spectral-content consumption. One can use rectangular pulses, raised cosine pulses, Gaussian pulses, sinc pulses, or some other waveform. To illustrate this type of AM, we consider below only rectangular pulse-trains. Figure 3.35 gives an example of PAM, in which we assume that each rectangular pulse is filled with a carrier signal. We notice that, in digital communications, low-frequency message signals may also be converted to digital format of PAM before modulation and transmission.

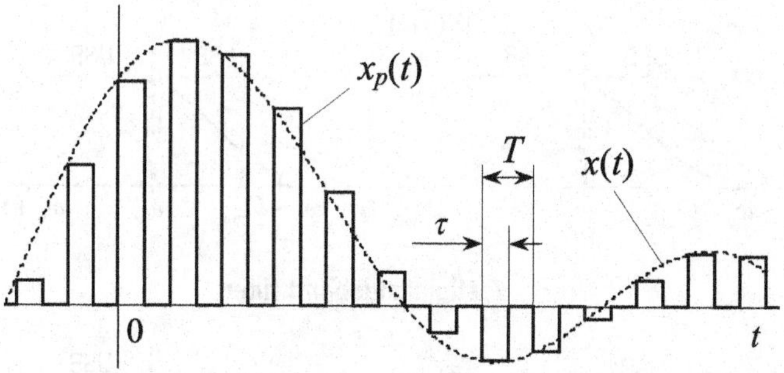

Fig. 3.35 An example of PAM.

Most commonly, a PAM signal may be written, by (3.12), as

$$y(t) = x_p(t)z(t)\,, \tag{3.71}$$

where the carrier signal is $z(t) = e^{j(\omega_0 t+\psi_0)}$ and a pulse-modulating signal $x_p(t)$ is performed using its analog origin $x(t)$ taken at discrete time $t = nT$ by a convolution

$$x_p(t) = \sum_{n=-\infty}^{\infty} x(nT)h(t-nT)\,, \tag{3.72}$$

in which an auxiliary function

$$h(t) = \begin{cases} 1, & 0 \leqslant t \leqslant \tau \\ 0, & \text{otherwise} \end{cases} \tag{3.73}$$

may be treated as the rectangular impulse response of a filter and $x(nT)$ is determined by

$$x(nT) = \sum_{n=-\infty}^{\infty} x(t_n)\delta(t-nT)\,, \tag{3.74}$$

where

$$\delta(n) = \begin{cases} 1, & n = 0 \\ 0, & n \neq 0 \end{cases}$$

is the Kronecker[3] symbol that is an analog of the Dirac delta function in the discrete time. By the convolution property, the Fourier transform of (3.72) becomes

$$X_p(j\omega) = X_n(j\omega)H(j\omega)\,, \tag{3.75}$$

where $X_p(j\omega) \overset{\mathcal{F}}{\Leftrightarrow} x_p(t)$, $X_n(j\omega) \overset{\mathcal{F}}{\Leftrightarrow} x(nT)$, and $H(j\omega) \overset{\mathcal{F}}{\Leftrightarrow} h(t)$ is the transfer function of a filter. An analog signal may be recovered using an LP filter with a cutoff frequency exceeding the maximum frequency of the modulating spectrum. Figure 3.36 gives a typical structure of PAM system.

[3] Leopold Kronecker, German mathematician, 7 December 1823–29 December 1891.

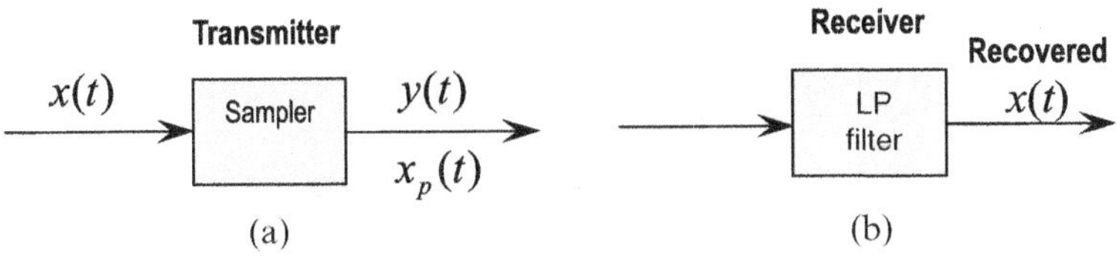

Fig. 3.36 PAM system: (**a**) transmitter and (**b**) receiver.

3.7.2 Amplitude Shift Keying

ASK is associated with digital coding. An example is the BASK. A BASK signal can be defined by

$$y(t) = A_0 m(t) \cos \omega_0 t\,, \tag{3.76}$$

where A_0 is a constant, $m(t) = 1$ or 0, $0 \leqslant t \leqslant T$, and T is the bit duration. The signal has a power $P = A^2/2$, so that $A_0 = \sqrt{2P} = \sqrt{2E/T}$, where $E = PT$ is the energy contained in a bit duration. Figure 3.37 shows an example of the BASK signal sequence generated by the binary sequence 0101001.

The Fourier transform of the BASK signal $y(t)$ is akin to that with the DSB-SC and is given by

$$\begin{aligned} Y(j\omega) &= \frac{A_0}{2}\int_{-\infty}^{\infty} m(t)\mathrm{e}^{-j(\omega-\omega_0)t}\mathrm{d}t + \frac{A_0}{2}\int_{-\infty}^{\infty} m(t)\mathrm{e}^{-j(\omega+\omega_0)t}\mathrm{d}t \\ &= \frac{A_0}{2}M(\omega-\omega_0) + \frac{A_0}{2}M(\omega+\omega_0)\,. \end{aligned} \tag{3.77}$$

As in DSB-SC, the effect of multiplication by the carrier signal $A_0 \cos \omega_0 t$ is simply to shift the spectral density of the modulating signal $m(t)$ to $\pm\omega_0$. The signal structure (Fig. 3.37) can be made more sophisticated by using several amplitude levels: for example, one can group the bits into groups of 2, i.e., 00, 01, 10, and 11, and have four different amplitude levels for each of these groups. This is referred to as *quadrature pulse amplitude modulation* (QPAM), which, in applications, is used for alphabets of size other than 4.

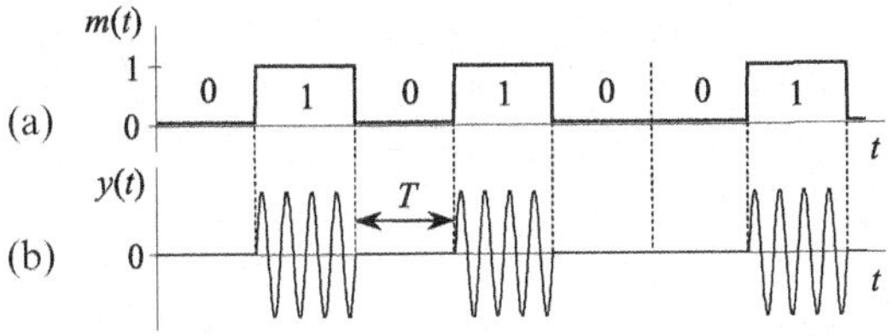

Fig. 3.37 BASK: (**a**) modulating signal and (**b**) modulated signal.

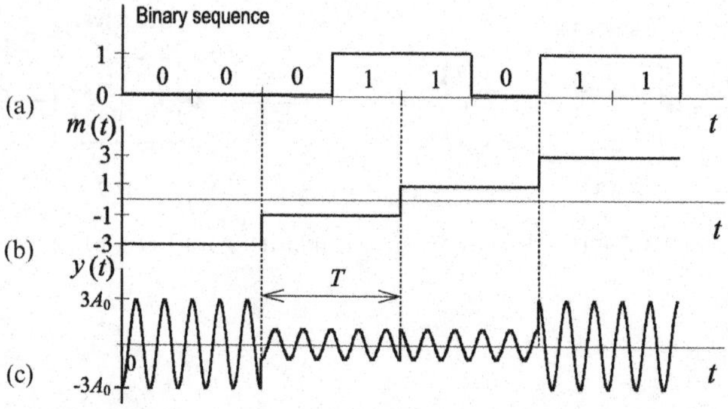

Fig. 3.38 4-ASK modulation: (**a**) binary sequence, (**b**) 4-ary signal, and (**c**) 4-ASK signal.

3.7.3 M-ary Amplitude Shift Keying

M-ASK signal can be defined by

$$y(t) = \begin{cases} A_0 m_i(t) \cos \omega_0 t, & 0 \leqslant t \leqslant T \\ 0, & \text{otherwise} \end{cases}, \tag{3.78}$$

where $m_i(t) = m[2i - (M-1)]$ for $i = 0, 1, \ldots, M-1$ with $M \geqslant 4$. The detector for M-ASK is an envelope detector. Figure 3.38 shows the 4-ASK signal sequence generated by the binary signal sequence 00011011.

3.8 Frequency Modulation

In line with AM, *angular modulation* has also gained wide currency in electronic systems and communications. This kind of modulation may be realized in two ways. We may straightforwardly modulate the carrier frequency in what is called FM. The modulating technique may also be organized to modulate the phase of a carrier signal in PM. In fact, both FM and PM are two kinds of angular modulation and therefore their separate explanation is conditional in a sense. Even so, in applications, FM and PM are often used as independent types of modulation. Therefore, we will consider them separately.

While thinking about FM, we imagine that frequency of some signal is varied by some other signal.

> ***Frequency modulation***: *FM* is a kind of angular modulation with which the frequency of oscillations is varied in accordance with the modulating signal.

□

If frequency variations are assumed, the angular measure is changed as in (3.5) and the modulated signal becomes

$$y(t) = A_0 \cos\left[\int_0^t \omega(t)\mathrm{d}t + \psi_0\right]$$
$$= A_0 \cos\left[\omega_0 t + k_\omega \int_0^t m(t)\mathrm{d}t + \psi_0\right], \quad (3.79)$$

where $m(t)$ is the message signal, $k_\omega = 2\pi k_f$, and k_f is the frequency sensitivity. Let us add that, with angular modulation, it is usually supposed that the amplitude A_0 is time-constant.

3.8.1 Simplest FM

In the simplest FM case, frequency of a carrier signal is varied by the harmonic message law

$$m(t) = m_0 \cos(\Omega t + \vartheta), \quad (3.80)$$

where m_0 is the peak amplitude, Ω is a modulation frequency, and ϑ is some constant phase. By (3.80), the function of the modulated frequency becomes

$$\omega(t) = \omega_0 + k_\omega m_0 \cos(\Omega t + \vartheta) = \omega_0 \left[1 + \frac{\Delta\omega}{\omega_0} \cos(\Omega t + \vartheta)\right], \quad (3.81)$$

where $\Delta\omega = k_\omega m_0$ is called the *frequency deviation*. Using both (3.80) and (3.81), the signal (3.79) may be transformed to

$$y(t) = A_0 \cos[\omega_0 t + \beta \sin(\Omega t + \vartheta) + \psi_0], \quad (3.82)$$

where

$$\beta = \frac{\Delta\omega}{\Omega} = \frac{k_\omega m_0}{\Omega},$$

also known as the *modulation index*. It is to be noticed now that the initial phase ϑ plays no significant role with FM and is usually set to be zero. If to permit this, then (3.82) will become the simplest FM oscillation

$$y(t) = A_0 \cos(\omega_0 t + \beta \sin \Omega t + \psi_0). \quad (3.83)$$

Figure 3.39 illustrates (3.83) graphically. Here we watch for the three principal processes represented by the modulating signal $x(t)$ (a), modulated frequency $\omega(t)$ (b), and modulated phase $\psi(t)$ (c). Accordingly, an FM signal $y(t)$ may be performed as in Fig. 3.39(d).

As it may be seen, the FM law inherently coincides with the modulating signal. The signal phase with FM is nonstationary with a drift $\omega_0 t$ and, contrary to the frequency, varies in time with a delay on $\pi/2$, by the modulating function $\beta \sin \Omega t$. It also follows that the peak phase deviation is equal to the modulation index that depends on both Ω and $\Delta\omega$. The latter is considered to be a disadvantage of PM caused by FM.

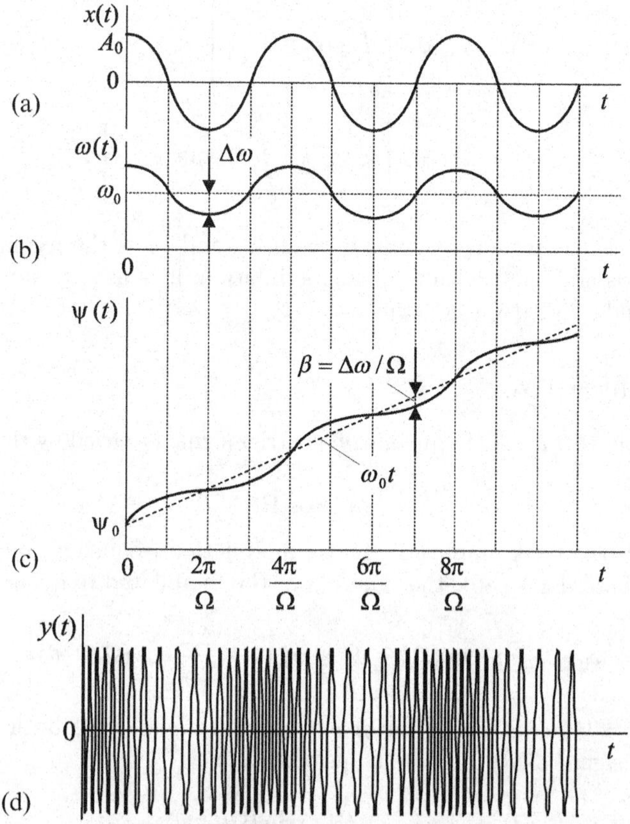

Fig. 3.39 Simplest FM: (**a**) modulating signal, (**b**) FM, (**c**) PM caused by FM, and (**d**) FM signal.

Spectrum of a signal (3.83) is derived by extending the right-hand side to the harmonic series. By simple manipulations, we first go to

$$y(t) = A_0 \cos(\omega_0 t + \psi_0) \cos(\beta \sin \Omega t) - A_0 \sin(\omega_0 t + \psi_0) \sin(\beta \sin \Omega t)\,, \quad (3.84)$$

where the functions $\cos(\beta \sin \Omega t)$ and $\sin(\beta \sin \Omega t)$ are performed through the infinite series of the components involving the Bessel[4] functions of the first kind and nth order,

$$\cos(\beta \sin \Omega t) = J_0(\beta) - 2J_2(\beta) \cos 2\Omega t + \ldots, \quad (3.85)$$

$$\sin(\beta \sin \Omega t) = 2J_1(\beta) \sin \Omega t + 2J_3(\beta) \sin 3\Omega t + \ldots \quad (3.86)$$

For the following convenience, several Bessel functions are sketched in Fig. 3.40. It is seen that, by a small modulation index, $\beta < 1$, the Bessel

[4] Friedrich Wilhelm Bessel, German scientist, 22 July 1784–17 March 1846.

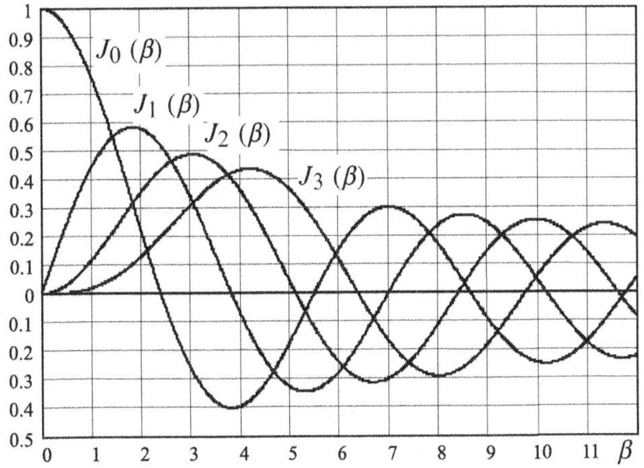

Fig. 3.40 Bessel functions of the first kind and nth order.

functions may approximately be calculated as

$$J_0(\beta) \approx 1.0\,, \quad J_1(\beta) \approx \beta/2\,, \quad \text{and} \quad J_{n>2}(\beta) \approx 0\,. \tag{3.87}$$

These approximations bring (3.84) to the harmonic series. For the simplicity, we allow $\psi_0 = 0$ and go to

$$\begin{aligned} y(t) &\cong A_0 \cos \omega_0 t - A_0 \beta \sin \omega_0 t \sin \Omega t \\ &\cong A_0 \cos \omega_0 t - \frac{A_0 \beta}{2} \cos(\omega_0 - \Omega)t + \frac{A_0 \beta}{2} \cos(\omega_0 + \Omega)t\,. \end{aligned} \tag{3.88}$$

Spectrum of a signal (3.88) is illustrated in Fig. 3.41. Even a quick look shows that the picture is similar to that featured to the simplest AM signal. The difference is in the lower side spectral line that has a negative amplitude (it is positive in the AM signal spectrum shown in Fig. 3.1a). This causes important changes in the vector diagram of FM signal (Fig. 3.42).

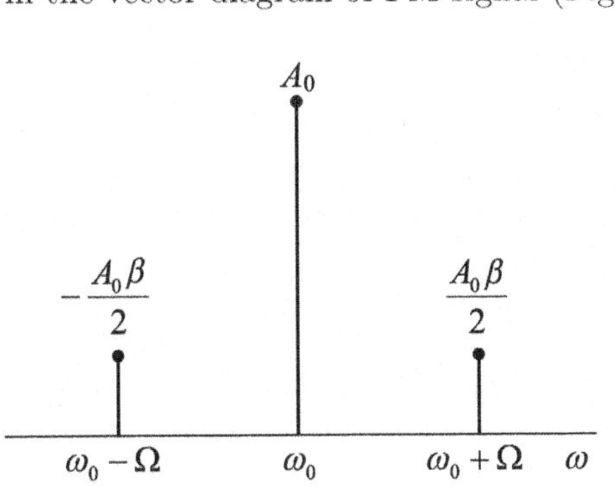

Fig. 3.41 Spectrum of a simplest FM signal.

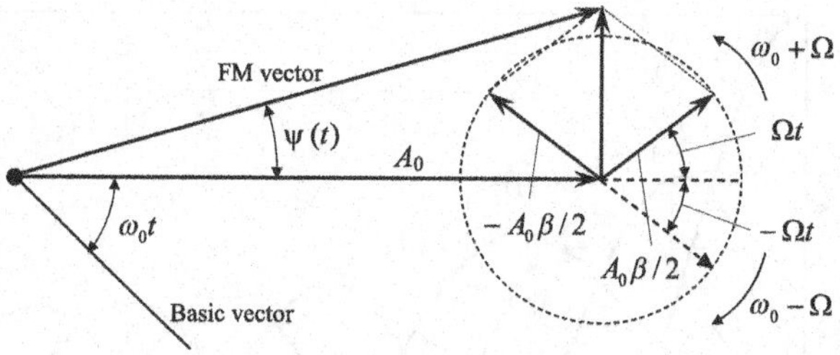

Fig. 3.42 Vector diagram of a simplest FM signal.

It follows (Fig. 3.42) that the resulting FM vector has a variable amplitude that certainly contradicts with the claim to keep the amplitude to be constant. However, if we assume that $\beta \ll 1$, then the side vectors become very short and variations in the amplitude almost negligible. This case is described by (3.88).

3.8.2 Complex FM

Generally, FM might be supposed to be accomplished with a complex modulating signal and large FM index $\beta > 1$. If so, spectrum of an FM signal will substantially differ from what is shown in Fig. 3.41. To describe such a signal, the modulating signal may be extended to the Fourier series

$$x(t) = \sum_{k=1}^{\infty} k_{\omega k} m_k \cos(k\Omega t + \vartheta_k)\,, \tag{3.89}$$

where $k_{\omega k} m_k = \Delta\omega_k$ is the frequency deviation of the kth harmonic of a modulating signal, ϑ_k is the relevant phase, $k_{\omega k} = 2\pi k_{fk}$, k_{fk} is the frequency sensitivity of the kth harmonic, and m_f is the peak message signal associated with the kth harmonic. Accordingly, the FM signal is performed as

$$\begin{aligned} y(t) &= A_0 \cos\left[\omega_0 t + \int_0^t \sum_{k=1}^{\infty} \Delta\omega_k \cos(k\Omega t + \vartheta_k)\mathrm{d}t + \psi_0\right] \\ &= A_0 \cos\left[\omega_0 t + \sum_{k=1}^{\infty} \beta_k \sin(k\Omega t + \vartheta_k) + \psi_0\right], \end{aligned} \tag{3.90}$$

where $\beta_k = \Delta\omega_k / k\Omega$ is the modulation index associated with the kth harmonic of a modulating signal.

As it may be seen, a signal (3.90) consists of an infinite number of spectral lines around the carrier frequency ω_0, whose amplitudes are proportional to the relevant modulation index. With small $\beta_k < 1$, only three terms are usually saved and we go to (3.88). If $\beta_k > 1$, the series becomes long and the spectral analysis entails difficulties. Figure 3.43 gives an idea about the spectral content of such a signal.

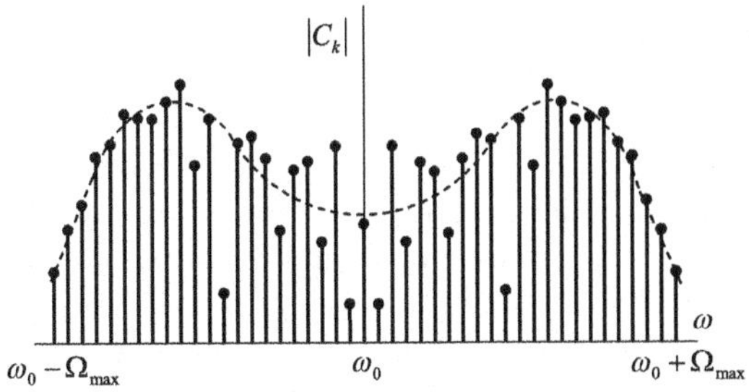

Fig. 3.43 Spectrum of FM signal with large FM index.

To evaluate the spectrum width of FM signal, the following approximate estimate may be used

$$\Delta\Omega_{\mathrm{FM}} \cong 2\Omega_{\max}(1+\beta+\sqrt{\beta})\,. \tag{3.91}$$

The formula suggests that, if $\beta \ll 1$, the spectrum width is determined by the doubled maximum modulation frequency $\Omega_{\max}$ as in Fig. 3.41. When $\beta = 1$, the width increases by the factor of 3 and is estimated as $\Delta\Omega_{\mathrm{FM}} \cong 6\Omega_{\max}$.

3.8.3 Analog Modulation of Frequency

A typical FM technique is an analog FM exploiting the voltage-controlled oscillator (VCO) to vary its frequency by the message signal converted to voltage.

Figure 3.44 sketches a typical scheme illustrating analog FM. Without modulation, the VCO generates a carrier signal, which frequency ω_0 is mostly determined by the quartz crystal resonator X_1 and the initial capacitance value of a nonlinear diode varactor D_1. With modulation, the message signal $m(t)$ is applied to the diode D_1, which plays the role of a voltage-controlled capacitor. Here the inductance L_1 allows passage of only the low-frequency message signal to the diode. The message signal $m(t)$ changes the reverse bias on the varactor diode, hence changing its capacitance and in turn the frequency of the oscillator. The output oscillator signal then becomes frequency-modulated.

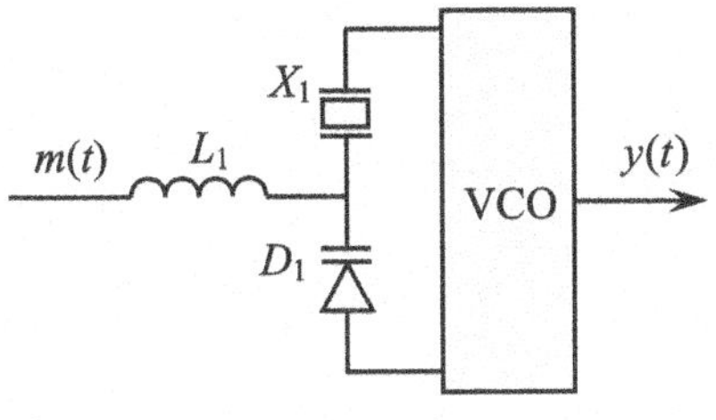

Fig. 3.44 Scheme of analog FM.

3.9 Linear Frequency Modulation

One of the combined modulation techniques is that used in radars and called LFM. Here impulse AM is accompanied with FM to suppress the pulse at receiver. The envelope of such an RF pulse may be arbitrary, e.g., rectangular, trapezium, triangle, or Gaussian. Further, we will learn the effect of pulse suppression with LFM in the time domain studying the correlation function. Here, let us be concerned only with the spectral content of such a signal, showing that its envelope in the time domain is almost fully recurred in the frequency domain.

3.9.1 Rectangular RF Pulse with LFM

Commonly, a carrier signal with simultaneous amplitude and angular modulation is written as

$$y(t) = A(t)\cos[\omega_0 t + \psi(t)]\,, \tag{3.92}$$

where the AM function $A(t)$ and PM function $\psi(t)$ may be either harmonic or complex. If AM is performed with a rectangular pulse and PM is due to FM, then the RF single pulse may be sketched as in Fig. 3.45a. Here, the frequency $\omega(t)$ is modulated with a linear law having a deviation $\pm\Delta\omega/2$ around the carrier frequency ω_0.

A generalized form of (3.92) is

$$y(t) = A(t)\cos\Psi(t)\,, \tag{3.93}$$

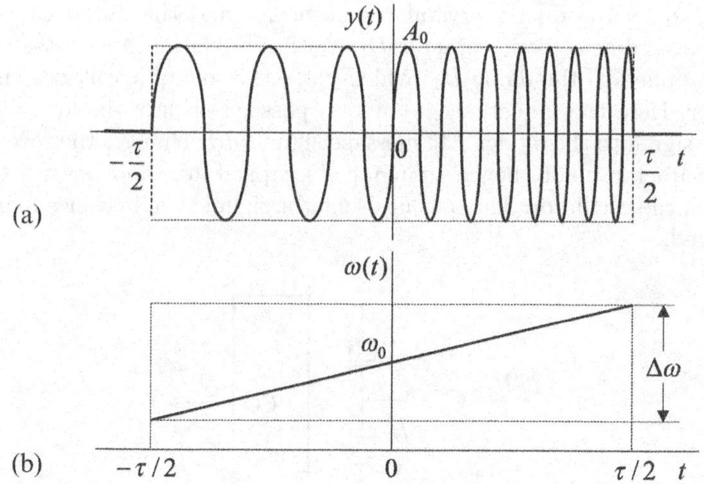

Fig. 3.45 Linear FM: (**a**) single LFM RF rectangular pulse and (**b**) linearly modulated frequency.

where the total phase $\Psi(t)$ is determined by the integral relation

$$\Psi(t) = \int \omega(t)\mathrm{d}t + \psi_0 \,, \tag{3.94}$$

in which $\omega(t)$ is given with the linear FM law

$$\omega(t) = \omega_0 + \alpha t \,, \tag{3.95}$$

where the coefficient α defines the rate of LFM.

Allowing $\psi_0 = 0$, the phase function (3.94) becomes

$$\Psi(t) = \int_0^t (\omega_0 + \alpha t)\mathrm{d}t = \omega_0 t + \frac{\alpha t^2}{2} \,, \tag{3.96}$$

where α is determined by the deviation of frequency $\Delta\omega$ and duration τ as

$$\alpha = \frac{\Delta\omega}{\tau} \,. \tag{3.97}$$

By (3.96), a rectangular LFM pulse may be written as

$$y(t) = \begin{cases} A_0 \cos\left(\omega_0 t + \frac{\alpha t^2}{2}\right), & -\frac{\tau}{2} \leqslant t \leqslant \frac{\tau}{2} \\ 0, & \text{otherwise} \end{cases} \,. \tag{3.98}$$

It has to be remarked now that, in applications, the envelope of a pulse (3.98) may not obligatory be rectangular. Trapezium, triangular, Gaussian and other waveforms are also used. To understand how the envelope affects the spectral density of an LFM pulse, we consider the following several particular cases.

3.9.2 Spectral Density of a Rectangular RF Pulse with LFM

Consider a rectangular RF pulse with LFM (Fig. 3.45a). By the Euler formula (1.11) and Fourier transform, the spectral density of (3.98) is first written as

$$\begin{aligned} Y(j\omega) &= A_0 \int_{-\tau/2}^{\tau/2} \cos\left(\omega_0 t + \frac{\alpha t^2}{2}\right) \mathrm{e}^{-j\omega t}\mathrm{d}t \\ &= \frac{A_0}{2} \int_{-\tau/2}^{\tau/2} \mathrm{e}^{-j\left[(\omega-\omega_0)t - \frac{\alpha t^2}{2}\right]}\mathrm{d}t + \frac{A_0}{2} \int_{-\tau/2}^{\tau/2} \mathrm{e}^{-j\left[(\omega+\omega_0)t + \frac{\alpha t^2}{2}\right]}\mathrm{d}t \end{aligned} \tag{3.99}$$

and we see that it consists of two sidebands placed at $\pm\omega_0$.

In practice, of importance is the case when the overlapping effect between the sidebands is small. Thus, the frequency deviation $\Delta\omega$ during the pulse duration τ is much smaller than the carrier, $\Delta\omega = \alpha\tau \ll \omega_0$. If so, only the first integral in (3.99) may be calculated as related to the physical positive carrier frequency.

By simple manipulations, the first integral in (3.77) transforms to

$$Y_+(j\omega) = \frac{A_0}{2} \mathrm{e}^{-j\frac{(\omega-\omega_0)^2}{2\alpha}} \int\limits_{-\tau/2}^{\tau/2} \mathrm{e}^{j\frac{\alpha}{2}\left(t-\frac{\omega-\omega_0}{\alpha}\right)^2} \mathrm{d}t\,. \tag{3.100}$$

Now, we change the variable to

$$\xi = \sqrt{\frac{\alpha}{\pi}}\left(t - \frac{\omega - \omega_0}{\alpha}\right)$$

and arrive at

$$Y_+(j\omega) = \frac{A_0}{2}\sqrt{\frac{\pi}{\alpha}} \mathrm{e}^{-j\frac{(\omega-\omega_0)^2}{2\alpha}} \int\limits_{-\nu_1}^{\nu_2} \mathrm{e}^{j\frac{\pi\xi^2}{2}} \mathrm{d}\xi, \tag{3.101}$$

where the integral bounds are determined as

$$\nu_1 = \frac{\alpha\tau + 2(\omega - \omega_0)}{2\sqrt{\alpha\pi}},$$

$$\nu_2 = \frac{\alpha\tau - 2(\omega - \omega_0)}{2\sqrt{\alpha\pi}}.$$

The integral in (3.101) may be expressed by the Fresnel[5] integrals

$$C(x) = \int\limits_0^x \cos\frac{\pi\xi^2}{2}\mathrm{d}\xi\,,$$

$$S(x) = \int\limits_0^x \sin\frac{\pi\xi^2}{2}\mathrm{d}\xi\,,$$

which functions are sketched in Fig. 3.46. The transform (3.101) then goes to its ultimate form of

$$\begin{aligned} Y_+(j\omega) &= \frac{A_0}{2}\sqrt{\frac{\pi}{\alpha}} \mathrm{e}^{-j\frac{(\omega-\omega_0)^2}{2\alpha}} \left[C(\nu_1) + C(\nu_2) + jS(\nu_1) + jS(\nu_2)\right] \\ &= |Y_+(j\omega)|\,\mathrm{e}^{-\varphi_y(\omega)}\,, \end{aligned} \tag{3.102}$$

defining the magnitude spectral density by

$$|Y_+(j\omega)| = \frac{A_0}{2}\sqrt{\frac{\pi}{\alpha}}\sqrt{[C(\nu_1) + C(\nu_2)]^2 + [S(\nu_1) + S(\nu_2)]^2}\,. \tag{3.103}$$

The phase spectral density associated with (3.102) may be performed by the sum of two terms, e.g., $\varphi_y(\omega) = \varphi_{y1}(\omega) + \varphi_{y2}(\omega)$. The first term is given

[5] Augustin Jean Fresnel, French mathematician, 10 May 1788–14 July 1827.

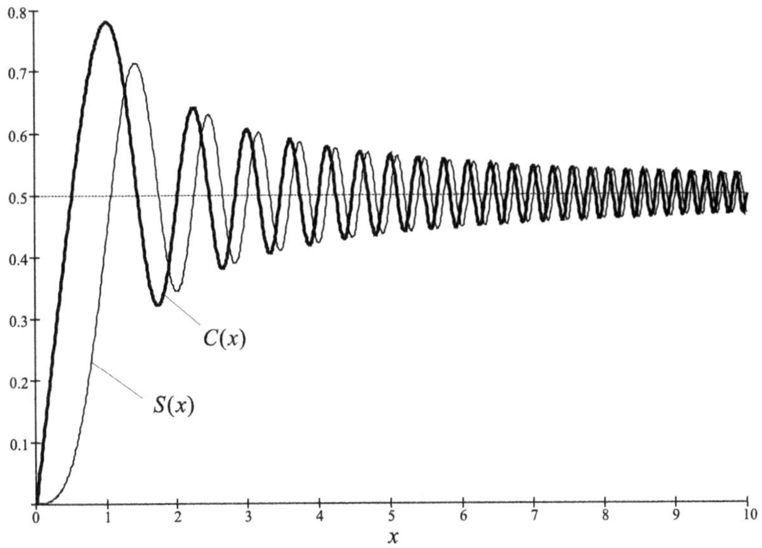

Fig. 3.46 Fresnel functions (integrals).

by the power of the exponential function in (3.102),

$$\varphi_{y1}(\omega) = \frac{(\omega - \omega_0)^2}{2\alpha}, \tag{3.104}$$

and the second one is calculated by the expression in brackets of (3.102) as

$$\varphi_{y2}(\omega) = \arctan \frac{S(\nu_1) + S(\nu_2)}{C(\nu_1) + C(\nu_2)}. \tag{3.105}$$

Setting aside the rigorous form (3.102), the following approximate expression is also used in applications,

$$Y_+(j\omega) \cong \begin{cases} A_0 K(\omega)\sqrt{\frac{\pi}{2\alpha}} e^{j\left[\frac{\pi}{4} - \frac{(\omega-\omega_0)^2}{2\alpha}\right]}, & \omega_0 - \frac{\Delta\omega}{2} \leqslant \omega \leqslant \omega_0 + \frac{\Delta\omega}{2} \\ 0, & \text{otherwise} \end{cases}, \tag{3.106}$$

where $K(\omega) \cong 1$ is some function with a poor dependence on frequency. Using (3.106), the magnitude and phase spectral densities of an RF LFM rectangular pulse are approximately given, respectively, by

$$Y_+(j\omega) \cong \begin{cases} A_0 K(\omega)\sqrt{\frac{\pi}{2\alpha}}, & \omega_0 - \frac{\Delta\omega}{2} \leqslant \omega \leqslant \omega_0 + \frac{\Delta\omega}{2} \\ 0, & \text{otherwise} \end{cases}, \tag{3.107}$$

$$\varphi_y(\omega) = \begin{cases} \frac{\pi}{4} - \frac{(\omega - \omega_0)^2}{2\alpha}, & \alpha > 0 \\ -\frac{\pi}{4} + \frac{(\omega - \omega_0)^2}{2\alpha}, & \alpha < 0 \end{cases}, \quad \omega_0 - \frac{\Delta\omega}{2} \leqslant \omega \leqslant \omega_0 + \frac{\Delta\omega}{2}. \tag{3.108}$$

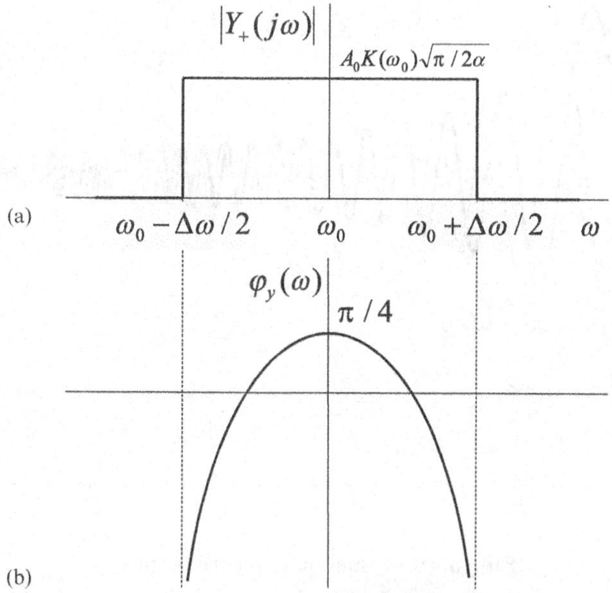

Fig. 3.47 Approximate spectral density of an LFM rectangular pulse with $\alpha > 0$: (**a**) magnitude and (**b**) phase.

Figure 3.47 illustrates (3.107) and (3.108) for $\alpha > 0$. It is seen that the rectangular envelope is saved in the magnitude spectral density and the phase density is quadratic keeping the value of $\pi/4$ at ω_0.

3.9.3 RF LFM Pulses with Large PCRs

The splendid property of a rectangular LFM pulse to save its envelope in the spectral density is also demonstrated by other waveforms. A diversity strongly depends on the parameter called the *pulse-compression ratio* (PCR),

$$\text{PCR} = \Delta f\tau = \frac{\alpha\tau^2}{2\pi}, \tag{3.109}$$

where the frequency deviation during a pulse is $\Delta f = \Delta\omega/2\pi$.

Depending on the PCR, the shape of the magnitude spectral density of a real LFM single pulse becomes closer (PCR $\gg 1$) or farther (PCR ~ 1) to the pulse envelope shape. Therefore, signals with small PCRs have not gained currency.

By PCR $\gg 1$, several important simplifications in (3.102) may be made, noting that the magnitude density (3.103) appears to be almost constant and its oscillations attenuate substantially. Indeed, by large $x \gg 1$, the Fresnel functions may approximately be calculated by

$$C(x)|_{x\gg 1} \cong \frac{1}{2} + \frac{1}{\pi x}\sin\frac{\pi x^2}{2}\,,$$
$$S(x)|_{x\gg 1} \cong \frac{1}{2} - \frac{1}{\pi x}\cos\frac{\pi x^2}{2}\,,$$

and then, by simple manipulations, the part in brackets of (3.102) becomes

$$\begin{aligned} C(\nu_1) + C(\nu_2) + jS(\nu_1) + jS(\nu_2) &\cong 1 + j + \frac{1}{\pi}\left(\frac{1}{\nu_1}e^{j\frac{\pi\nu_1^2}{2}} + \frac{1}{\nu_2}e^{j\frac{\pi\nu_2^2}{2}}\right) \\ &\cong 1 + j\,. \end{aligned}$$

Owing to this, (3.103) simplifies to

$$|Y_+(j\omega_0)| \cong \begin{cases} A_0\sqrt{\frac{\pi}{2\alpha}}, & \omega_0 - \frac{\Delta\omega}{2} \leqslant \omega \leqslant \omega_0 + \frac{\Delta\omega}{2} \\ 0, & \text{otherwise} \end{cases}\,. \tag{3.110}$$

Employing (3.110), energy of the LFM pulse with large PCR calculates within the width of the spectral density, approximately, by the constant value

$$|Y_+(j\omega)|^2 \cong \frac{A_0^2\pi}{2\alpha}\,.$$

Note that rigorous analytical analysis of LFM pulses with different envelopes (include the rectangular one) entails large burden and a superior way is to do it numerically. We examine numerically several LFM pulses shaped with the rectangular, trapezium, triangular, and Gaussian waveforms in the following.

Example 3.8. Given a rectangular RF LFM pulse with PCR = 20 and 200. Figure 3.48 shows the magnitude density for each of the cases. To compare, the magnitude density for PCR = 0 is also given. It may be concluded that increasing the PCR results in the following:

- The spectral envelope becomes closer to the pulse envelope.
- Oscillations in the spectral envelope smooth.

We notice that intensive oscillations in the magnitude spectral density (Fig. 3.48) are due to sharp sides of the rectangular pulse. This disadvantage is efficiently overcome in the trapezium LFM pulse that we examine below.

□

Example 3.9. Given a trapezium LFM pulse (Fig. 3.49a) with PCR = 20 and 100. Figures 3.49b and c show the relevant magnitude spectral densities. It follows that, unlike the rectangular pulse case, oscillations in the spectral envelope are substantially suppressed here.

□

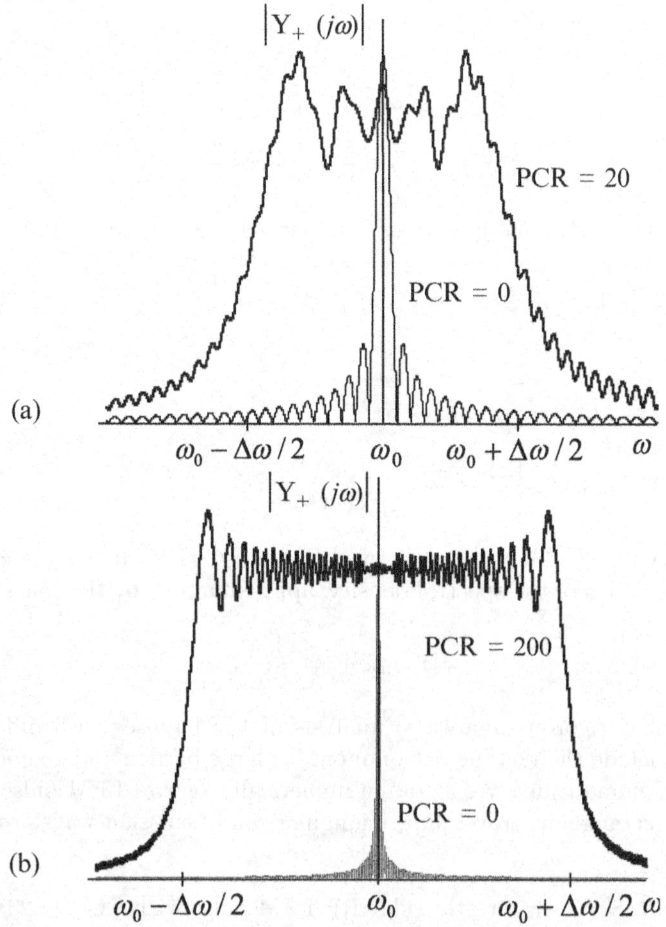

Fig. 3.48 Magnitude density of a rectangular LFM pulse: (**a**) PCR = 20 and (**b**) PCR = 200.

Example 3.10. Given a triangular LFM pulse (Fig. 3.50) with PCR = 20 and 100. Figure 3.51 shows the relevant magnitude spectral densities. It is seen that the spectral shape looks almost triangular even by PCR = 20 and no visible oscillations are fixed.

□

Example 3.11. Given a Gaussian LFM pulse (Fig. 3.52) with PCR = 20 and 100. The magnitude spectral densities are shown in Fig. 3.53. One concludes that the Gaussian shape is saved here without oscillations for arbitrary PCRs.

□

The most common properties of RF pulses with LFM may now be outlined:

- By large PCR, its waveform, spectral density, and magnitude density have close shapes.

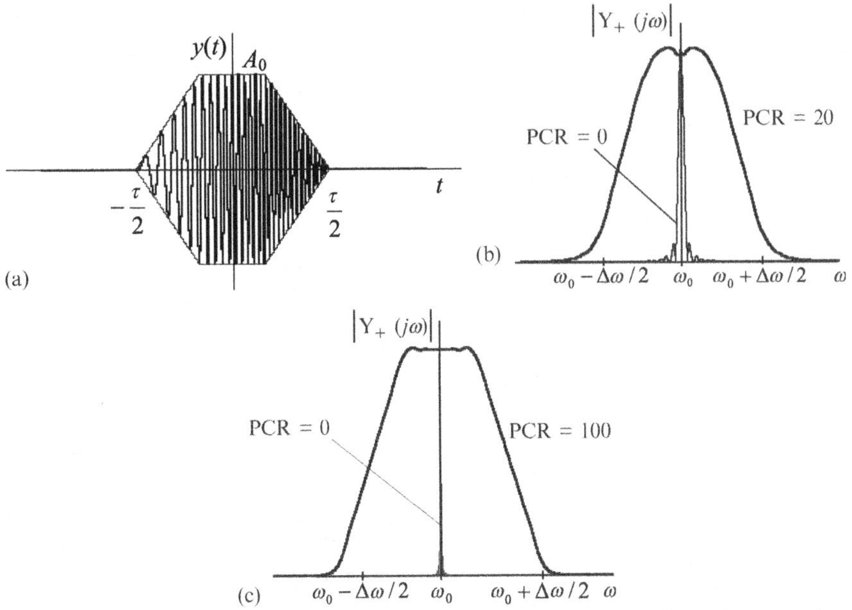

Fig. 3.49 Trapezium LFM pulse: (**a**) time presentation, (**b**) magnitude density for PCR = 20, and (**b**) magnitude density for PCR = 100.

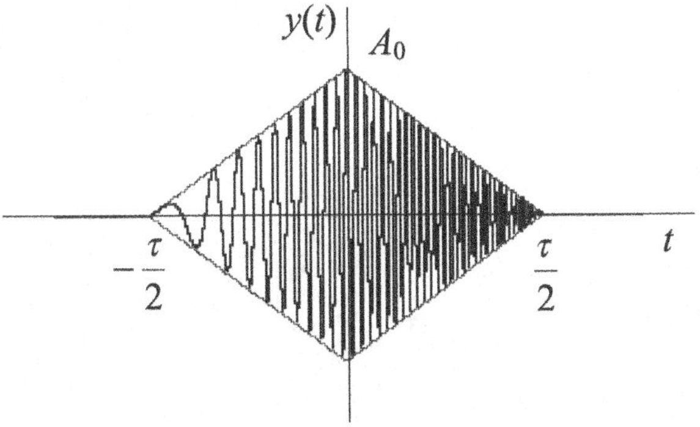

Fig. 3.50 A triangular LFM pulse.

- Its spectrum width is defined by the frequency deviation,

$$\Delta\omega_{\mathrm{L}FM} = \Delta\omega\,. \tag{3.111}$$

- Small oscillations in its spectral density are guaranteed by smoothed sides in the waveform.

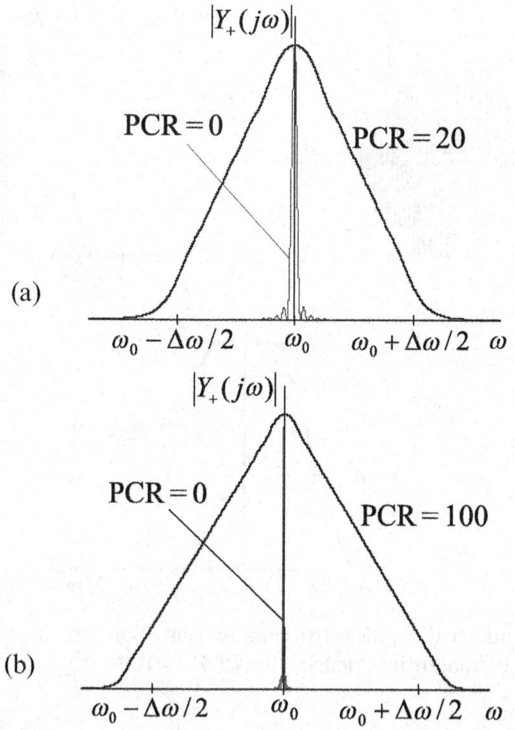

Fig. 3.51 Magnitude spectral density of a triangular LFM pulse: (**a**) PCR = 20 and (**b**) PCR = 100.

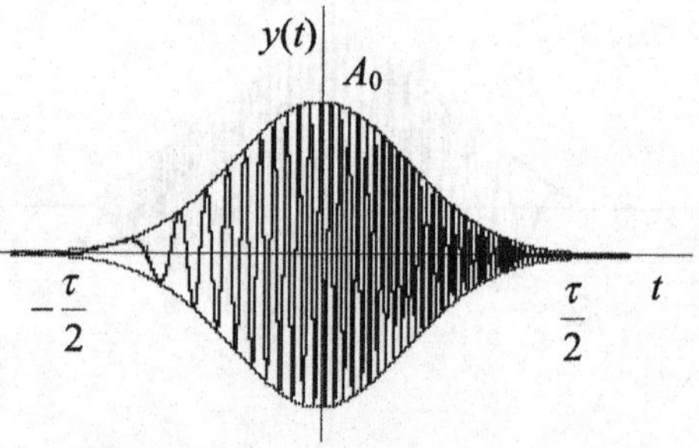

Fig. 3.52 A Gaussian LFM pulse.

3.10 Frequency Shift Keying

In analog communications with FM, a modulating signal is used to shift the frequency of an oscillator at an audio rate. In digital communications with FM, another form of modulation called the *frequency shift keying* (FSK) is used that is somewhat similar to continuous wave keying (or ASK) in AM

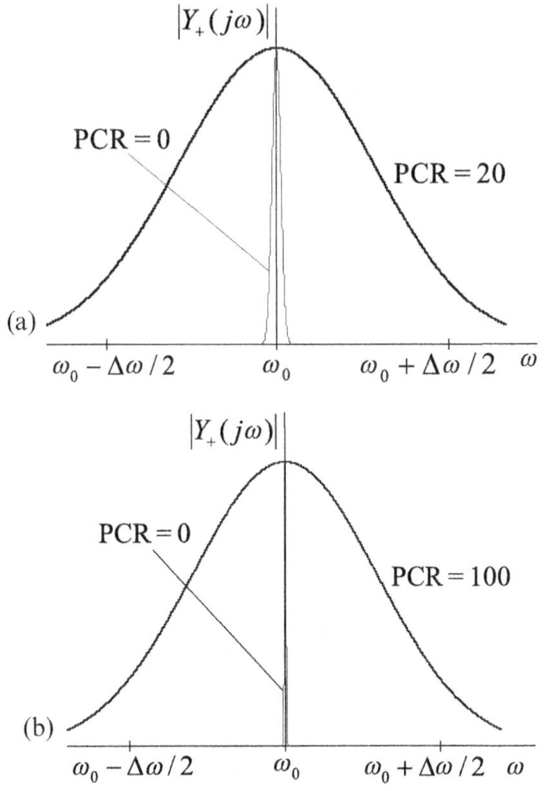

Fig. 3.53 Magnitude spectral density of a Gaussian LFM pulse: (**a**) PCR = 20 and (**b**) PCR = 100.

transmission. Several types of FSK are recognized. Among them the *binary frequency shift keying* (BFSK) and *multifrequency shift keying* or *M-ary frequency shift keying* (MFSK) are most widely used.

3.10.1 Binary Frequency Shift Keying

In BFSK, the frequency of a fixed amplitude carrier signal is abruptly changed between two differing frequencies by opening and closing the key according to the binary stream of data to be transmitted. Figure 3.54 shows an example of the BFSK signal sequence generated, e.g., by the same binary sequence 0101001 ("1" is a mark and "0" is a space) as that used in Fig. 3.37 to illustrate the BASK signal. For illustrative purposes, the spacing frequency is shown here as double the marking frequency.

A block diagram for the BFSK transmitter can be realized as it is shown in Fig. 3.55. Two carrier signals $z_1(t) = e^{j(\omega_{01}t+\psi_{01})}$ and $z_2(t) = e^{j(\omega_{02}t+\psi_{02})}$ are formed separately at different frequencies $\omega_{01} = 2\pi f_{01}$ and $\omega_{02} = 2\pi f_{02}$.

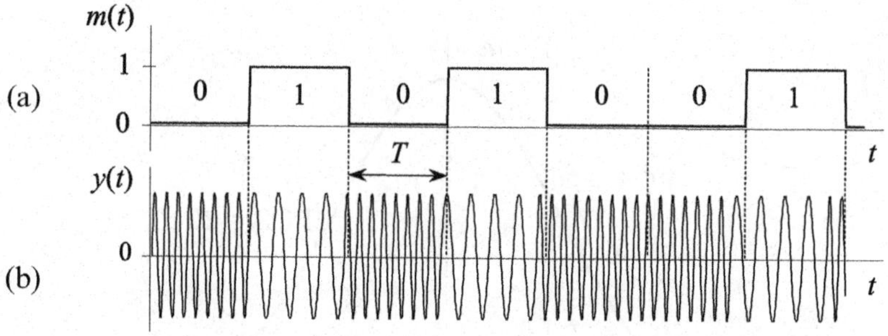

Fig. 3.54 Frequency shift keying: (**a**) binary modulating signal and (**b**) BFSK signal.

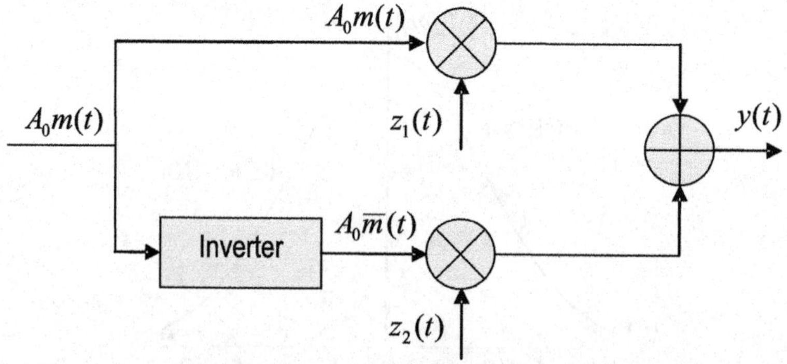

Fig. 3.55 Block diagram for BFSK transmitter.

The binary FSK wave is then formed as

$$y(t) = A_0[m(t)z_1(t) + \bar{m}(t)z_2(t)] = \begin{cases} A_0 z_1(t), & m(t) = 1 \\ A_0 z_2(t), & m(t) = 0 \end{cases}. \tag{3.112}$$

In such a structure, the BFSK signal has constant amplitude A_0 and the difference $f_{02} - f_{01}$ is usually less than 1,000 Hz, even when operating at several megahertz. To receive the BFSK signal, the receiver is designed to have a structure shown in Fig. 3.56. Its logical part is designed with two correlators and a comparator. So, if the first correlator produces a signal that is larger than that of the second correlator, the receiver outputs "1" and vice versa.

On the whole, FSK is used primarily in low-speed modems capable of transmitting data up to 300 bits/s. So, this is a relatively slow data rate. To achieve higher speeds, other forms of modulation are used, including phase-shift keying modulation.

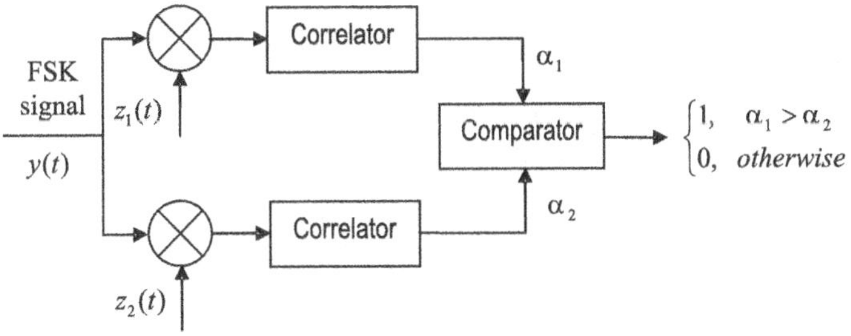

Fig. 3.56 Block diagram for BFSK receiver.

3.10.2 Multifrequency Shift Keying

To increase the speed of data transmitting, MFSK is used. MFSK is bandwidth-efficient, fast, and relatively simple to encode. This kind of coding is akin to that in M-ASK. It implies using several tones in the carrier with a number of frequencies multiple to 2. For example, MFSK4 allows for high-speed (up to 2,400 baud) data transmission. Another coding is known as Hadamard MFSK, which advanced modulation scheme is used to minimize the effects of frequency-dependent fading. This scheme also allows the system to operate at a lower signal-to-noise ratio (SNR) by working reliably at lower transmit power levels.

3.11 Phase Modulation

Usually, the terms angular modulation, FM, and PM are used interchangeably. Indeed, if the modulating function varies proportionally the signal angle, it is PM; and if the derivative of the modulating signal affects this angle, then it is FM. Thus, a transition from FM to PM and backward may be done by preprocessing the modulating signal with the integrator and differentiator, respectively.

> ***Phase modulation:*** *PM* is a kind of angular modulation with which the phase of oscillations is varied in accordance with the modulating signal.

□

In PM, the modulated signal phase (3.6) may be performed by

$$\Psi(t) = \omega_0 t + k_p m(t) + \psi_0 , \tag{3.113}$$

where $k_p m(t) = \Delta\psi(t)$ is the *phase deviation*, $m(t)$ is the message signal, and k_p is the *phase sensitivity*.

3.11.1 Simplest PM

By the harmonic message signal, $m(t) = m_0 \cos(\Omega t + \vartheta)$, the angle (3.113) is rewritten as

$$\Psi(t) = \omega_0 t + \Delta\psi \cos(\Omega t + \vartheta) + \psi_0 \,, \tag{3.114}$$

where $\Delta\psi = k_\mathrm{p} m_0$ is the *peak phase deviation* caused by the modulating signal. The PM signal then becomes

$$y(t) = A_0 \cos[\omega_0 t + \Delta\psi \cos(\Omega t + \vartheta) + \psi_0] \,. \tag{3.115}$$

The instantaneous frequency associated with (3.114) calculates

$$\omega(t) = \frac{\mathrm{d}\Psi(t)}{\mathrm{d}t} = \omega_0 - \Delta\psi\Omega \sin(\Omega t + \vartheta) \tag{3.116}$$

and hence the frequency deviation caused by PM is

$$\Delta\omega = -\Delta\psi\Omega = -k_\mathrm{p} m_0 \Omega \,. \tag{3.117}$$

Figure 3.57 exhibits the modulation processes associated with PM, i.e., the modulating signal $x(t)$ (a), frequency variations $\omega(t)$ caused by PM (b), PM function $\psi(t)$ (c), and modulated PM signal $y(t)$ (d). It is seen that the instantaneous phase is nonstationary having a drift $\omega_0 t$ and, unlike the FM case, the PM law inherently coincides here with the modulating signal $x(t)$. The frequency varies by the differentiated modulating signal, thus obeying the function $-\Delta\psi\Omega \sin(\Omega t + \vartheta)$. Inherently, the frequency function is shifted on $\pi/2$ for the modulating signal. Yet, the peak phase deviation $\Delta\psi\Omega = k_\mathrm{p} m_0 \Omega$ depends not only on the phase sensitivity k_p and peak value of the message signal m_0 but also on the modulation frequency Ω. The latter is treated as a disadvantage of PM.

Spectrum of simplest PM demonstrates quite some similarity with that of FM if the phase deviation $\Delta\psi$ is small. By zero constant phases $\vartheta = \psi_0 = 0$ and $|\Delta\psi| \ll \pi$, (3.115) becomes

$$\begin{aligned} y(t) &= A_0 \cos(\omega_0 t + \Delta\psi \cos \Omega t) \\ &= A_0 \cos(\omega_0 t) \cos(\Delta\psi \cos \Omega t) - A_0 \sin(\omega_0 t) \sin(\Delta\psi \cos \Omega t) \,. \end{aligned} \tag{3.118}$$

The functions $\cos(\Delta\psi \sin \Omega t)$ and $\sin(\Delta\psi \sin \Omega t)$ are extended to the Bessel functions of the first kind and nth order, (3.85) and (3.86), respectively. By $|\Delta\psi| < 1$ and (3.87), (3.118) reduces to the series

$$y(t) \cong A_0 \cos \omega_0 t - \frac{A_0 \Delta\psi}{2} \cos(\omega_0 - \Omega)t + \frac{A_0 \Delta\psi}{2} \cos(\omega_0 + \Omega)t \tag{3.119}$$

and we see that (3.119) and (3.88) are consistent. Therefore, spectrum of a signal with simplest PM will be performed by Fig. 3.41 if to substitute β with $\Delta\psi$. And, like the FM case, increasing $\Delta\psi$ increases a number of spectral lines in the signal spectrum.

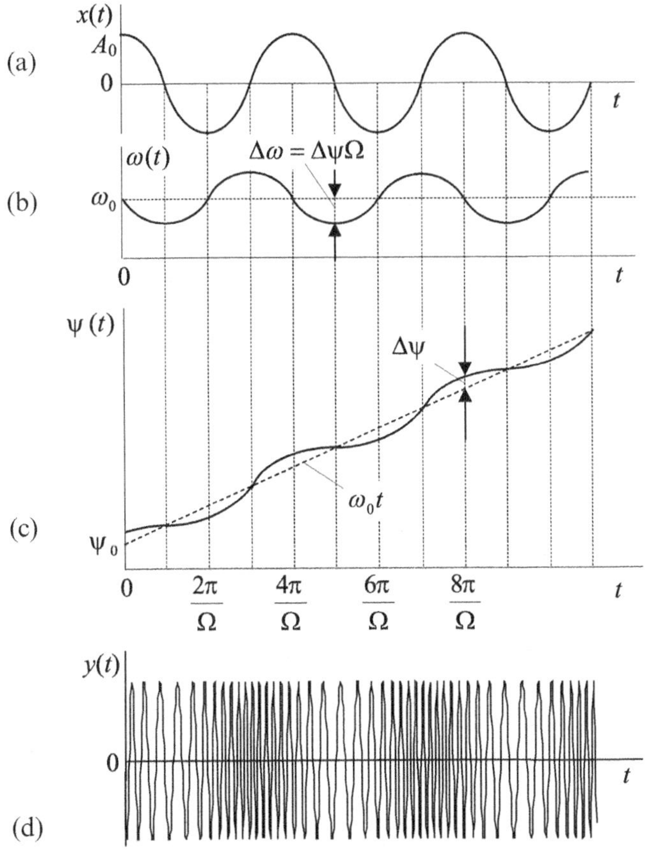

Fig. 3.57 Simplest PM: (**a**) modulating signal, (**b**) FM caused by PM, (**c**) PM, and (**d**) PM signal.

3.11.2 Spectrum with Arbitrary Angle Deviation

A generalized form of (3.118) can be chosen to be

$$y(t) = \mathrm{Re}\left[A_0 e^{j\omega_0 t} e^{j\Delta\psi \cos \Omega t}\right], \tag{3.120}$$

where no assumption about the phase deviation $\Delta\psi$ is made.

We may then extend the modulating term to the harmonic series by the Bessel functions of the first kind and nth order (3.85) and (3.86), and write

$$e^{j\Delta\psi \cos \Omega t} = \sum_{n=-\infty}^{\infty} j^n J_n(\Delta\psi) e^{jn\Omega t}\,. \tag{3.121}$$

By substituting $j^n = e^{jn\pi/2}$ and invoking (3.121), (3.120) transforms to

$$
\begin{aligned}
y(t) &= \mathrm{Re}\left[A_0 e^{j\omega_0 t} \sum_{n=-\infty}^{\infty} J_n(\Delta\psi) e^{jn\Omega t + j\frac{n\pi}{2}}\right] \\
&= \mathrm{Re}\left[A_0 \sum_{n=-\infty}^{\infty} J_n(\Delta\psi) e^{j\left(\omega_0 t + n\Omega t + \frac{n\pi}{2}\right)}\right] \\
&= A_0 \sum_{n=-\infty}^{\infty} J_n(\Delta\psi) \cos\left[(\omega_0 + n\Omega)t + \frac{n\pi}{2}\right]. \qquad (3.122)
\end{aligned}
$$

The Fourier transform of (3.122), using the Euler formula, may now be performed by

$$
\begin{aligned}
Y(j\omega) &= \frac{A_0}{2} \sum_{n=-\infty}^{\infty} J_n(\Delta\psi) \mathcal{F}\left[e^{j(\omega_0 + n\Omega)t + j\frac{n\pi}{2}} + e^{-j(\omega_0 + n\Omega)t - j\frac{n\pi}{2}}\right] \\
&= \frac{A_0}{2} \sum_{n=-\infty}^{\infty} J_n(\Delta\psi) \left[e^{j\frac{n\pi}{2}} \mathcal{F} e^{j(\omega_0 + n\Omega)t} + e^{-j\frac{n\pi}{2}} \mathcal{F} e^{-j(\omega_0 + n\Omega)t}\right].
\end{aligned}
\qquad (3.123)
$$

By the shift property of the delta function,

$$
\delta(z - z_0) = \frac{1}{2\pi} \int_{-\infty}^{\infty} e^{\pm j(z - z_0)u} du, \qquad (3.124)
$$

(3.123) finally goes to

$$
\begin{aligned}
Y(j\omega) = &\sum_{n=-\infty}^{\infty} \pi A_0 J_n(\Delta\psi) \delta(\omega + \omega_0 + n\Omega) e^{-j\frac{n\pi}{2}} \\
&+ \sum_{n=-\infty}^{\infty} \pi A_0 J_n(\Delta\psi) \delta(\omega - \omega_0 - n\Omega) e^{j\frac{n\pi}{2}}, \qquad (3.125)
\end{aligned}
$$

where the first and second sums represent the PM spectrum around the negative and positive carrier frequencies, respectively.

Relation (3.125) gives a generalized solution of an analysis problem for signals with angular modulation. It is a straightforward solution for PM and it holds true for FM if to substitute the phase deviation $\Delta\psi$ with the FM modulation index β. In either case, the calculus requires high orders of the Bessel functions if $\Delta\psi$ is large. This thereby increases a number of spectral lines, which amplitudes are calculated by the Bessel functions. To illustrate, Fig. 3.58 sketches two plots of the magnitude spectrums with angular modulation for $\Delta\psi = \beta = 2$ and $\Delta\psi = \beta = 5$.

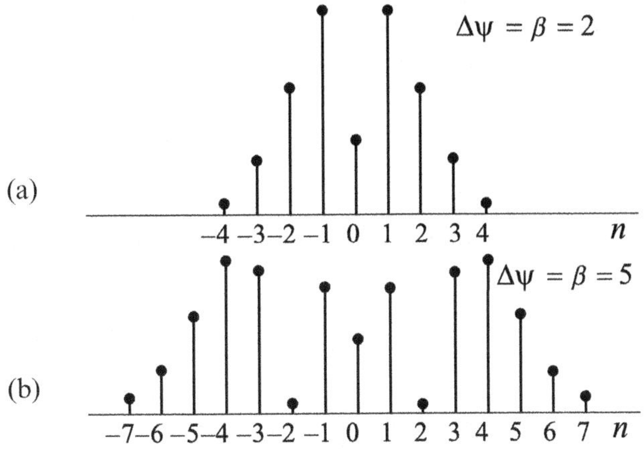

Fig. 3.58 Magnitude spectrum with angular modulation: (**a**) $\Delta\psi = \beta = 2$ and (**b**) $\Delta\psi = \beta = 5$.

3.11.3 Signal Energy with Angular Modulation

Energy plays an important role when signals are transmitted to a distance and then received and processed at receiver. Two major measures of energy are of prime interest: an average energy during the period of a carrier and modulating signal.

With simplest PM (3.119), an instantaneous power of an electric current i_{PM} may be assumed to be generated on some real resistance R such that

$$P(t) = i_{\mathrm{PM}}^2 R = \frac{v_{\mathrm{PM}}^2}{R} = \frac{A_0^2}{R}\cos^2(\omega_0 t + \Delta\psi\cos\Omega t)\,, \tag{3.126}$$

where v_{PM} is an electric voltage induced by i_{PM} on R.

An average power over the period of a carrier signal $T_\omega = 2\pi/\omega_0$ is given by

$$\begin{aligned}
\langle P\rangle_\omega &= \frac{1}{T_\omega}\int_{-T_\omega/2}^{T_\omega/2} P(t)\mathrm{d}t = \frac{A_0^2}{RT_\omega}\int_{-T_\omega/2}^{T_\omega/2}\cos^2(\omega_0 t + \Delta\psi\cos\Omega t)\mathrm{d}t \\
&= \frac{A_0^2}{2RT_\omega}\int_{-T_\omega/2}^{T_\omega/2}[1+\cos 2(\omega_0 t + \Delta\psi\cos\Omega t)]\,\mathrm{d}t \\
&= \frac{A_0^2}{2R} + \frac{A_0^2}{2RT_\omega}\int_{-T_\omega/2}^{T_\omega/2}[\cos 2\omega_0 t\cos(2\Delta\psi\cos\Omega t) - \sin 2\omega_0 t\sin(2\Delta\psi\cos\Omega t)]\,\mathrm{d}t\,.
\end{aligned} \tag{3.127}$$

Typically, angular modulation is performed with $\Omega \ll \omega_0$ and the functions $\cos(2\Delta\psi \cos \Omega t)$ and $\sin(2\Delta\psi \cos \Omega t)$ become almost constants as compared to the carrier frequency terms. An average power therefore may approximately be calculated by

$$\langle P \rangle_\omega \cong \frac{A_0^2}{2R} \,. \tag{3.128}$$

In contrast to AM (3.21), there is no modulating terms in (3.128) and thus this is also an average power over the period of the modulating frequency,

$$\langle P \rangle_\Omega \cong \langle P \rangle_\omega \cong \frac{A_0^2}{2R} \,. \tag{3.129}$$

This peculiarity of angular modulation (PM and FM) speaks in favor of its application in communications. Indeed, signals with time-constant energy are the best candidates for efficient matching with system channels.

3.12 Phase Shift Keying

In line with LFM, there are several other types of complex modulation of practical importance. One of them is perfected with the impulse AM and phase manipulation, also known as the PSK or the binary PSK (BPSK). In BPSK modulation, the phase of the RF carrier is shifted on $\pm\pi$ in accordance with a digital bit stream. Therefore, the BPSK signal may be performed as the RF pulse-burst or pulse-train with a unit period-to-pulse duration ratio, $q = 1$. Phase in each pulse is manipulated by a digital code. In communications, the periodic series of BPSK bit pulses is used. In radars, they exploit the pulse-bursts to compress the correlation function of a signal at the receiver.

3.12.1 RF Signal with Phase Keying

Most generally, the *phase keying (manipulation)* signal is the one which phase changes in accordance with some impulse modulating signal that may be periodic or it may be the burst.

In practical applications, the widest currency has gained the signal which phase is manipulated on 0 or π with some periodicity. Examples are the BPSK in communications and the binary phase coding in radars. Herewith, other phase shifts may also be used in communications, in particular.

To learn a spectral content of such signals, it is in order considering them as the series of single pulses with $q = 1$ and phases manipulated in accordance with some law. The envelope of elementary pulses might be arbitrary and, in some cases, coded.

Since the burst duration is calculated by

$$\tau_B = N\tau \,, \tag{3.130}$$

where τ is an elementary pulse duration and N is a number of pulses in the burst, the phase-manipulated signal of the rectangular envelope with amplitude A_0 may be performed by

$$y(t) = \begin{cases} A_0 \cos[\omega_0 t + k_\mathrm{p} m(t) + \psi_0], & 0 \leqslant t \leqslant \tau_B \\ 0 & \text{otherwise} \end{cases}, \tag{3.131}$$

where the message signal $m(t)$ makes the phase shift $\Delta\psi(t) = k_\mathrm{p} m(t)$ to be 0 or π in each of elementary pulses. If we let $\psi_0 = 0$ and suppose that the ratio τ/T_0 is integer, where $T_0 = 2\pi/\omega_0$ is the period of the carrier, then the signal phase will take a constant value of either

$$\Psi(t) = \omega_0 t \quad \text{or} \quad \Psi(t) = \omega_0 t + \pi$$

and (3.131) will be performed by a superposition of the elementary pulses,

$$y(t) = \begin{cases} A_0 \sum\limits_{n=0}^{N-1} \cos(\omega_0 t + \Delta\psi_n), & n\tau \leqslant t \leqslant (n+1)\tau \\ S0 & \text{otherwise} \end{cases}, \tag{3.132}$$

where N may be arbitrary in general and it is determined by the code length, in particular. The phase shift $\Delta\psi_n$ takes the values either 0 or π induced by the digital code. Figure 3.59 gives an example of a PSK signal in question.

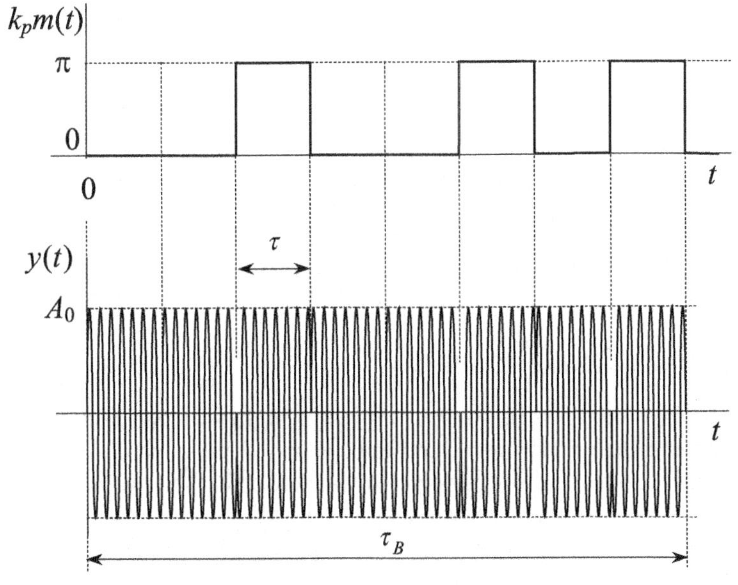

Fig. 3.59 PSK signal.

3.12.2 Spectral Density of RF Signal with BPSK

An analysis problem of (3.132) for finite N may be solved using the properties of the Fourier transform. We thus write

$$Y(j\omega) = \sum_{n=1}^{N} X_n(j\omega)\,, \tag{3.133}$$

where $X_n(j\omega)$ is the spectral density of the nth elementary shifted RF pulse.

Referring to Fig. 3.59, we note that a center of the nth elementary RF pulse is shifted on $(n-0.5)\tau$ and hence its spectral density, by the time-shifting property, is

$$X_n(j\omega) = X(j\omega)\mathrm{e}^{-j\left[(\omega-\omega_0)\left(n+\frac{1}{2}\right)\tau-\Delta\psi_n\right]}\,, \tag{3.134}$$

where $X(j\omega)$ is the spectral density of an elementary unshifted RF pulse. By (3.134), a density (3.133) becomes

$$Y(j\omega) = X(j\omega)\sum_{n=1}^{N} \mathrm{e}^{-j\left[(\omega-\omega_0)\left(n+\frac{1}{2}\right)\tau-\Delta\psi_n\right]}\,. \tag{3.135}$$

The main problem with an analysis of (3.135) is coupled with the value of the phase $\Delta\psi_n$ that is assigned by the code. The phase $\Delta\psi_n$ may take different values in a common case and it is just 0 or π in the BPSK. Fortunately, $X(j\omega)$ is not involved in the analysis and is accounted as a multiplier.

Earlier, we derived spectral densities of several elementary single RF pulses. If we restrict ourself with using only the rectangular waveform, which spectral density at ω_0 is

$$X(j\omega) = A_0\tau\frac{\sin\left[(\omega-\omega_0)\,\tau_0/2\right]}{(\omega-\omega_0)\,\tau_0/2}\,, \tag{3.136}$$

then (3.135) will attain the form of

$$Y(j\omega) = A_0\tau\frac{\sin\left[(\omega-\omega_0)\,\tau_0/2\right]}{(\omega-\omega_0)\,\tau_0/2}\sum_{n=1}^{N} \mathrm{e}^{-j\left[(\omega-\omega_0)\left(n-\frac{1}{2}\right)\tau-\Delta\psi_n\right]}\,. \tag{3.137}$$

Further transformations of (3.137) can be provided only for the particular types of phase codes, among which the Barker codes are most widely used.

3.12.3 PSK by Barker Codes

Many radars operate with short compressed pulses at the receiver. An example of pulse-compression radar is phase-coded pulse compression. As we have already mentioned, in pulse-coded waveform the long pulse (Fig. 3.59) is subdivided into N short subpulses of equal duration τ. Each is then transmitted with a particular phase in accordance with a phase code (usually binary coding). Phase of the transmitted signal alternates between 0 and π in accordance with the sequence of elements in the phase code. The phase code

Table 3.1 Barker codes of length N and the side-lobe level achieved in the signal power spectral density

Code length N	Code elements	Side-lobe level (dB)
$N2$	01 or 00	−6.0
$N3$	001	−9.5
$N4$	0001 or 0010	−13.0
$N5$	00010	−14.0
$N7$	0001101	−16.9
$N11$	00011101101	−20.8
$N13$	0000011001010	−23.3

used is generally a standard code, which has proved to provide the best resolution and least ambiguity in determining the target parameters. The codes used can be either Barker or some form of pseudorandom code. Table 3.1 gives Barker codes along with the side-lobe level achieved in the signal power spectral density (PSD).

The known Barker codes are limited to 13 bits in length. To reduce the probability of a high correlation with the message bits, longer codes are desirable. To meet this need, another set of codes called Willard sequences have been developed. These can be longer than Barker codes, but the performance is not as good.

Examples of spectral densities of signals, which phases are manipulation by Barker codes, are given below.

Example 3.12. Given a rectangular pulse-burst, $q = 1$, with phase manipulation by the Barker code of length $N = 3$. Figure 3.60 shows this signal along with the Barker code $N3$.

By (3.135) and $e^{j\pi} = -1$, the spectral density of a signal is written as

$$\begin{aligned} Y(j\omega) &= X(j\omega)\left\{e^{-j(\omega-\omega_0)\frac{\tau}{2}} + e^{-j(\omega-\omega_0)\frac{3\tau}{2}} + e^{-j[(\omega-\omega_0)\frac{5\tau}{2}-\pi]}\right\} \\ &= X(j\omega)e^{-j(\omega-\omega_0)\frac{3\tau}{2}}\left[e^{j(\omega-\omega_0)\tau} + 1 - e^{-j(\omega-\omega_0)\tau}\right] \\ &= X(j\omega)e^{-j(\omega-\omega_0)\frac{3\tau}{2}}\left[1 + 2j\sin(\omega-\omega_0)\tau\right] . \end{aligned} \quad (3.138)$$

By using (3.32) and (3.33), we go to the USB of the spectral density

$$\begin{aligned} Y_+(j\omega) &= \frac{A_0\tau}{2}\frac{\sin[(\omega-\omega_0)\tau/2]}{(\omega-\omega_0)\tau/2}\sqrt{1+4\sin^2(\omega-\omega_0)\tau} \\ &\quad \times e^{-j\left\{(\omega-\omega_0)\frac{3\tau}{2}+\arctan[2\sin(\omega-\omega_0)\tau]\right\}} = |Y_+(j\omega)|\, e^{-j\Psi_y(\omega)} , \end{aligned} \quad (3.139)$$

by which the magnitude and phase spectral densities are calculated to be, respectively,

$$|Y_+(j\omega)| = \frac{A_0\tau}{2}\left|\frac{\sin[(\omega-\omega_0)\tau/2]}{(\omega-\omega_0)\tau/2}\right|\sqrt{1+4\sin^2(\omega-\omega_0)\tau} , \quad (3.140)$$

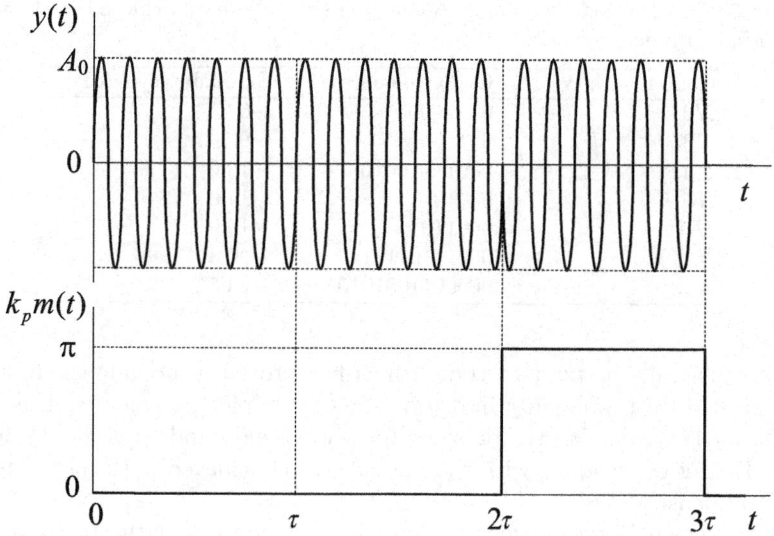

Fig. 3.60 Phase-manipulated signal with the Barker code $N3$.

$$\varphi_y(\omega) = \begin{cases} \frac{3\tau(\omega-\omega_0)}{2} + \arctan[2\sin(\omega-\omega_0)\tau] \pm 2\pi n, & \frac{\sin(\omega-\omega_0)\tau/2}{(\omega-\omega_0)\tau/2} \geqslant 0 \\ \frac{3\tau(\omega-\omega_0)}{2} + \arctan[2\sin(\omega-\omega_0)\tau] \pm \pi \pm 2\pi n, & \frac{\sin(\omega-\omega_0)\tau/2}{(\omega-\omega_0)\tau/2} < 0 \end{cases} . \tag{3.141}$$

Figure 3.61 shows the magnitude density (3.140) along with that associated with the rectangular RF single pulse of the same duration. It is neatly seen that the spectral waveform of the coded burst originates from that of the rectangular pulse.

□

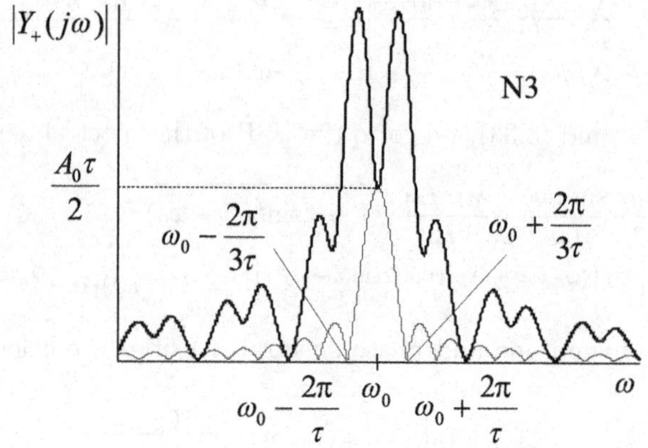

Fig. 3.61 Magnitude spectral density of an RF pulse-burst with the phase manipulated by the Barker code $N3$ (sold) and RF rectangular pulse (light).

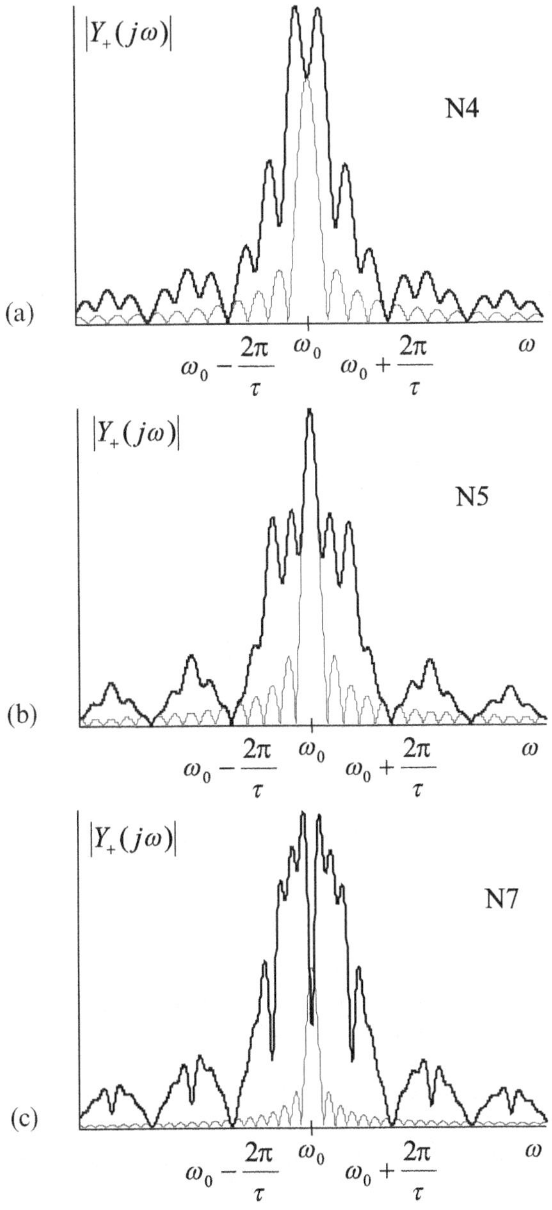

Fig. 3.62 Magnitude spectral densities of the pulse-bursts with the Barker phase-coding (sold) and RF rectangular pulse (light): **(a)** $R4$, **(b)** $R5$, **(c)** $R7$, **(d)** $R11$, and **(e)** $R13$.

Example 3.13. Given a rectangular pulse-burst with phase-coding by the Barker codes $N4$, $N5$, $N7$, $N11$, and $N13$. Figure 3.62 sketches the magnitude spectral densities associated with all these codes. For the comparison, each

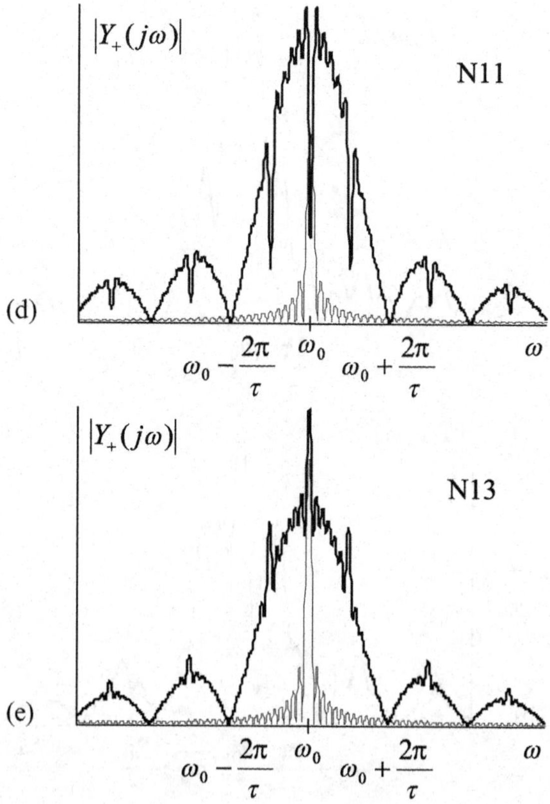

Fig. 3.62 *Continued*

plot is accompanied by the magnitude density of a single RF pulse of the same duration. □

The most important practical features of BPSK are the following:

- Noise immunity
- The channel utilization efficiency is low because of using one phase shift to represent single bit
- The need for carrier recovery circuit, which make the circuit complex and expensive

Although the noise immunity, the BPSK is not widely used owing to the low channel utilization efficiency and necessity of carrier recovery. For synchronous data transmission, it is essential to enforce a phase change from the previously transmitted phase to help the receiver detect the beginning of every bit. This technique is called *differential phase shift keying* (DPSK).

3.12.4 Differential PSK

In DPSK, at the end of every bit, the value of the next bit in the input data stream is as follows: if the bit is binary 1, insert $+90°$ phase shift; if the bit is binary 0, insert $-90°$ phase shift. The main problem with BPSK and DPSK is that the speed of data transmission is limited in a given bandwidth. One way to increase the binary data rate while not increasing the bandwidth required for the signal transmission is to encode more than 1 bit per phase change. One commonly used system for doing this is known as *quadrature phase shift keying* (QPSK or 4-PSK). More efficient use of the bandwidth can be achieved if we squeeze more bits per second into a given bandwidth. For example, instead of a phase shift of 180°, as in PSK, this modulation technique allows phase shifts of multiples of 90°.

In QPSK, the incoming bit stream is broken into pairs of bits (called dibits). Hence, there are only four possible values of these pairs (dibits) in any bit stream; 00, 01, 10, 11. These pairs change the carrier's phase in accordance with the digits, values.

3.13 Summary

Modulation is a tool to transfer useful signals to a distance with highest efficiency. In spite of a number of practical techniques, modulation is obtained by varying the signal amplitude, phase, and/or frequency and this is independent on the system type, operation principle, and solved problems. The following notations are of importance:

- AM signal has poor jamming immunity because of noise and environmental, which directly affect its amplitude and thus vary its energy.
- AM spectrum is narrower than that featured to angular modulation.
- FM signal has nice jamming immunity as its amplitude may be stabilized with a limiter.
- FM spectrum is wider than that of AM signal and the SSB transmitting cannot be used in FM.
- FM signal is accompanied with PM whose phase deviation varies inversely to the modulation frequency.
- PM signal is accompanied with FM whose frequency is changed proportionally to the modulating frequency.
- An initial signal phase is insignificant for FM and may be crucial for analog PM.
- Signals with LFM have almost equal time and spectral waveforms.
- Phase coding is used to achieve high noise immunity and channel utilization efficiency.

3.14 Problems

3.1 (Harmonic modulation). Explain the principle difference between AM and FM in the time and frequency domains.

3.2. Given a signal

$$A(t)\cos\left[\int \omega(t)\mathrm{d}t + \psi_0\right],$$

where $A(t) = A_0[1+\mu\cos(\Omega t+\vartheta_0)]$ and $\omega(t) = \omega_0 + k_\omega m_0\cos(\Omega t+\theta_0)$. Show the vector diagram of this AM–FM signal. Derive the spectrum and determine the conditions under which the spectrum becomes SSB.

3.3. Solve Problem 3.2 for

1. $A(t) = A_0[1+\mu\cos(\Omega_1 t+\vartheta_0)]$
   $\omega(t) = \omega_0 + k_\omega m_0\cos(\Omega_2 t+\theta_0)$
2. $A(t) = A_0[1+\mu_1\cos(\Omega_1 t+\vartheta_{01}) + \mu_2\cos(\Omega_2 t+\vartheta_{02})]$
   $\omega(t) = \omega_0$
3. $A(t) = A_0$
   $\omega(t) = \omega_0 + k_\omega[m_{01}\cos(\Omega_1 t+\theta_{01}) + m_{02}\cos(\Omega_2 t+\theta_{02})]$
4. $A(t) = A_0[1+\mu_1\cos(\Omega_1 t+\vartheta_{01}) + \mu_2\cos(\Omega_2 t+\vartheta_{02})]$
   $\omega(t) = \omega_0 + k_\omega m_0\cos(\Omega t+\theta_0)$

3.4. The following signals are given in their spectral forms. Write the signals in compact forms and determine the type of modulation in each of those:

1. $y(t) = 0.1\cos(1.8\times 10^6\pi t + \pi/4) + 0.1\cos(2.2\times 10^6\pi t + \pi/4)$
   $+4\cos(2\times 10^6\pi t + \pi/4)$
2. $y(t) = 0.1\cos(1.8\times 10^6\pi t + 5\pi/4) + 0.1\cos(2.2\times 10^6\pi t + \pi/4)$
   $+4\cos(2\times 10^6\pi t + \pi/4)$
3. $y(t) = 0.1\cos(2.2\times 10^6\pi t + \pi/4) + 4\cos(2\times 10^6\pi t + \pi/4)$
4. $y(t) = 0.1\cos(1.8\times 10^6\pi t + 5\pi/4) + 4\cos(2\times 10^6\pi t + \pi/4)$
5. $y(t) = \cos(2.2\pi f_0 t) + 4\cos(2\pi f_0 t + \pi/2)$
6. $y(t) = 0.1\cos(1.8\pi f_0 t + \pi/2) + 4\cos(2\pi f_0 t)$

3.5. Assume the signal (Problem 3.4) to be an electric current inducing a voltage on a real resistance of 1 ohm. Determine

1. An instantaneous power $P(t)$ given by (3.20)
2. An average power $\langle P\rangle_\omega$ over the period of a carrier (3.21)
3. An average power $\langle P\rangle_\Omega$ over the period of a modulated frequency (3.25)

3.6. Given the spectra (Fig. 3.63). Investigate, how the coefficient k affects the modulated signal. What type of modulation is it? Consider a general case of $k \neq 1$ and a particular situation with $k = 1$.

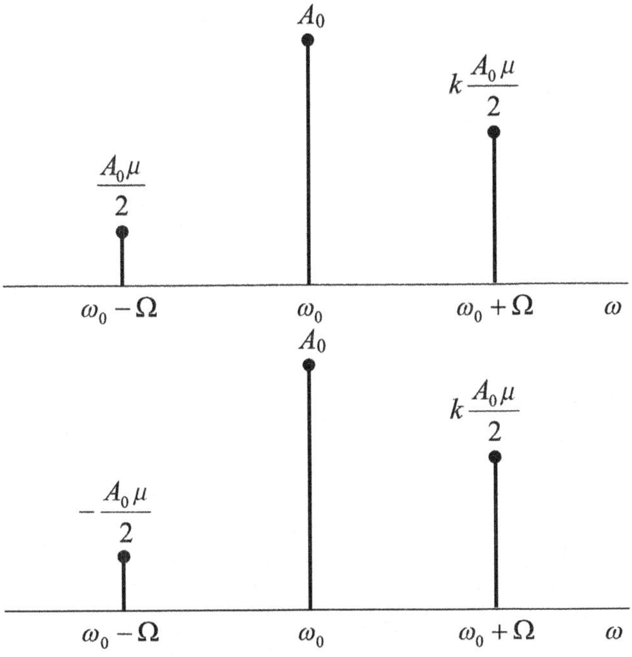

Fig. 3.63 Spectra of modulated signals.

3.7 (Impulse AM). Given a signal of a waveform (Fig. 2.64) that provides AM of a signal $A(t)\cos(\omega_0 t + \psi_0)$. The Fourier transform for the given waveform was found in Problem 2.13. Now show the spectral density of the relevant AM signal.

3.8. Given a signal with a rectangular impulse AM. Determine the spectral density of a signal, using the properties of the Fourier transforms:

1. $y(t) = \begin{cases} 4\cos(2\pi f_0 t + \pi/2), & -2/f_0 \le t \le 10/f_0 \\ 0, & \text{otherwise} \end{cases}$

2. $y(t) = \begin{cases} 0.1\sin(1.8\pi f_0 t), & -10/f_0 \le t \le 4/f_0 \\ 0, & \text{otherwise} \end{cases}$

3. $y(t) = \begin{cases} \cos(2.2\pi f_0 t) + 4\cos(2\pi f_0 t + \pi/2), & -6/f_0 \le t \le 6/f_0 \\ 0, & \text{otherwise} \end{cases}$

4. $y(t) = \begin{cases} 0.1\sin(1.8\pi f_0 t) + 4\cos(2\pi f_0 t), & -10/f_0 \le t \le 10/f_0 \\ 0, & \text{otherwise} \end{cases}$

3.9. Consider Problem 3.7 assuming a burst of two such pulses with $q = 2$. Calculate and plot the spectral density of the relevant modulated signal.

3.10. Consider Problem 3.9 assuming a periodic pulse-burst with $q_1 = 2$. Calculate and plot the spectrum of the relevant modulated signal.

3.11. Analyze Fig. 3.17 and realize what will happen with the spectrum if period T will tend toward infinity?

3.12 (Types of analog AM). Consider the following kinds of analog AM, namely DSB-LC, DSB-SC, and DSB-RC, and answer on the question: which type is most band-efficient and which is most energy-efficient?

3.13. Compare DSB, SSB, and VSB modulations. Which type is most band-efficient and which is energy-efficient?

3.14 (Amplitude shift keying). Given an ASK signal with $N = 7$ (Fig. 3.37). Using the properties of the Fourier transforms, derive the spectral density of this pulse-burst.

3.15. Given an ASK signal (Fig. 3.38). Derive the spectral density of this pulse-burst. Compare the density derived with that featured to the signal (Fig. 3.37). Based on these two spectral densities, formulate the advantages and disadvantages of BASK and MASK.

3.16 (Frequency modulation). Given a signal $y(t) = A_0 \cos(\omega_0 t + \beta \sin \Omega t)$ with simplest FM. Derive the spectrum of this signal for $\beta < 2$, neglecting the Bessel functions, which values are lesser than 0.1. Analyze the result and deduce how the spectrum will be changed, by increasing β ?

3.17. Plot the vector diagram of the spectrum of a FM signal (Problem 3.16, Case $\beta < 2$). Explain, how this diagram will change if β increases?

3.18 (Frequency shift keying). Given an ASK signal (Fig. 3.54). Using the properties of the Fourier transforms, derive the spectral density of this pulse-burst.

3.19 (Phase modulation). Explain the principle difference between PM and FM, their advantages and disadvantages.

3.20 (Phase shift keying). Following Example 3.12.1, derive the spectral densities of PSK signals for the Barker codes $N2$(01 and 10) and $N4$(0001 and 0010).

3.21. Derive the spectral density of the PSK signal for your own number M represented by the binary code (xxxx). Compare with the spectral density corresponding to the Barker code $N4$. Make the conclusions.

4

Signal Energy and Correlation

4.1 Introduction

Signals can be represented not only by their waveforms and spectra, but also by energy, power, and correlation performance. One of the most impressive examples of practical use of signal's energy may be found in radars. Here the transmitted interrogating pulse is also delayed at the receiver. When the time delay becomes equal to that in the received pulse, the joint energy of two signals attains a maximum. Through the identified time delay, the distance between the interrogator and the target is measured. Signal energy in the frequency domain is also widely used. For example, the signal energy spectrum may be restricted in communication channels with the essential bandwidth within which the signal contains of about 90–99% of its total energy. To possess the necessary energy properties, a signal must be of certain waveform that is discussed in correlation analysis.

4.2 Signal Power and Energy

The *instantaneous signal power* of a signal $x(t)$ is calculated by

$$P_x(t) = x(t)x^*(t) = |x(t)|^2 \,. \tag{4.1}$$

If a signal $x(t)$ is the voltage, it is tacitly assumed that the power (4.1) is distributed across the resistance of 1 ohm. The total signal energy is then defined by integrating the product of a multiplication of the signal $x(t)$ and its complex conjugate $x^*(t)$ over the infinite bounds. The result may be written as the inner product

$$E_x = \langle x, x \rangle = \int_{-\infty}^{\infty} x(t)x^*(t)\mathrm{d}t = \int_{-\infty}^{\infty} |x(t)|^2 \,\mathrm{d}t. \tag{4.2}$$

We may also suppose that there are two different signals $x(t)$ and $y(t)$. Then their joint energy will be determined analogously,

$$E_{xy} = \langle x, y \rangle = \int_{-\infty}^{\infty} x(t)y^*(t)\mathrm{d}t. \tag{4.3}$$

It is important that all real nonperiodic signals have finite energy. In turn, all periodic signals have infinite energy, therefore cannot be described in terms of energy. An average power is an appropriate measure for such signals. These key properties separate all signals onto the *energy signals* and *power signals*.

4.2.1 Energy Signals

There is an important class of signals, which energy is finite; thus the integral in (4.2) converges. Nonperiodic real signals belong to this class.

Energy signal: A signal is the energy signal or finite energy signal if its energy is finite over the infinite time bounds. All finite energy signals vanish at infinity.

□

Example 4.1. Given two signals,

$$x(t) = \begin{cases} A, & -\tau/2 \leqslant t \leqslant \tau/2 \\ 0, & \text{otherwise} \end{cases},$$

$$y(t) = A_0 \cos \omega_0 t.$$

The first signal $x(t)$ has a finite energy

$$E_x = A^2 \int_{-\tau/2}^{\tau/2} \mathrm{d}t = A^2\tau < \infty ,$$

hence this signal is the energy signal. The second signal $y(t)$ has an infinite energy

$$E_y = A_0^2 \int_{-\infty}^{\infty} \cos^2 \omega_0 t\mathrm{d}t = A_0^2\infty$$

and hence is not an energy signal. □

The other form of signal energy is allowed by the Rayleigh theorem. In fact, supposing that the transforms are known, $x(t) \overset{\mathcal{F}}{\Leftrightarrow} X(j\omega)$ and $y(t) \overset{\mathcal{F}}{\Leftrightarrow} Y(j\omega)$, we specify the joint signal energy, by (2.62), as the inner product,

$$E_{xy} = \frac{1}{2\pi}\langle X, Y\rangle = \frac{1}{2\pi}\int_{-\infty}^{\infty} X(j\omega)Y^*(j\omega)\mathrm{d}\omega. \tag{4.4}$$

One may also suppose that $X = Y$ and then (4.4) will describe the signal energy

$$\begin{aligned} E_x &= \frac{1}{2\pi}\langle X, X\rangle = \frac{1}{2\pi}\int_{-\infty}^{\infty} |X(j\omega)|^2 \mathrm{d}\omega \\ &= \frac{1}{\pi}\int_{0}^{\infty} |X(j\omega)|^2 \mathrm{d}\omega. \end{aligned} \tag{4.5}$$

Comparing the Fourier transform of a signal and its energy performance, we notice that

- Signal energy is calculated by integrating the real numbers (4.5) that may easily be done; but, the inverse Fourier transform of $X(j\omega)$ requires operating with complex numbers.

□

- By studying signals through the quantity $|X(j\omega)|^2$, we inevitably lose information about the signal phase; but, the Fourier transform does not allow losing this information.

□

The quantity $G_x(\omega) = |X(j\omega)|^2$ has a sense of the *energy spectral density* (ESD) of the signal $x(t)$, once the integral in (4.5) gives the energy contained in the infinite frequency range. Thus, the function

$$G_{xy}(j\omega) = X(j\omega)Y^*(j\omega) \tag{4.6}$$

represents the *cross energy spectral density* (cross ESD) of the signals $x(t)$ and $y(t)$.

Several important observations may be done regarding the cross ESD:

- By substituting X and Y with their real and imaginary components, i.e., with $X = X_\mathrm{r} + jX_\mathrm{i}$ and $Y^* = Y_\mathrm{r} - jY_\mathrm{i}$, respectively, we go to

$$\begin{aligned} G_{xy} &= (X_\mathrm{r} + jX_\mathrm{i})(Y_\mathrm{r} - jY_\mathrm{i}) = X_\mathrm{r}Y_\mathrm{r} + X_\mathrm{i}Y_\mathrm{i} + j(X_\mathrm{i}Y_\mathrm{r} - X_\mathrm{r}Y_\mathrm{i}) \\ &= \mathrm{Re}\, XY^* + j\,\mathrm{Im}\, XY^* \end{aligned}$$

and conclude that the cross ESD of two different signals is a complex function.

□

- The $\operatorname{Re} G_{xy}(j\omega)$ is an even and $\operatorname{Im} G_{xy}(j\omega)$ is an odd function of frequency.
- The following property holds true

$$G_{xy} = G^*_{yx}. \tag{4.7}$$

- Only a real component of (4.6) contributes to the integral (4.5), therefore

$$E_{xy} = \frac{1}{2\pi}\int_{-\infty}^{\infty} \operatorname{Re} G_{xy}(j\omega)\mathrm{d}\omega = \frac{1}{\pi}\int_{0}^{\infty} \operatorname{Re} G_{xy}(j\omega)\mathrm{d}\omega. \tag{4.8}$$

If $G_{xy} \neq 0$, then two signals are nonorthogonal and, if $G_{xy} = 0$, they are orthogonal. To go from the nonorthogonal signals to the orthogonal ones, it needs to suppress their cross ESD function in the range where their joint energy is maximum.

Let us notice that the relation (4.8) is basic to analyze the joint energy of two signals in the frequency domain.

Example 4.2. Given two shifted rectangular single pulses of duration τ and amplitude $A = 1/\tau$ (Fig. 4.1). By (2.80) and the time-shifting property, the spectral densities of $x(t)$ and $y(t) = x(t-\theta)$ become, respectively,

$$X(jf) = \frac{\sin(\pi f\tau)}{\pi f\tau}, \tag{4.9}$$

$$Y(jf) = \frac{\sin(\pi f\tau)}{\pi f\tau}\mathrm{e}^{-j2\pi f\theta}, \tag{4.10}$$

The cross ESD function of these signals is determined, by (4.6), to be

$$G_{xy}(j2\pi f) = \frac{\sin^2(\pi f\tau)}{(\pi f\tau)^2}\mathrm{e}^{j2\pi f\theta}, \tag{4.11}$$

having a real component

$$\operatorname{Re} G_{xy}(j2\pi f) = \frac{\sin^2(\pi f\tau)}{(\pi f\tau)^2}\cos 2\pi f\theta. \tag{4.12}$$

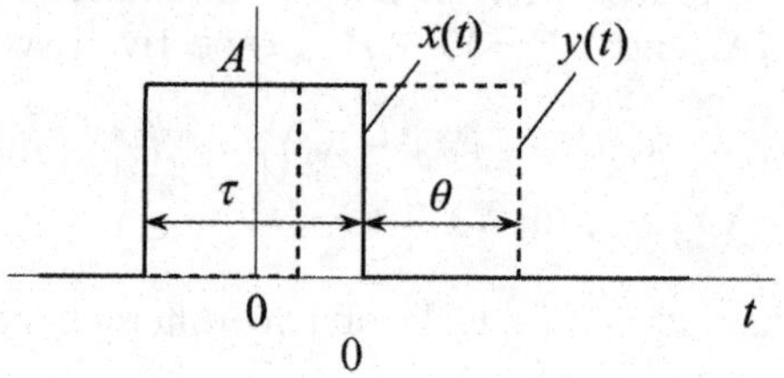

Fig. 4.1 Two shifted rectangular pulses.

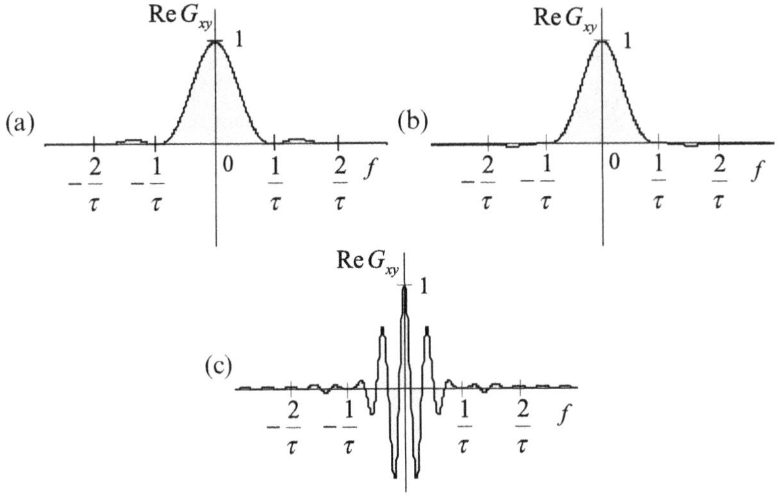

Fig. 4.2 ESD functions of two shifted rectangular pulses: (**a**) $\theta = 0$, (**b**) $\theta = 0.22\tau$, and (**c**) $\theta = 2.5\tau$.

Figure 4.2 illustrates (4.12) for three particular values of the time shift θ. By $\theta = 0$, the function (a) represents the energy density of the rectangular pulse in the frequency domain. The case of $\theta = 0.22\tau$ shows (b) that the side lobes of the cross ESD function may be suppressed almost to zero. If a time shift overcomes the pulse duration, $\theta = 2.5\tau$, then G_{xy} oscillates around zero owing to the cosine multiplier.

□

Example 4.3. Example 4.2 determines the cross ESD function of two rectangular pulses and illustrates what will happen if the pulses are shifted in time. Now consider the case of $\theta = 0$ (Fig. 4.2a) that corresponds to the ESD function of a single rectangular pulse, specified by (2.80),

$$G_x(\omega) = A^2\tau^2 \frac{\sin^2(\omega\tau/2)}{(\omega\tau/2)^2}. \tag{4.13}$$

The function (4.13) is real, so that $G_x = \operatorname{Re} G_x$. Figure 4.3 demonstrates a normalized version of (4.13) in the positive frequency range. It follows that the major part of the signal energy is concentrated in the main lobe and that $G_x(f)$ attenuates with the slope f^{-2}, whereas the spectral density of this pulse attenuates by f^{-1}.

The side lobes do not contribute too much to the signal energy and thus, practically, there is no big sense in accounting the spectral components beyond the main lobe. In fact, by (4.5), the energy of the rectangular pulse calculates

$$E_x = \frac{1}{\pi}\int_0^\infty X^2(\omega)\mathrm{d}\omega = \frac{A^2\tau^2}{\pi}\int_0^\infty \frac{\sin^2(\omega\tau/2)}{(\omega\tau/2)^2}\mathrm{d}\omega = A^2\tau \tag{4.14}$$

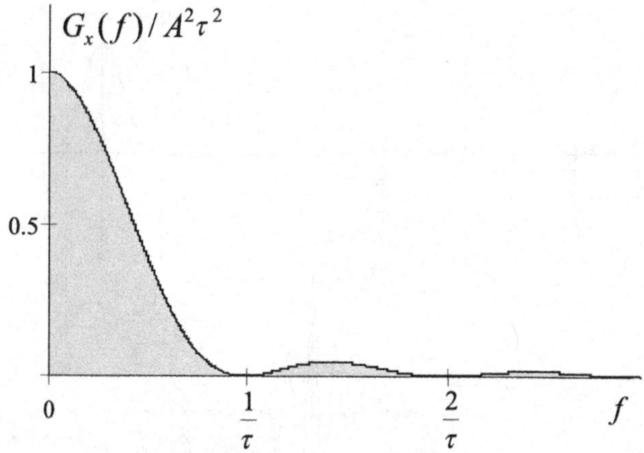

Fig. 4.3 ESD function of the rectangular pulse.

and we notice that the same result is obtained elementary in the time domain by integrating the squared rectangular waveform (Fig. 4.1).

Aimed at evaluating the energy contained only in several side lobes of the ESD function, we rewrite (4.14) as follows:

$$E_{x1} = \frac{2A^2\tau}{\pi}\int_0^{k\pi}\frac{\sin^2\vartheta}{\vartheta^2}\mathrm{d}\vartheta\,, \tag{4.15}$$

where $\vartheta = \pi f\tau$ and the number k is integer. This integral can be calculated analytically but has no simple closed form. Numerical integration shows that

- If $k = 1$, then $E_{x1}/E_x \cong 0.903$.
- If $k = 2$, then $E_{x1}/E_x \cong 0.95$.
- If $k = 3$, then $E_{x1}/E_x \cong 0.966$.

It becomes obvious that if to provide filtering of the rectangular pulse with an ideal LP filter having the cutoff frequency $f_c = 1/\tau$ then, at the output, we will get a signal, which energy consists of 90.3% of the input energy. It also follows that accounting for the second lobe contributes to the output energy with only 4.8%. On the other hand, by extending the filter bandwidth twice with $k = 2$, the energy of the uniformly distributed influencing fluctuations will also be increased twice at the output that is undesirable.

□

Example 4.4. Given a symmetric triangular pulse with amplitude A and duration τ (Fig. 2.21a) described by (2.85). Using (4.2) and (2.90), its energy calculates

$$E_x = \frac{A^2\tau}{3}. \tag{4.16}$$

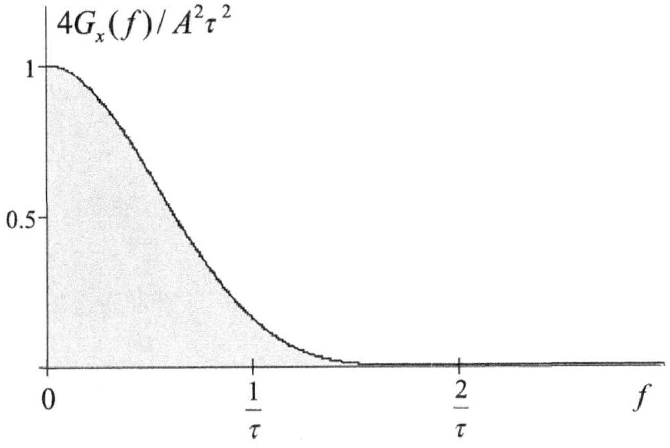

Fig. 4.4 ESD function of a symmetric triangular pulse.

The spectral density of this pulse (2.90) specifies its ESD function

$$G_x(f) = \frac{A^2\tau^2}{4} \frac{\sin^4(\pi f\tau/2)}{(\pi f\tau/2)^4} \tag{4.17}$$

and the integration of (4.17), by (4.5), leads to (4.16). Figure 4.4 illustrates (4.17) and we realize that, by integrating similarly to (4.15),

- If $k = 1$, then $E_{x1}/E_x \cong 0.95$.
- If $k = 2$, then $E_{x1}/E_x \cong 0.997$.
- If $k = 3$, then $E_{x1}/E_x \cong 0.999$.

Thus, practically, there is no big sense in transmitting the side lobes of this pulse, since the main lobe consists of 95% of its energy.

□

Example 4.5. Given a signal $x(t) = \mathrm{e}^{-\alpha t}u(t)$. Its energy is calculated, by (4.2), to be

$$E_x = \int_0^\infty x^2(t)\mathrm{d}t = \int_0^\infty \mathrm{e}^{-2\alpha t}\mathrm{d}t = \frac{1}{2\alpha}. \tag{4.18}$$

The spectral density of this signal is given by (2.66) as $X(j\omega) = 1/(\alpha+j\omega)$. Using (4.5), we arrive at (4.18) in an alternative way,

$$E_x = \frac{1}{\pi}\int_0^\infty \frac{\mathrm{d}\omega}{\alpha^2+\omega^2} = \frac{1}{\alpha\pi}\arctan\frac{\omega}{\alpha}\Big|_0^\infty = \frac{1}{2\alpha}. \tag{4.19}$$

To find, e.g., the 95% containment BW, we solve

$$\frac{1}{\pi}\int_{0}^{\mathrm{BW}} \frac{\mathrm{d}\omega}{\alpha^2+\omega^2} = \frac{0.95}{2\alpha} \tag{4.20}$$

for BW and get $\mathrm{BW} = \alpha \tan(0.95/2)$.

□

An important measure of the ESD function is the *logarithmic ESD* (in decibel units),

$$G(f)_{\mathrm{dB}} = 10 \log_{10} G(f).$$

If the ESD occupies a wide frequency range of several decades, the logarithmic measure of frequency is also preferable. The display of $G(f)_{\mathrm{dB}}$ vs. $\log(f)$ is known as the *Bode plot.* A benefit of a Bode plot is that (1) it allows evaluating the particular parts of the energy spectra in terms of the function rate in dB/decade and (2) the ESD function may by bounded with lines that, in many cases, lead to useful generalizations.

Example 4.6. Given the ESDs of the rectangular pulse (Examples 4.2 and 4.3) and triangular pulse (Example 4.4). The Bode plots of both are shown in Fig. 4.5.

It is seen that the nearest zero of the triangular pulse appears at $2/\tau$. This pulse contains more energy (95%) in the range limited with $1/\tau$ than the rectangular pulse (90.2%). Such a tendency saves if to invoke more lobes: with $k = 2$, there is 99.7% against 95%; and with $k = 3$, we have 99.9% against 96.7%, respectively.

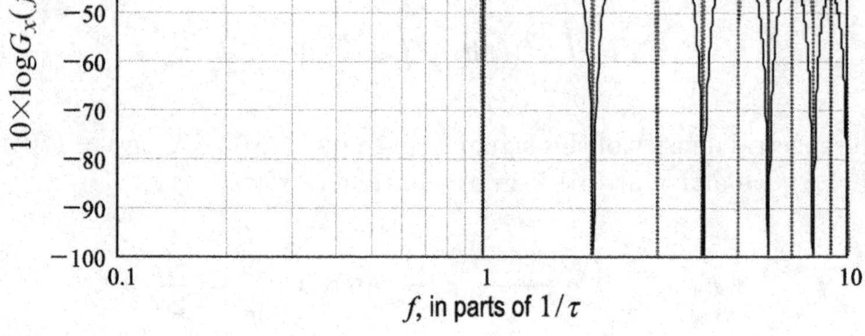

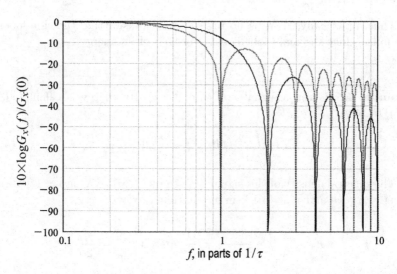

Fig. 4.5 Bode plots of ESD functions: rectangular pulse (doted) and symmetric triangular pulse (sold).

The width of the main lobe of the triangular pulse is twice larger than that of the rectangular pulse. The side lobes of the triangular pulse attenuate with −40 dB/decade and in the rectangular pulse with −20 dB/decade. Therefore, the triangular pulse has a wider BW at the level of −10 dB and narrower BW at −20 dB and below.

□

4.2.2 Power Signals

Signals of the other class of practical importance have infinite total energy. In Example 4.1, we examined one of such signals, the periodic harmonic wave.

To characterize the signals, which energy is not finite, the other measure is used, i.e., the *average power* over the period of repetition,

$$P_x = \left\langle |x(t)|^2 \right\rangle = \lim_{T\to\infty} \frac{1}{2T} \int_{-T}^{T} |x(t)|^2 \,\mathrm{d}t. \tag{4.21}$$

It follows from (4.21) that the average power is determined by the signal energy calculated in the infinite time interval T and then divided by T.

> ***Power signal***: A signal, which average power is finite is said to be the power signal or finite power signal.

□

A special subclass of power signals is represented by periodic signals, which typically have finite energy in one period $T = 1/F = 2\pi/\Omega$, where F is the fundamental frequency. Owing to this periodicity, the average energy remains the same if to average a signal over N periods T. Thus, instead of (4.21), for periodic signals we may write

$$P_x = \left\langle |x(t)|^2 \right\rangle = \frac{1}{T} \int_{-T/2}^{T/2} |x(t)|^2 \,\mathrm{d}t. \tag{4.22}$$

To evaluate P_x in the frequency domain, a periodic signal $x(t)$ may be represented by the Fourier series,

$$x(t) = \sum_{k=-\infty}^{\infty} C_k \mathrm{e}^{jk\Omega t}, \tag{4.23}$$

where the complex amplitude is defined, by (2.15), to be

$$C_k = \frac{1}{T} \int_{-T/2}^{T/2} x(t)\mathrm{e}^{-jk\Omega t}\mathrm{d}t. \tag{4.24}$$

By substituting (4.23) to (4.22) and making the transformations, we go to

$$\begin{aligned}
P_x &= \frac{1}{T}\int_{-T/2}^{T/2} |x(t)|^2\,\mathrm{d}t \\
&= \frac{1}{T}\int_{-T/2}^{T/2} \sum_{k=-\infty}^{\infty} C_k \mathrm{e}^{jk\Omega t} \sum_{l=-\infty}^{\infty} C_l^* \mathrm{e}^{-jl\Omega t}\mathrm{d}t \\
&= \frac{1}{T}\sum_{k=-\infty}^{\infty} C_k \sum_{l=-\infty}^{\infty} C_l^* \int_{-T/2}^{T/2} \mathrm{e}^{j(k-l)\Omega t}\mathrm{d}t \\
&= \frac{1}{T}\sum_{k=-\infty}^{\infty} C_k \sum_{l=-\infty}^{\infty} C_l^* \frac{\mathrm{e}^{j(k-l)\pi} - \mathrm{e}^{-j(k-l)\pi}}{j(k-l)\Omega} \\
&= \sum_{k=-\infty}^{\infty} C_k \sum_{l=-\infty}^{\infty} C_l^* \frac{\sin(k-l)\pi}{(k-l)\pi}.
\end{aligned}$$

Since $\sin(k-l)\pi = 0$ with any unequal k and l and the sinc function is unity by $k = l$, then the result transforms to the addition of the products of $C_k C_k^* = |C_k|^2$. We thus arrive at

$$P_x = \frac{1}{T}\int_{-T/2}^{T/2} |x(t)|^2\,\mathrm{d}t = \sum_{k=-\infty}^{\infty} |C_k|^2 \tag{4.25}$$

that is stated by the Parseval theorem (2.24) for the periodic (power) signals.

Example 4.7. Given a rectangular pulse-train with the amplitude A, pulse duration τ, and period T. Its average power is calculated, by (4.22) and $q = T/\tau$, to be

$$P_x = \frac{A^2}{T}\int_{-\tau/2}^{\tau/2} \mathrm{d}t = \frac{A^2}{q}. \tag{4.26}$$

The coefficients of the Fourier series are provided, by (2.122), to be

$$C_k = \frac{A}{q}\frac{\sin(k\pi/q)}{k\pi/q}$$

and the average power may thus alternatively be calculated by (4.25). Indeed, by an identity

$$\sum_{k=1}^{\infty} \frac{\sin^2(k\pi/q)}{k^2} = \frac{\pi^2(q-1)}{2q^2},$$

we arrive at (4.26),

$$P_x = \frac{A^2}{q^2}\sum_{k=-\infty}^{\infty}\frac{\sin^2(k\pi/q)}{(k\pi/q)^2} = \frac{A^2}{q^2} + \frac{2A^2}{\pi^2}\sum_{k=1}^{\infty}\frac{\sin^2(k\pi/q)}{k^2}$$

$$= \frac{A^2}{q^2} + \frac{A^2}{q^2}(q-1) = \frac{A^2}{q}. \tag{4.27}$$

This example demonstrates that the calculus of P_x may be elementary in the time domain (4.26) and take efforts in the frequency domain (4.27), or vice versa.

4.3 Signal Autocorrelation

In line with the spectral presentation of signal's energy and power, their time functions are also very often used in what is called the *correlation analysis.*

> ***Correlation***: Correlation is a relation between two or more signals such that changes in the value of one signal are accompanied by changes in the other.

The correlation analysis allows for evaluating the relationship between two or more (same or different) signals and hereby gives an idea about their energy and power avoiding decomposing the signals into the harmonic components. If correlation is evaluated between two shifted versions of the same signal, the relevant relation is called the *autocorrelation function.* If it is evaluated for different signals, the function is said to be the *cross-correlation function.*

It is of high importance that the autocorrelation function is calculated with the signal energy or PSD. If the energy signal is analyzed, the *energy autocorrelation function* is used coupled with the signal *ESD function* by the pair of the Fourier transform. If the power signal is studied, the *power autocorrelation function* is exploited. The Fourier transform couples this function with the signal PSD.

4.3.1 Monopulse Radar Operation

An example of the use of energy autocorrelation function is in the pulse radar operation. Figure 4.6 sketches a picture explaining the principle of operation. A short pulse $x(t)$ is transmitted to the target (car), and its reflected copy $x(t-\theta)$ is received being delayed on a propagation time θ that is proportional to the double distance between the antenna and the target. The time delay θ is then measured at the receiver.

To evaluate a distance with high accuracy, the structure of the receiver is designed as in Fig. 4.7. The processing system consists of N elements, each of which delays the transmitted interrogating signal $x(t)$ on one of the fixed

Fig. 4.6 Operation principle of pulse radars.

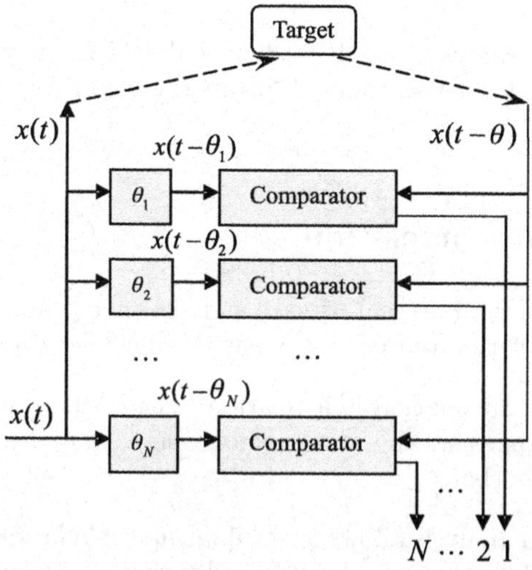

Fig. 4.7 Structure of the pulse radar processing system.

time intervals θ_1, θ_2, ..., θ_N. The delayed signals go to the first input of the relevant comparator. A received signal appears at the second input of each comparator. The comparator compares two inputs and its output appears only if the delayed signal is an unshifted copy of the received signal. Thus, ideally, only one signal may be fixed at the output of some nth channel. If this channel is identified, the time delay $\theta \cong \theta_n$ is measured and the distance to the target is evaluated.

Obviously the accuracy of such measurements is higher as more difference is fixed between the interrogating pulse and its delayed copy around $\theta = \theta_n$. The "goodness" of the used signal is therefore determined by its shape and structure.

How to compare two same shifted signals? Even intuitively, one may suggest calculating their joint energy. If it attains a maximum, the time shift is zero and thus the copy is equal to the original. The joint energy, in turn, is calculated by integrating the product of a multiplication of two signals. We thus need introducing a new important characteristic, the *energy autocorrelation function.*

4.3.2 Energy Autocorrelation Function

Consider the inner product (4.3) of two signals $x(t)$ and $y(t)$, each of which has finite energy. Let us assume that $y(t)$ is just a shifted version of $x(t)$ such that $y(t) = x(t - \theta) = x_\theta(t)$. By definition, the inner product of a signal $x(t)$ and its shifted copy $x_\theta(t)$ is said to be the *energy autocorrelation function.* This function represents the joint energy of the original signal and its shifted copy and is written as

$$\begin{aligned}\phi_x(\theta) = \langle x, x_\theta \rangle &= \lim_{T\to\infty} \int_{-T}^{T} x(t)x^*(t-\theta)\mathrm{d}t \\ &= \int_{-\infty}^{\infty} x(t)x^*(t-\theta)\mathrm{d}t. \qquad (4.28)\end{aligned}$$

We notice that the energy of $x(t)$ must be finite to avoid divergency of the integral.

An example of the integrand function $x(t)x(t-\theta)$ is given in Fig. 4.8. Here the energy autocorrelation function (4.28) is represented by the shaded area for a given θ. A simple analysis shows that if $x(t)$ is to be displaced relative to itself by an amount of θ, then the integral of the product will be the same whether the shift θ is positive or negative. So, the autocorrelation function is an even function.

Example 4.8. Given a rectangular pulse of duration τ and amplitude A. By (4.28), its energy autocorrelation function is calculated to be

$$\phi_x(\theta) = \begin{cases} A^2 \int\limits_0^{\tau-|\theta|} \mathrm{d}t = A^2\tau\left(1 - \frac{|\theta|}{\tau}\right), & |\theta| \leqslant \tau \\ 0, & |\theta| > \tau \end{cases}. \qquad (4.29)$$

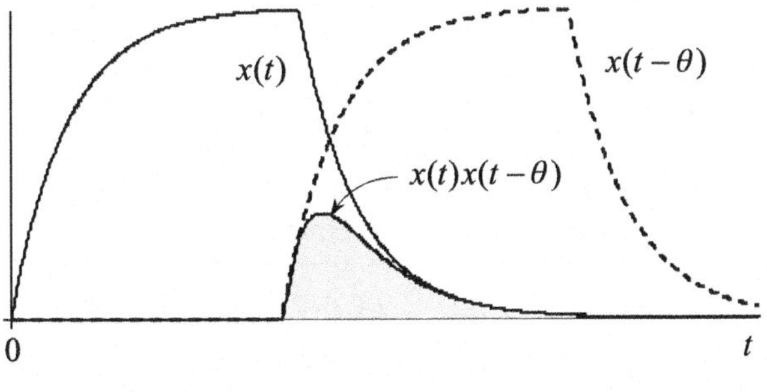

Fig. 4.8 A single value of the autocorrelation function represented by the shaded area.

It is seen that (4.29) is symmetric and triangular. The value $\theta = 0$ makes $\phi_x(\theta)$ to be identically the signal energy (4.14),

$$\phi_x(0) = E_x = A^2\tau. \tag{4.30}$$

A geometrical interpretation of (4.29) is given in Fig. 4.9 for three different values of θ. By the advance time shift, $-\theta < -\tau$, the joint energy of two pulses is zero and thus the correlation is zero as well. If $-\tau < -\theta_1 < 0$, then some joint energy occurs (b) and hence $\phi_x(-\theta_1) > 0$ (e). With zero time shift, $\theta > 0$, the joint energy attains a maximum (c) and the correlation function reaches the signal energy $\phi_x(0) = E_x$ (e). The time delay $0 < \theta_1$ (d) reduces the joint energy (e) so that, with $\tau < \theta$, both the joint energy and the autocorrelation function become identically zero.

It is seen that the autocorrelation function of a rectangular pulse is triangular and symmetric with a twice larger duration. □

Example 4.9. Given a rectangular RF pulse of duration τ, amplitude A_0, and carrier frequency $\omega_0/2\pi = f_0 \gg 1/\tau$. The pulse is described by

$$x(t) = \begin{cases} A_0 \cos \omega_0 t, & -\tau/2 \leqslant t \leqslant \tau/2 \\ 0, & \text{otherwise} \end{cases}. \tag{4.31}$$

The energy autocorrelation function (4.28) is calculated for this pulse in the range of $|\theta| \leqslant \tau$ by

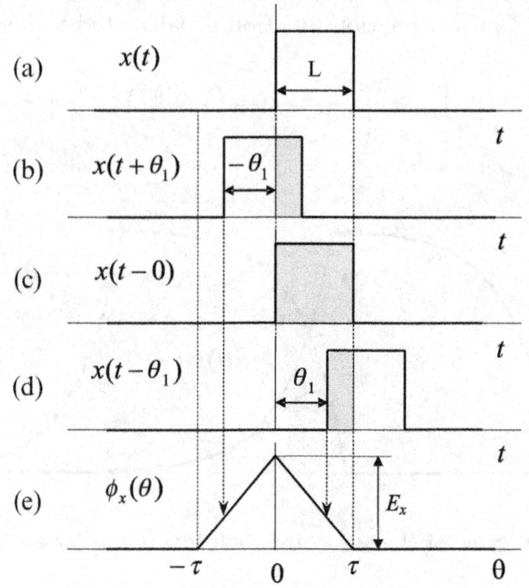

Fig. 4.9 Energy autocorrelation function of the rectangular pulse.

$$\phi_x(\theta) = A_0^2 \int_0^{\tau-|\theta|} \cos\omega_0 t \cos\omega_0(t-\theta)\mathrm{d}t$$
$$= \frac{A_0^2(\tau-|\theta|)}{2}\left[\cos\omega_0\theta + \frac{\sin\omega_0(\tau-|\theta|)}{\omega_0(\tau-|\theta|)}\cos\omega_0(\tau-|\theta|-\theta)\right]. \tag{4.32}$$

By setting $\theta = 0$, we arrive at the signal energy

$$E_x = \phi_x(0) = \frac{A_0^2\tau}{2}\left(1 + \frac{\sin 2\omega_0\tau}{2\omega_0\tau}\right). \tag{4.33}$$

Figure 4.10 illustrates (4.32). It follows that, unlike the rectangular video pulse case (Fig. 4.9), this function oscillates. Moreover, its energy (4.33) is lower than that (4.30) of the rectangular waveform. □

4.3.3 Properties of the Energy Autocorrelation Function

Even though we illustrated the energy autocorrelation function (4.28) with only two examples associated with the rectangular video and RF pulses, we already pointed out several of its important properties. These and other properties of importance are outlined as follows:

- The function is even, since a sign of the time shift does not change the joint energy. Thus,

$$\phi_x(\theta) = \phi_x(-\theta)$$
$$= \int_{-\infty}^{\infty} x(t)x^*(t-\theta)\mathrm{d}t = \int_{-\infty}^{\infty} x(t)x^*(t+\theta)\mathrm{d}t. \tag{4.34}$$

□

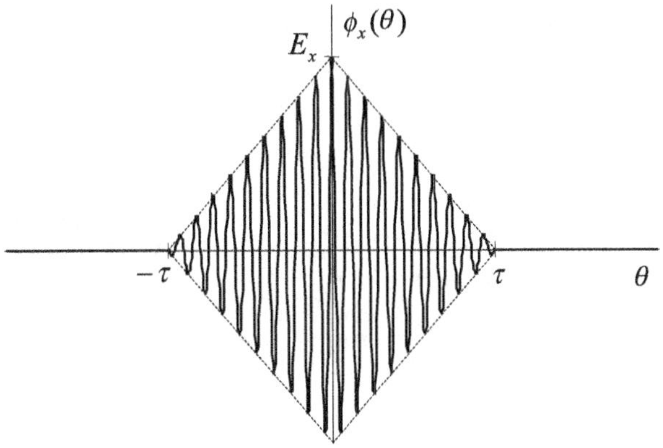

Fig. 4.10 Energy autocorrelation function of a rectangular RF pulse.

- Increasing θ results in diminishing the joint energy and correlation. When θ occurs to be more than the pulse duration, $\theta > \tau$, the function becomes identically zero.

$\square$

- With $\theta = 0$, the function reaches the signal energy,

$$\phi_x(0) = \int_{-\infty}^{\infty} |x(t)|^2 \, \mathrm{d}t = E_x. \tag{4.35}$$

$\square$

- By $|\theta| > 0$, the function value does not overcome the signal energy,

$$\phi_x(\theta) = \langle x, x_\theta \rangle \leqslant \phi_x(0) = E_x. \tag{4.36}$$

This fact follows straightforwardly from the Cauchy–Bunyakovskii inequality

$$|\langle x, x_\theta \rangle| \leqslant \|x\| \cdot \|x_\theta\| = E_x \,, \tag{4.37}$$

where $\|x\| \equiv \|x\|_2 = \sqrt{\int_{-\infty}^{\infty} x^2(t)\mathrm{d}t}$ is the 2-norm of $x(t)$ and $\|x_\theta\| \equiv \|x_\theta\|_2 = \sqrt{\int_{-\infty}^{\infty} x^2(t-\theta)\mathrm{d}t}$.

$\square$

- The function has the dimension

$$[\text{signal dimension}]^2 \times \text{s}.$$

$\square$

- The normalized measure of the function is dimensionless and called the *energy autocorrelation coefficient* $\gamma_x(\theta)$. This coefficient exists from -1 to $+1$ and is defined by

$$-1 \leqslant \gamma_x(\theta) = \frac{\phi_x(\theta)}{\phi_x(0)} = \frac{\phi_x(\theta)}{E_x} \leqslant 1. \tag{4.38}$$

$\square$

- The function duration is equal to the doubled duration of the pulse,

$$\tau_\phi = 2\tau.$$

$\square$

- *Narrowness.* The function width W_{xx} at the level of half of energy $E_x/2$ cannot exceed the pulse duration τ,

$$W_{xx} \leqslant \tau.$$

The maximum value $W_{xx} = \tau$ corresponds to the rectangular pulse (Figs. 4.9 and 4.10). Complex RF pulses can obtain $W_{xx} \ll \tau$.

$\square$

Narrowness, as the most appreciable applied property of $\phi_x(\theta)$, is exploited widely in signals detection and identification. This is because $\phi_x(\theta)$ reaches a maximum when the signals are unshifted and, for some complex signals, it may diminish toward zero rapidly even though having a negligible time shift θ.

At the early electronics age, the complex signals (LFM and PSK) had not been required as the systems (e.g., communications, navigation, radars) were relatively simple and the signals were simple as well. A situation was changed cardinally several decades ago, when use of complex signals had become motivated not only by technical facilities but also primarily by possibilities of optimum task solving. The role of correlation analysis had then been gained substantially.

4.3.4 Power Autocorrelation Function of a Signal

So far, we were concerned with autocorrelation of the signals, which energy is finite. A periodic signal $x(t) = x(t + kT)$ has infinite energy and, therefore, (4.28) cannot be applied. If such a signal has finite power, the *power autocorrelation function* is used as the average of the energy autocorrelation function over the infinite time bounds. This function is written as

$$R_x(\theta) = \lim_{T\to\infty} \frac{1}{2T} \int_{-T}^{T} x(t)x^*(t-\theta)\mathrm{d}t \tag{4.39}$$

and it reaches the signal average power with $\theta = 0$,

$$R_x(0) = \lim_{T\to\infty} \frac{1}{2T} \int_{-T}^{T} |x(t)|^2 \,\mathrm{d}t = P_x\,, \tag{4.40}$$

having a dimension of [(signal dimension)^2]. An example is a simplest harmonic signal.

Example 4.10. Given a real harmonic signal

$$x(t) = A\cos\omega_0 t\,, \tag{4.41}$$

with constant amplitude A_0 and natural frequency ω_0. The signal exists in infinite time bounds.

To determine the power autocorrelation function, the signal (4.41) may be assumed to be a rectangular RF pulse (Examples 4.2 and 4.3), which duration 2τ tends toward infinity. The function (4.39) is then calculated by

$$R_x(\theta) = \lim_{\tau\to\infty} \frac{1}{2\tau} \int_{-\tau}^{\tau} x(t)x^*(t-\theta)\mathrm{d}t = \lim_{\tau\to\infty} \frac{1}{2\tau}\phi_x(\theta)\,, \tag{4.42}$$

where $\phi_x(\theta)$ is determined by (4.32). Setting $\tau \to \infty$ to (4.32), we arrive at

$$R_x(\theta) = \frac{A^2}{2}\cos\omega_0\theta. \tag{4.43}$$

The average power of a signal is calculated by (4.43) at $\theta = 0$,

$$P_x = R_x(0) = \frac{A^2}{2}.$$

Thus, the power autocorrelation function of a periodic harmonic signal is also a harmonic function (4.43) having a peak value that is equal to the average power.

□

Alternatively, the power autocorrelation function of a periodic signal may be calculated by averaging over the period $T = 2\pi/\omega_0$. The modified relation (4.39) is

$$R_x(\theta) = \frac{1}{T}\int\limits_{-T/2}^{T/2} x(t)x^*(t-\theta)\mathrm{d}t. \tag{4.44}$$

Example 4.11. Given a signal (4.41). By (4.44), we get

$$\begin{aligned} R_x(\theta) &= \frac{A^2}{T}\int\limits_{-T/2}^{T/2} \cos\omega_0 t\cos\omega_0(t-\theta)\mathrm{d}t \\ &= \frac{A^2}{2T}\left[\int\limits_{-T/2}^{T/2}\cos\omega_0(2t-\theta)\,\mathrm{d}t + \int\limits_{-T/2}^{T/2}\cos\omega_0\theta\mathrm{d}t\right]. \end{aligned}$$

Here, the first integral produces zero and the second one leads to (4.43).

□

4.3.5 Properties of the Power Autocorrelation Function

The following important properties of the power autocorrelation function (4.39) may now be outlined:

- The function is even,

$$R_x(\theta) = R_x(-\theta). \tag{4.45}$$

- By $\theta = 0$, it is the average power of a signal,

$$R_x(0) = \lim_{T\to\infty}\frac{1}{2T}\int\limits_{-T}^{T}|x(t)|^2\,\mathrm{d}t = P_x. \tag{4.46}$$

- The function is periodic and it does not exceed the signal average power,

$$R_x(\theta) \leqslant P_x(0).$$

- The Cauchy–Schwarz inequality claims that

$$R_x(\theta) = \langle x, x_\theta \rangle \leqslant \sqrt{\langle x, x \rangle}\sqrt{\langle x_\theta, x_\theta \rangle} = P_x. \tag{4.47}$$

- The function has a dimension

$$[\text{signal dimension}]^2 .$$

- The normalized function is called the *power autocorrelation coefficient*

$$r_x(\theta) = \frac{R_x(\theta)}{R_x(0)} = \frac{R_x(\theta)}{P_x}, \tag{4.48}$$

which absolute value does not exceed unity, $-1 \leqslant r_x \leqslant 1$.

4.4 Energy and Power Spectral Densities

Usually, neither the autocorrelation function nor its spectral analog can solely help solving system's problems satisfactorily. Frequently, both functions are required. In fact, an optimum signal waveform and structure in radars is derived through the correlation function. On the contrary, an optimum system structure for the desired signal is easily obtained in the frequency domain by the relevant spectral performance. Thus, the translation rule between the signal autocorrelation function and its spectral performance is important not only from the standpoint of the signals theory but also from the motivation by practical needs.

4.4.1 Energy Spectral Density

To determine a correspondence between the signal energy autocorrelation function $\phi_x(\theta)$ and ESD function $G_x(\omega)$, we apply the direct Fourier transform to $\phi_x(\theta)$, employ (4.28), and go to

$$\begin{aligned}\mathcal{F}\{\phi_x(\theta)\} &= \int_{-\infty}^{\infty} \phi_x(\theta)\mathrm{e}^{-j\omega\theta}\mathrm{d}\theta \\ &= \int_{-\infty}^{\infty}\int_{-\infty}^{\infty} x(t)x^*(t-\theta)\mathrm{e}^{-j\omega\theta}\mathrm{d}t\mathrm{d}\theta \\ &= \int_{-\infty}^{\infty}\int_{-\infty}^{\infty} x(t)\mathrm{e}^{-j\omega t}x^*(t-\theta)\mathrm{e}^{j\omega(t-\theta)}\mathrm{d}t\mathrm{d}\theta\end{aligned}$$

$$= \left[\int_{-\infty}^{\infty} x(t) e^{-j\omega t} dt \right] \left[\int_{-\infty}^{\infty} x(\lambda) e^{-j\omega\lambda} d\lambda \right]^*$$

$$= X(j\omega)X^*(j\omega) = |X(j\omega)|^2 = G_x(\omega). \tag{4.49}$$

The inverse relation, of course, also holds true,

$$\mathcal{F}^{-1}\{G_x(\omega)\} = \frac{1}{2\pi} \int_{-\infty}^{\infty} G_x(\omega) e^{j\omega\theta} d\omega = \phi_x(\theta), \tag{4.50}$$

and we conclude that the energy autocorrelation function $\phi_x(\theta)$ and the ESD function $G_x(\omega)$ are coupled by the Fourier transforms,

$$\phi_x(\theta) \overset{\mathcal{F}}{\Leftrightarrow} G_x(\omega). \tag{4.51}$$

We may also denote $X_\theta(j\omega) = X(j\omega)e^{-j\omega\theta}$ and $X_\theta^*(j\omega) = X^*(j\omega)e^{j\omega\theta}$, and rewrite (4.50) as the inner product

$$\phi_x(\theta) = \frac{1}{2\pi} \int_{-\infty}^{\infty} G_x(\omega) e^{j\omega\theta} d\omega = \frac{1}{2\pi} \int_{-\infty}^{\infty} X(j\omega) X^*(j\omega) e^{j\omega\theta} d\omega$$

$$= \frac{1}{2\pi} \int_{-\infty}^{\infty} X(j\omega) X_\theta^*(j\omega) d\omega = \frac{1}{2\pi} \langle X, X_\theta \rangle . \tag{4.52}$$

Three important applied meaning of the transformation (4.51) may now be mentioned:

- It is a tool to define the ESD function through the energy autocorrelation function, and vice versa.
- It gives a choice in measuring either the ESD or the energy autocorrelation function. After one of these functions is measured, the other one may be recovered without measurements.
- It gives two options in determining both the ESD function and the energy autocorrelation function. Indeed, $\phi_x(\theta)$ may be defined either by the signal time function (4.28) or by the ESD function, by (4.50). In turn, $G_x(\omega)$ may be defined either by the signal spectral density (4.6) or by the autocorrelation function, by (4.49).

So, there are two options in defining the ESD and energy autocorrelation functions. Which way is simpler and which derivation routine is shorter? The answer depends on the signal waveform that we demonstrate in the following examples.

Example 4.12. Suppose a rectangular video pulse with amplitude A and duration τ. By (4.49) and (2.80), we go to the ESD function

$$G_x(j\omega) = |X(j\omega)|^2 = A^2\tau^2 \frac{\sin^2(\omega\tau/2)}{(\omega\tau/2)^2}. \tag{4.53}$$

By a larger burden, the same result (4.53) appears if to apply the direct Fourier transform (4.49) to the autocorrelation function of this pulse (4.29).

Relation (4.29) shows that the energy autocorrelation function of the pulse is defined in an elementary way. Alternatively, one may apply the inverse Fourier transform to (4.53) and go to (4.29), however, with lower burden. Thus, unlike the direct calculus, the Fourier transform is less efficient to produce both $G_x(\omega)$ and $\phi_x(\theta)$ of the rectangular pulse. Figure 4.11 sketches both functions.

□

Example 4.13. Suppose a sinc-shaped video pulse (2.91). By (2.93), its ESD function is basically defined to be

$$G_x(\omega) = |X(j\omega)|^2 = \begin{cases} A^2\pi^2/\alpha^2, & |\omega| \leqslant \alpha \\ 0, & |\omega| > \alpha \end{cases} . \tag{4.54}$$

Applying (4.50) to (4.54), we go to the energy autocorrelation function in an elementary way as well,

$$\phi_x(\theta) = \frac{A^2\pi}{2\alpha^2} \int_{-\alpha}^{\alpha} e^{j\omega\theta} d\omega = \frac{A^2\pi\omega}{\alpha^2} \frac{\sin \omega\theta}{\omega\theta} , \tag{4.55}$$

and conclude that shapes of the sinc pulse and its autocorrelation function coincide.

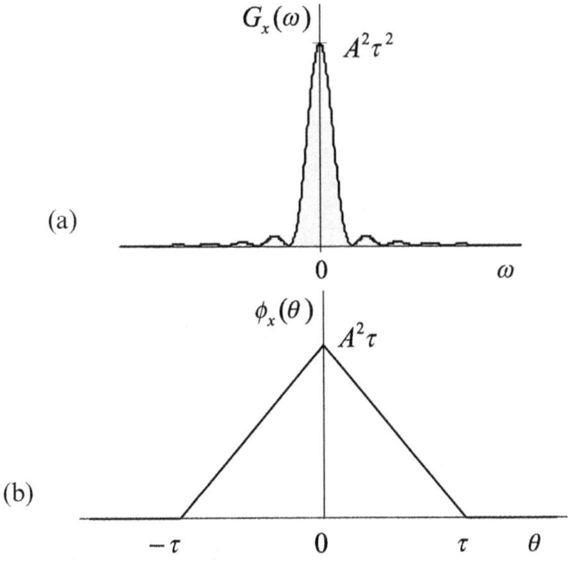

Fig. 4.11 Energy functions of a rectangular pulse: (**a**) ESD and (**b**) autocorrelation.

One can also derive $\phi_x(\theta)$ by integrating the shifted versions of a sinc-function (2.91). In turn, $G_x(\omega)$ may be obtained from $\phi_x(\theta)$ employing the direct Fourier transform. In doing so, one realizes that the derivation burden rises substantially. Thus, the Fourier transform is efficient in (4.55) and not efficient in deriving $G_x(\omega)$ by $\phi_x(\theta)$. Figure 4.12 illustrates both functions.

Example 4.14. Suppose a Gaussian video pulse (2.95), whose spectral density is specified by (2.96). By (4.49), its ESD function is written as

$$G_x(\omega) = |X(j\omega)|^2 = \frac{A^2\pi}{\alpha^2}\mathrm{e}^{-\frac{\omega^2}{2\alpha^2}}. \tag{4.56}$$

The autocorrelation function may be derived using either (4.28) or (4.50). In the former case, we employ an identity $\int\limits_{-\infty}^{\infty} \mathrm{e}^{-pz^2-qz}\mathrm{d}z = \sqrt{\frac{\pi}{p}}\mathrm{e}^{q^2/4p}$ and obtain

$$\begin{aligned}
\phi_x(\theta) &= A^2 \int\limits_{-\infty}^{\infty} \mathrm{e}^{-\alpha^2 t^2}\mathrm{e}^{-\alpha^2(t+\theta)^2}\mathrm{d}t \\
&= A^2\mathrm{e}^{-\alpha^2\theta^2}\int\limits_{-\infty}^{\infty} \mathrm{e}^{-2\alpha^2(t^2+t\theta)}\mathrm{d}t \\
&= \frac{A^2}{\alpha}\sqrt{\frac{\pi}{2}}\mathrm{e}^{-\frac{\alpha^2\theta^2}{2}}.
\end{aligned} \tag{4.57}$$

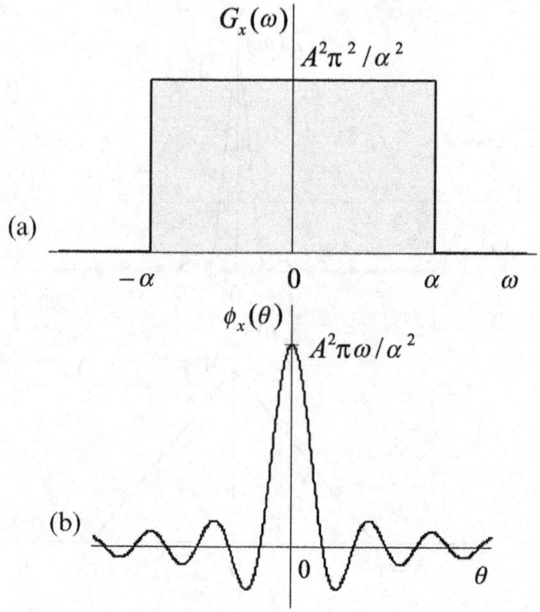

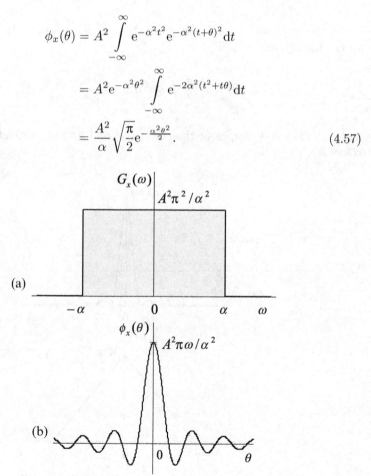

Fig. 4.12 Energy functions of a sinc-shaped pulse: (**a**) ESD and (**b**) autocorrelation.

In the latter case, we first write

$$\phi_x(\theta) = \frac{1}{2\pi}\frac{A^2\pi}{\alpha^2}\int\limits_{-\infty}^{\infty} e^{-\frac{\omega^2}{2\alpha^2}+j\theta\omega}d\omega.$$

Then recall that $\phi_x(\theta) = \phi_x(-\theta)$, use the above-mentioned integral identity, and go to (4.57).

If to derive the ESD function by (4.49) using (4.57), then the derivation routine will have almost the same burden. Figure 4.13 illustrates both functions.

Resuming, we first recall the splendid property of the Gaussian pulse that is its ESD and autocorrelation functions have the same shape. We then notice that its ESD (4.56) and correlation function (4.57) are Gauss-shaped as well. Of importance also is that different ways in deriving the energy functions of this pulse offer almost the same burden.

□

It may be deduced from the above-given examples that the energy autocorrelation functions of simple single waveforms are not sufficiently narrow. The question then is if such a property is featured to the pulse-bursts. To answer, we first consider an example of the rectangular one.

Example 4.15. Suppose a rectangular video pulse-burst with $N = 4$ and $q = T/\tau = 2$. By (2.145) and (4.49), its ESD function becomes

$$G_x(\omega) = |X(j\omega)|^2 = A^2\tau^2\frac{\sin^2(\omega\tau/2)}{(\omega\tau/2)^2}\frac{\sin^2(\omega NT/2)}{\sin^2(\omega T/2)}. \quad (4.58)$$

The energy autocorrelation function may be derived by (4.28) in analogy to the single pulse (4.29). Figure 4.14 illustrates both $G_x(\omega)$ and $\phi_x(\theta)$.

□

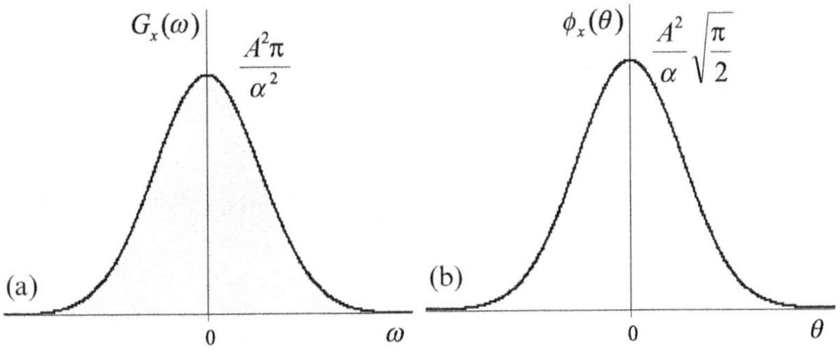

Fig. 4.13 Energy functions of the Gaussian pulse: (**a**) ESD and (**b**) autocorrelation.

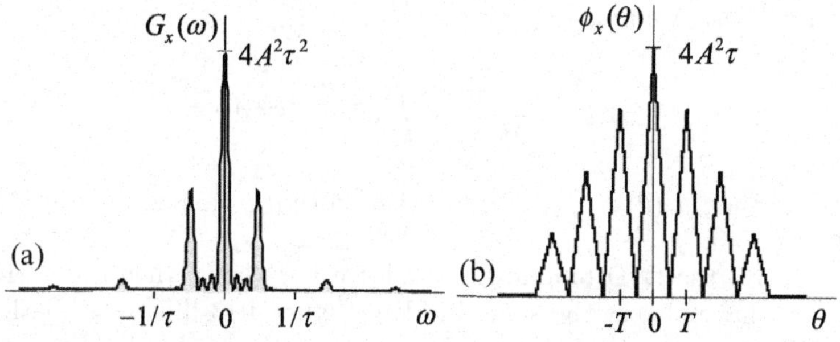

Fig. 4.14 Energy functions of the rectangular pulse-burst: (**a**) ESD and (**b**) autocorrelation.

Figure 4.14b demonstrates that the peak value of the main lobe of the autocorrelation function $\phi_x(\theta)$ of the rectangular pulse is increased and its width reduced by the number N of pulses in the burst. This is certainly an advantage of the latter. However, the main lobe is accompanied with the multiple side lobes. This nuisance effect is of course a disadvantage. To overcome this, the pulse-burst needs to be modified somehow that we consider in further.

4.4.2 Power Spectral Density

In line with the energy autocorrelation function and ESD, the power autocorrelation function and the associated PSD function have also gained wide currency in applications. To derive the PSD function, we apply the direct Fourier transform to $R_x(\theta)$, use (4.39), and go to

$$\mathcal{F}\{R_x(\theta)\} = \int_{-\infty}^{\infty} R_x(\theta)\mathrm{e}^{-j\omega\theta}\mathrm{d}\theta$$

$$= \int_{-\infty}^{\infty} \lim_{T\to\infty}\frac{1}{2T}\int_{-T}^{T} x(t)x^*(t-\theta)\mathrm{e}^{-j\omega\theta}\mathrm{d}t\,\mathrm{d}\theta$$

$$= \lim_{T\to\infty}\frac{1}{2T}\int_{-T}^{T} x(t)\mathrm{e}^{-j\omega t}\int_{-T}^{T} x^*(t-\theta)\mathrm{e}^{j\omega(t-\theta)}\mathrm{d}\theta\,\mathrm{d}t$$

$$= \lim_{T\to\infty}\frac{1}{2T}\int_{-T}^{T} x(t)\mathrm{e}^{-j\omega t}\mathrm{d}t\int_{-T}^{T} x^*(\lambda)\mathrm{e}^{j\omega\lambda}\mathrm{d}\lambda$$

$$= \lim_{T\to\infty} \frac{1}{2T} \left[\int_{-T}^{T} x(t)\mathrm{e}^{-j\omega t}\mathrm{d}t \right] \left[\int_{-T}^{T} x(\lambda)\mathrm{e}^{-j\omega\lambda}\mathrm{d}\lambda \right]^*$$

$$= \lim_{T\to\infty} \frac{1}{2T} |X(j\omega)|^2 = \lim_{T\to\infty} \frac{1}{2T} G_x(\omega). \tag{4.59}$$

The spectral function

$$S_x(\omega) = \lim_{T\to\infty} \frac{1}{2T} G_x(\omega) = \int_{-\infty}^{\infty} R_x(\theta)\mathrm{e}^{-j\omega\theta}\mathrm{d}\theta \tag{4.60}$$

has a meaning of the PSD of a signal $x(t)$. The inverse transform thus produces

$$R_x(\theta) = \frac{1}{2\pi} \int_{-\infty}^{\infty} S_x(\omega)\mathrm{e}^{j\omega\theta}\mathrm{d}\theta \tag{4.61}$$

to mean that, in line with $\phi_x(\theta)$ and $G_x(\omega)$, the power autocorrelation function $R_x(\theta)$ and PSD function $S_x(\omega)$ are also subjected to the pair of the Fourier transforms,

$$S_x(\omega) \overset{\mathcal{F}}{\Leftrightarrow} R_x(\theta). \tag{4.62}$$

Example 4.16. Suppose a real periodic harmonic signal $x(t) = A\cos\omega_0 t$, whose power autocorrelation function is defined by (4.43). The PSD function is obtained, by (4.60), to be

$$S_x(\omega) = \frac{A^2}{2} \int_{-\infty}^{\infty} \cos\omega_0\theta\mathrm{e}^{-j\omega\theta}\mathrm{d}\theta = \frac{A^2}{2} \int_{-\infty}^{\infty} \frac{\mathrm{e}^{j\omega_0\theta} + \mathrm{e}^{-j\omega_0\theta}}{2} \mathrm{e}^{-j\omega\theta}\mathrm{d}\theta$$

$$= \frac{A^2}{4} \int_{-\infty}^{\infty} \mathrm{e}^{-j(\omega-\omega_0)\theta}\mathrm{d}\theta + \frac{A^2}{4} \int_{-\infty}^{\infty} \mathrm{e}^{-j(\omega+\omega_0)\theta}\mathrm{d}\theta$$

$$= \frac{A^2}{4} \left[\delta(f - f_0) + \delta(f + f_0)\right]. \tag{4.63}$$

Figure 4.15 shows the PSD function of this signal.

□

For periodic signals, the PSD function is calculated by the coefficients of the Fourier series (2.14). To go to the relevant formula, we apply the Fourier

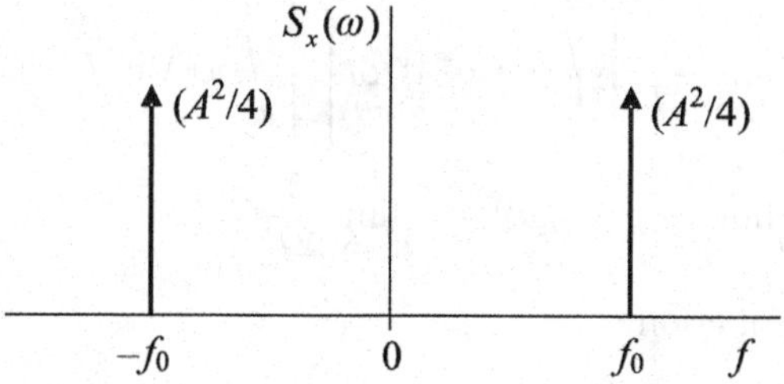

Fig. 4.15 PSD function of a harmonic signal $x(t) = A\cos 2\pi f_0 t$.

transform to (4.44), use (2.14), and arrive at

$$\begin{aligned}
\mathcal{F}\{R_x(\theta)\} &= \int_{-\infty}^{\infty} R_x(\theta)\mathrm{e}^{-j\omega\theta}\mathrm{d}\theta \\
&= \int_{-\infty}^{\infty} \frac{1}{T}\int_{-T/2}^{T/2} x(t)x^*(t-\theta)\mathrm{e}^{-j\omega\theta}\mathrm{d}t\mathrm{d}\theta \\
&= \int_{-\infty}^{\infty} \frac{1}{T}\int_{-T/2}^{T/2} \sum_{k=-\infty}^{\infty} C_k \mathrm{e}^{jk\Omega t} \sum_{l=-\infty}^{\infty} C_k^* \mathrm{e}^{-jl\Omega(t-\theta)}\mathrm{e}^{-j\omega\theta}\mathrm{d}t\mathrm{d}\theta \\
&= \sum_{k=-\infty}^{\infty}\sum_{l=-\infty}^{\infty} C_k C_l^* \frac{1}{T}\int_{-T/2}^{T/2} \mathrm{e}^{j(k-l)\Omega t}\mathrm{d}t \int_{-\infty}^{\infty} \mathrm{e}^{-j(\omega-l\Omega)\theta}\mathrm{d}\theta \\
&= 2\pi \sum_{k=-\infty}^{\infty}\sum_{l=-\infty}^{\infty} C_k C_l^* \frac{\sin(k-l)\pi}{(k-l)\pi}\delta(\omega - l\Omega).
\end{aligned}$$

As may be seen, the transform is not equal to zero only if $k = l$. We then get

$$S_x(\omega) = \mathcal{F}\{R_x(\theta)\} = 2\pi \sum_{k=-\infty}^{\infty} |C_k|^2 \delta(\omega - k\Omega) \tag{4.64}$$

that is the other option to calculate the signal PSD function.

Example 4.17. Suppose a signal $x(t) = A\cos\omega_0 t$ with period $T = 2\pi/\omega_0$. The coefficients of the Fourier series are calculated by (2.15) to be

$$\begin{aligned}
C_k &= \frac{1}{T}\int_{-T/2}^{T/2} x(t)\mathrm{e}^{-jk\Omega t}\mathrm{d}t \\
&= \frac{A}{T}\int_{-T/2}^{T/2} \cos\omega_0 t\,\mathrm{e}^{-jk\Omega t}\mathrm{d}t = \frac{A}{T}\int_{-T/2}^{T/2} \frac{\mathrm{e}^{j\omega_0 t} + \mathrm{e}^{-j\omega_0 t}}{2}\mathrm{e}^{-jk\Omega t}\mathrm{d}t
\end{aligned}$$

$$= \frac{A}{2T}\int\limits_{-T/2}^{T/2} e^{j(\omega_0 - k\Omega)t}dt + \frac{A}{2T}\int\limits_{-T/2}^{T/2} e^{-j(\omega_0 + k\Omega)t}dt$$

$$= \frac{A}{2}\frac{\sin \pi(f_0 - kF)T}{\pi(f_0 - kF)T} + \frac{A}{2}\frac{\sin \pi(f_0 + kF)T}{\pi(f_0 + kF)T}.$$

The PSD function is calculated, by (4.64), to be

$$S_x(\omega) = 2\pi \sum_{k=-\infty}^{\infty} |C_k|^2 \delta(\omega - k\Omega)$$

$$= \frac{\pi A^2}{2}\sum_{k=-\infty}^{\infty} \frac{\sin^2(\omega_0 - k\Omega)T/2}{[(\omega_0 - k\Omega)T/2]^2}\delta(\omega - k\Omega)$$

$$+ \frac{\pi A^2}{2}\sum_{k=-\infty}^{\infty} \frac{\sin^2(\omega_0 + k\Omega)T/2}{[(\omega_0 + k\Omega)T/2]^2}\delta(\omega - k\Omega)$$

$$+ \frac{2\pi A^2}{2}\sum_{k=-\infty}^{\infty} \frac{\sin(\omega_0 - k\Omega)T/2}{(\omega_0 - k\Omega)T/2}\frac{\sin(\omega_0 + k\Omega)T/2}{(\omega_0 + k\Omega)T/2}\delta(\omega - k\Omega).$$

In this series, the first and second sums are not zero only if $k\Omega = \omega_0$ and $k\Omega = -\omega_0$, respectively. Yet, the last sum is zero. By such a simplification, we arrive at the same relation (4.63),

$$S_x(\omega) = \frac{\pi A^2}{2}\left[\delta(\omega - \omega_0) + \delta(\omega + \omega_0)\right]$$
$$= \frac{A^2}{4}\left[\delta(f - f_0) + \delta(f + f_0)\right].$$

□

Observing Examples 4.16 and 4.17, one may conclude that the transform (4.60) allows for a lesser routine if a periodic signal is harmonic. Fortunately, this conclusion holds true for many other periodic signals.

4.4.3 A Comparison of Energy and Power Signals

We have now enough to compare the properties of the energy and power signals. First, we notice that a signal cannot be both energy and power signals. If we suppose that $P_x > 0$ leads to $E_x = \infty$ (one needs to multiply the finite power with the infinite time duration), the signal is only a power signal. In the other case, if we assume that $E_x < \infty$, then $P_x = 0$ (one needs to divide the finite energy with the infinite time duration), the signal is an energy signal.

This substantial difference between two types of signals requires different approaches to calculate their performance in the time and frequency domains. Table 4.1 generalizes the relationships for the energy and power signals and Table 4.2 supplements it for the power periodic signals with period T. On the basis of the above-given analysis of these tables, one may arrive at several important conclusions:

- The wider the ESD function, the narrower the energy correlation function, and vice versa.
- A signal waveform is critical to derive the ESD and energy correlation functions with minimum burden. Here, one has two options: a direct derivation or the Fourier transform. The only signal, the Gaussian waveform, allows for the same routine in deriving either $G_x(\omega)$ or $\phi_x(\theta)$.
- The functions $G_x(\omega)$ and $\phi_x(\theta)$ are coupled by the Fourier transform. Therefore, frequently, one of these functions is calculated through the other

Table 4.1 Relationships for energy and power signals

Energy signal	Power signal
$E_x = \lim_{T\to\infty} \int_{-T}^{T} x(t)x^*(t)\mathrm{d}t$	$P_x = \lim_{T\to\infty} \frac{1}{2T} \int_{-T}^{T} x(t)x^*(t)\mathrm{d}t$
$= \frac{1}{2\pi} \int_{-\infty}^{\infty} X(j\omega)X^*(j\omega)\mathrm{d}\omega$	$= \lim_{T\to\infty} \frac{E_x}{2T}$
$= \frac{1}{2\pi} \int_{-\infty}^{\infty} G_x(\omega)\mathrm{d}\omega$	$= \frac{1}{2\pi} \int_{-\infty}^{\infty} S_x(\omega)\mathrm{d}\omega$
$\phi_x(\theta) = \lim_{T\to\infty} \int_{-T}^{T} x(t)x^*(t-\theta)\mathrm{d}t$	$R_x(\theta) = \lim_{T\to\infty} \frac{1}{2T} \int_{-T}^{T} x(t)x^*(t-\theta)\mathrm{d}t$
$= \frac{1}{2\pi} \int_{-\infty}^{\infty} G_x(\omega)\mathrm{e}^{j\omega\theta}\mathrm{d}\omega$	$= \lim_{T\to\infty} \frac{\phi_x(\theta)}{2T}$
	$= \frac{1}{2\pi} \int_{-\infty}^{\infty} S_x(\omega)\mathrm{e}^{j\omega\theta}\mathrm{d}\omega$
$G_x(\omega) = \int_{-\infty}^{\infty} \phi_x(\theta)\mathrm{e}^{-j\omega\theta}\mathrm{d}\theta$	$S_x(\omega) = \int_{-\infty}^{\infty} R_x(\theta)\mathrm{e}^{-j\omega\theta}\mathrm{d}\theta$
	$= \lim_{T\to\infty} \frac{1}{2T} G_x(\omega)$

Table 4.2 Additional relationships for power periodic signals

Power periodic signal	
$P_x = \frac{1}{T} \int_{-T/2}^{T/2} \lvert x(t)\rvert^2 \,\mathrm{d}t$	$R_x(\theta) = \frac{1}{T} \int_{-T/2}^{T/2} x(t)x^*(t-\theta)\mathrm{d}t$
$= \sum_{k=-\infty}^{\infty} \lvert C_k\rvert^2$	$S_x(\omega) = 2\pi \sum_{k=-\infty}^{\infty} \lvert C_k\rvert^2 \,\delta(\omega - k\Omega)$

one. It is important to remember that $\phi_x(\theta)$, by definition, has a peak value at $\theta = 0$, meaning that its shape cannot be supposed to be rectangular, for example. Indeed, a supposition

$$\phi_x(\theta) = \begin{cases} B, & |\theta| \leqslant \theta_\mathrm{c} \\ 0, & \text{otherwise} \end{cases}$$

leads to a wrong result, and we have

$$G_x(j\omega) = B \int_{-\theta_\mathrm{c}}^{\theta_\mathrm{c}} \mathrm{e}^{-j\omega\theta} \mathrm{d}\theta = 2B\theta_\mathrm{c} \frac{\sin \omega\theta_\mathrm{c}}{\omega\theta_\mathrm{c}}$$

and thus the derived ESD function is real but oscillates about zero. However, it cannot be negative-valued, by definition, whereas the obtained result can.

4.5 Single Pulse with LFM

We have shown in Section 4.6 that the LFM rectangular pulse has almost a rectangular spectral density about the carrier frequency. Therefore, its ESD function is almost rectangular as well. Such a peculiarity of the LFM pulse makes it to be easily detected and the measurements associated with such signals may be provided with high accuracy. Below, we learn the autocorrelation function of a single rectangular LFM pulse and study its properties in detail.

4.5.1 ESD Function of a Rectangular LFM Pulse

The ESD function of a rectangular LFM pulse $x(t)$ is readily provided by squaring the relevant amplitude spectral density (3.103). This function is

$$\begin{aligned} G_x(\omega) &= |X(j\omega)|^2 \\ &= \frac{A_0^2\pi}{4\alpha}\left\{[C(\nu_1) + C(\nu_2)]^2 + [S(\nu_1) + S(\nu_2)]^2\right\}, \end{aligned} \tag{4.65}$$

where the Fresnel functions, $C(\nu_1)$, $C(\nu_2)$, $S(\nu_1)$, and $S(\nu_2)$, were introduced in Section 3.9.2 and the auxiliary coefficients are specified by

$$\nu_1 = \frac{\alpha\tau + 2(\omega - \omega_0)}{2\sqrt{\alpha\pi}} \quad \text{and} \quad \nu_2 = \frac{\alpha\tau - 2(\omega - \omega_0)}{2\sqrt{\alpha\pi}}.$$

Figure 4.16a sketches (4.65) for a variety of pulse-compression ratios $\mathrm{PCR} = \Delta f\tau = \alpha\tau^2/2\pi$, in which the frequency deviation is calculated by $\Delta\omega = 2\pi\Delta f = \alpha\tau = 2\pi/\tau\mathrm{PCR}$. It follows that, in line with the spectral density of this pulse, the ESD function occupies a wider frequency range if

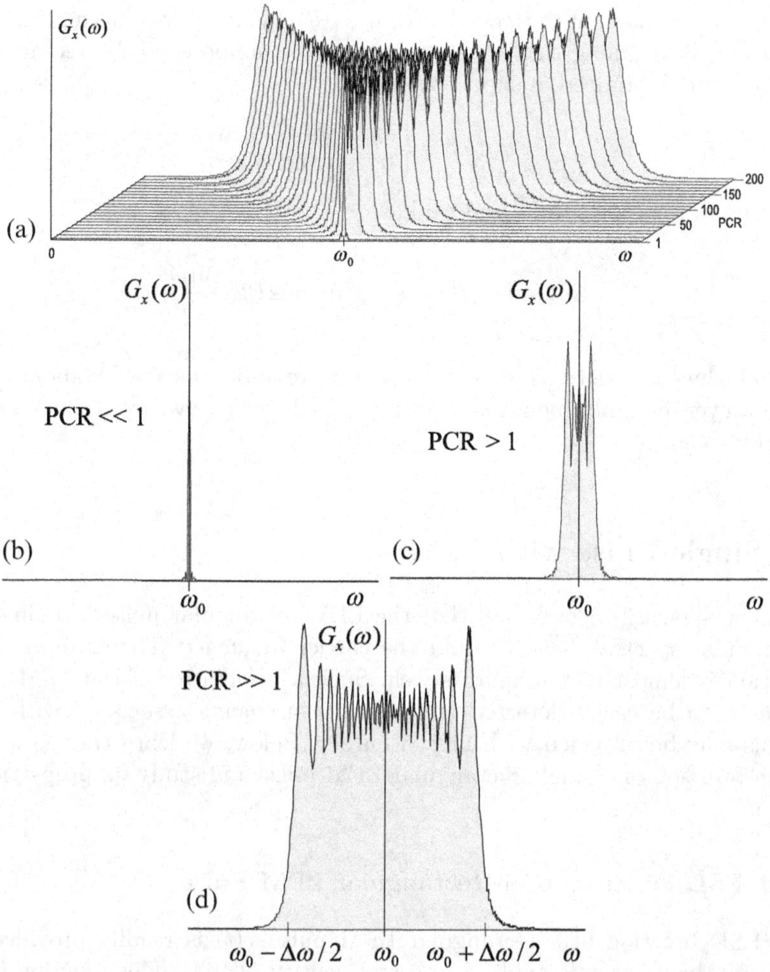

Fig. 4.16 ESD function of a rectangular LFM pulse: (**a**) $1 \leqslant \text{PCR} \leqslant 200$; (**b**) PCR $\ll 1$; (**c**) PCR > 1; and (**d**) PCR $\gg 1$.

PCR increases. With very low values of PCR $\ll 1$, the ESD function behaves closer to that corresponding to the rectangular RF pulse (Fig. 4.16b). With PCR > 1, the function occupies a wider frequency range and tends to be rectangular (Fig. 4.16c). With PCR $\gg 1$, it is almost rectangular if to neglect oscillations within a bandwidth (Fig. 4.16d).

Example 4.18. Given a rectangular LFM pulse with the amplitude $A_0 = 15\,\text{V}$, carrier frequency $f_0 = 9 \times 10^9$ Hz, pulse duration $\tau = 2 \times 10^{-6}$ s, and frequency deviation $\Delta f = 1 \times 10^8$ Hz. The PCR of this pulse calculates $\text{PCR} = \Delta f \tau = 200$. The LFM rate is $\alpha = 2\pi\Delta f/\tau = 2\pi \times 10^8/(2 \times 10^{-6}) = \pi{\times}10^{14}/\text{s}^4$. The average value of the ESD function around the carrier is defined to be $G_x(0) = A_0^2\pi/2\alpha = 225\pi/(2\pi \times 10^{14}) = 1.125 \times 10^{-12}\,\text{V}^2\ \text{s}^4$. Having a large PCR $= 200$, the signal spectrum is concentrated within a frequency range from $f_0 - \Delta f/2 = 8.95 \times 10^9$ Hz to $f_0 + \Delta f/2 = 9.05 \times 10^9$ Hz.

□

4.5.2 Autocorrelation Function of a Rectangular LFM Pulse

Basically, the autocorrelation function of a rectangular LFM pulse may be derived using the inverse Fourier transform applied to (4.65). The way is not short and we prefer starting with the pulse time function and then exploiting (4.28).

A single rectangular RF pulse with LFM is described by

$$x(t) = \begin{cases} A_0 \cos\left(\omega_0 t + \frac{\alpha t^2}{2}\right), & -\frac{\tau}{2} \leqslant t \leqslant \frac{\tau}{2} \\ 0, & \text{otherwise} \end{cases}. \tag{4.66}$$

Taking into account the symmetry property of the autocorrelation function, $\phi_x(\theta) = \phi_x(-\theta)$, we may apply (4.28) to (4.66) and specify $\phi_x(\theta)$ in the positive time range by

$$\phi_x(\theta) = A_0^2 \int\limits_{-\tau/2}^{(\tau/2)-\theta} \cos\left(\omega_0 t + \frac{\alpha t^2}{2}\right) \cos\left[\omega_0(t+\theta) + \frac{\alpha(t+\theta)^2}{2}\right] \mathrm{d}t. \tag{4.67}$$

By the Euler formula, we go to

$$\phi_x(\theta) = \frac{A_0^2}{4} \int\limits_{-\tau/2}^{(\tau/2)-\theta} \left[\mathrm{e}^{j\left(\omega_0 t + \frac{\alpha t^2}{2}\right)} + \mathrm{e}^{-j\left(\omega_0 t + \frac{\alpha t^2}{2}\right)}\right]$$
$$\times \left\{\mathrm{e}^{j\left[\omega_0(t+\theta) + \frac{\alpha(t+\theta)^2}{2}\right]} + \mathrm{e}^{-j\left[\omega_0(t+\theta) + \frac{\alpha(t+\theta)^2}{2}\right]}\right\} \mathrm{d}t,$$

provide the manipulations, and obtain

$$\phi_x(\theta) = \frac{A_0^2}{4} \mathrm{e}^{j\left(\omega_0\theta + \frac{\alpha\theta^2}{2}\right)} \int\limits_{-\tau/2}^{(\tau/2)-\theta} \mathrm{e}^{j\left[(2\omega_0+\alpha\theta)t + \alpha t^2\right]} \mathrm{d}t$$
$$+ \frac{A_0^2}{4} \mathrm{e}^{-j\left(\omega_0\theta + \frac{\alpha\theta^2}{2}\right)} \int\limits_{-\tau/2}^{(\tau/2)-\theta} \mathrm{e}^{-j\alpha\theta t} \mathrm{d}t$$
$$+ \frac{A_0^2}{4} \mathrm{e}^{j\left(\omega_0\theta + \frac{\alpha\theta^2}{2}\right)} \int\limits_{-\tau/2}^{(\tau/2)-\theta} \mathrm{e}^{j\alpha\theta t} \mathrm{d}t$$
$$+ \frac{A_0^2}{4} \mathrm{e}^{-j\left(\omega_0\theta + \frac{\alpha\theta^2}{2}\right)} \int\limits_{-\tau/2}^{(\tau/2)-\theta} \mathrm{e}^{-j\left[(2\omega_0+\alpha\theta)t + \alpha t^2\right]} \mathrm{d}t. \tag{4.68}$$

The second and third terms in (4.68) are now easily transformed as in the following,

$$
\begin{aligned}
&\frac{A_0^2}{4}\mathrm{e}^{-j\left(\omega_0\theta+\frac{\alpha\theta^2}{2}\right)}\int_{-\tau/2}^{(\tau/2)-\theta}\mathrm{e}^{-j\alpha\theta t}\mathrm{d}t+\frac{A_0^2}{4}\mathrm{e}^{j\left(\omega_0\theta+\frac{\alpha\theta^2}{2}\right)}\int_{-\tau/2}^{(\tau/2)-\theta}\mathrm{e}^{j\alpha\theta t}\mathrm{d}t\\
&=-\frac{A_0^2}{4j\alpha\theta}\mathrm{e}^{-j\left(\omega_0\theta+\frac{\alpha\theta^2}{2}\right)}\left[\mathrm{e}^{-j\alpha\theta\left(\frac{\tau}{2}-\theta\right)}-\mathrm{e}^{j\alpha\theta\frac{\tau}{2}}\right]\\
&\quad+\frac{A_0^2}{4j\alpha\theta}\mathrm{e}^{j\left(\omega_0\theta+\frac{\alpha\theta^2}{2}\right)}\left[\mathrm{e}^{j\alpha\theta\left(\frac{\tau}{2}-\theta\right)}-\mathrm{e}^{-j\alpha\theta\frac{\tau}{2}}\right]\\
&=\frac{A_0^2}{4j\alpha\theta}\mathrm{e}^{-j\left(\omega_0\theta+\frac{\alpha\theta^2}{2}\right)}\left[\mathrm{e}^{j\alpha\theta\left(\frac{\tau}{2}-\frac{\theta}{2}\right)}-\mathrm{e}^{-j\alpha\theta\left(\frac{\tau}{2}-\frac{\theta}{2}\right)}\right]\mathrm{e}^{j\alpha\frac{\theta^2}{2}}\\
&\quad+\frac{A_0^2}{4j\alpha\theta}\mathrm{e}^{j\left(\omega_0\theta+\frac{\alpha\theta^2}{2}\right)}\left[\mathrm{e}^{j\alpha\theta\left(\frac{\tau}{2}-\frac{\theta}{2}\right)}-\mathrm{e}^{-j\alpha\theta\left(\frac{\tau}{2}-\frac{\theta}{2}\right)}\right]\mathrm{e}^{-j\alpha\frac{\theta^2}{2}}\\
&=\frac{A_0^2\sin\left[\alpha\theta(\tau-\theta)/2\right]}{2\alpha\theta}\mathrm{e}^{-j(\omega_0\theta)}+\frac{A_0^2\sin\left[\alpha\theta(\tau-\theta)/2\right]}{2\alpha\theta}\mathrm{e}^{j(\omega_0\theta)}\\
&=\frac{A_0^2}{2}\frac{\sin\left[\alpha\theta(\tau-\theta)/2\right]}{\alpha\theta(\tau-\theta)/2}(\tau-\theta)\cos\omega_0\theta. \qquad (4.69)
\end{aligned}
$$

To find a closed form for the integral in the first term of (4.68), we first bring the power of the exponential function in the integrand to the full square and arrive at

$$
\begin{aligned}
I_1&=\int_{-\tau/2}^{(\tau/2)-\theta}\mathrm{e}^{j\left[(2\omega_0+\alpha\theta)t+\alpha t^2\right]}\mathrm{d}t\\
&=\mathrm{e}^{-j\frac{(2\omega_0+\alpha\theta)^2}{4\alpha}}\int_{-\tau/2}^{(\tau/2)-\theta}\mathrm{e}^{j\alpha\left(\frac{2\omega_0+\alpha\theta}{2\alpha}+t\right)^2}\mathrm{d}t.
\end{aligned}
$$

A new variable $\nu=\sqrt{\frac{2\alpha}{\pi}}\left(\frac{2\omega_0+\alpha\theta}{2\alpha}+t\right)$ then leads to

$$
\begin{aligned}
I_1&=\sqrt{\frac{\pi}{2\alpha}}\mathrm{e}^{-j\frac{(2\omega_0+\alpha\theta)^2}{4\alpha}}\int_{\nu_2(\theta)}^{\nu_1(\theta)}\mathrm{e}^{j\frac{\pi\xi^2}{2}}\mathrm{d}\xi\\
&=\sqrt{\frac{\pi}{2\alpha}}\mathrm{e}^{-j\frac{(2\omega_0+\alpha\theta)^2}{4\alpha}}\left(\int_{\nu_2(\theta)}^{\nu_1(\theta)}\cos\frac{\pi\xi^2}{2}\mathrm{d}\xi+j\int_{\nu_2(\theta)}^{\nu_1(\theta)}\sin\frac{\pi\xi^2}{2}\mathrm{d}\xi\right)\\
&=\sqrt{\frac{\pi}{2\alpha}}\mathrm{e}^{-j\frac{(2\omega_0+\alpha\theta)^2}{4\alpha}}
\end{aligned}
$$

$$\times\left(\int\limits_{\nu_2(\theta)}^{0}\cos\frac{\pi\xi^2}{2}\mathrm{d}\xi+\int\limits_{0}^{\nu_1(\theta)}\cos\frac{\pi\xi^2}{2}\mathrm{d}\xi+j\int\limits_{\nu_2(\theta)}^{0}\sin\frac{\pi\xi^2}{2}\mathrm{d}\xi+j\int\limits_{0}^{\nu_1(\theta)}\sin\frac{\pi\xi^2}{2}\mathrm{d}\xi\right)$$

$$=\sqrt{\frac{\pi}{2\alpha}}\mathrm{e}^{-j\frac{(2\omega_0+\alpha\theta)^2}{4\alpha}}\left[C(\nu_3)-C(\nu_4)+jS(\nu_3)-jS(\nu_4)\right], \tag{4.70}$$

where the integration bounds for the Fresnel functions, $C(\nu)$ and $S(\nu)$, are given by

$$\nu_3=\sqrt{\frac{2\alpha}{\pi}}\left(\frac{2\omega_0+\alpha\theta}{2\alpha}+\frac{\tau}{2}-\theta\right), \tag{4.71}$$

$$\nu_4=\sqrt{\frac{2\alpha}{\pi}}\left(\frac{2\omega_0+\alpha\theta}{2\alpha}-\frac{\tau}{2}\right). \tag{4.72}$$

Reasoning along similar lines, we find a solution for the integral in the fourth term in (4.68),

$$I_2=\sqrt{\frac{\pi}{2\alpha}}\mathrm{e}^{j\frac{(2\omega_0+\alpha\theta)^2}{4\alpha}}\left[C(\nu_3)-C(\nu_4)-jS(\nu_3)+jS(\nu_4)\right]. \tag{4.73}$$

Substituting (4.69), (4.70), and (4.73) into (4.68) yields

$$\begin{aligned}\phi_x(\theta)=&\frac{A_0^2}{2}\frac{\sin\left[\alpha\theta(\tau-\theta)/2\right]}{\alpha\theta(\tau-\theta)/2}(\tau-\theta)\cos\omega_0\theta\\&+\frac{A_0^2}{2}\sqrt{\frac{\pi}{2\alpha}}\left\{\left[C(\nu_3)-C(\nu_4)\right]\cos\frac{4\omega_0^2-\alpha^2\theta^2}{4\alpha}\right.\\&\qquad\left.+\left[S(\nu_3)-S(\nu_4)\right]\sin\frac{4\omega_0^2-\alpha^2\theta^2}{4\alpha}\right\}\end{aligned}$$

that finally may be written as

$$\begin{aligned}\phi_x(\theta)=&\frac{A_0^2}{2}\frac{\sin\left[\alpha\theta(\tau-\theta)/2\right]}{\alpha\theta(\tau-\theta)/2}(\tau-\theta)\cos\omega_0\theta\\&+\frac{A_0^2}{2}\sqrt{\frac{\pi}{2\alpha}}H(\theta)\cos\left[\frac{4\omega_0^2-\alpha^2\theta^2}{4\alpha}-\psi(\theta)\right],\end{aligned} \tag{4.74}$$

where

$$H=\sqrt{\left[C(\nu_3)-C(\nu_4)\right]^2+\left[S(\nu_3)-S(\nu_4)\right]^2}, \tag{4.75}$$

$$\psi=\arctan\frac{S(\nu_3)-S(\nu_4)}{C(\nu_3)-C(\nu_4)}. \tag{4.76}$$

It may be shown that the second term in (4.74) contributes insignificantly. Furthermore, as $\phi_x(\theta)=\phi_x(-\theta)$, one may substitute $|\theta|$ instead of θ. The correlation function of a single rectangular RF LFM pulse then attains the approximate form of

$$\phi_x(\theta) \cong \frac{A_0^2}{2} \frac{\sin[\alpha\theta(\tau - |\theta|)/2]}{\alpha\theta(\tau - |\theta|)/2}(\tau - |\theta|) \cos\omega_0\theta\,, \tag{4.77}$$

allowing for a comparative analysis of the signal power performance in the time domain with that obtained in the frequency domain (4.65).

Several particular situations may now be observed.

4.5.2.1 Negligible PCR

By PCR $\ll 1$, we let $\alpha\theta(\tau - |\theta|)/2 \cong 0$ and thus the sinc function in (4.77) is almost unity. The correlation function then becomes

$$\phi_x(\theta)_{\mathrm{PCR}\ll 1} \cong \frac{A_0^2}{2}(\tau - |\theta|) \cos\omega_0\theta. \tag{4.78}$$

This is of course an isolated case for the LFM pulse, as its performance becomes as that of the relevant rectangular RF pulse. The ESD function for this case is shown in Fig. 4.16a and Fig. 4.17 shows the relevant autocorrelation function.

4.5.2.2 Large PCR and $\alpha(\tau - |\theta|)/2 \ll \omega_0$

This is a typical case of the LFM pulse that is associated with the sinc-shaped envelope filled by harmonic oscillations of the carrier. Figure 4.18 shows the correlation function for this case. A disadvantage is a large level of the side lobes that reaches about 21% of the signal energy.

4.5.2.3 Large PCR and $\alpha(\tau - |\theta|)/2 \simeq \omega_0$

To increase narrowness of the function, the carrier frequency needs to be reduced and PCR increased. If we obtain $\alpha(\tau - |\theta|)/2 \simeq \omega_0$ with PCR $\gg 1$, we go to Fig. 4.19 that exhibits an excellent performance. Moreover, the function

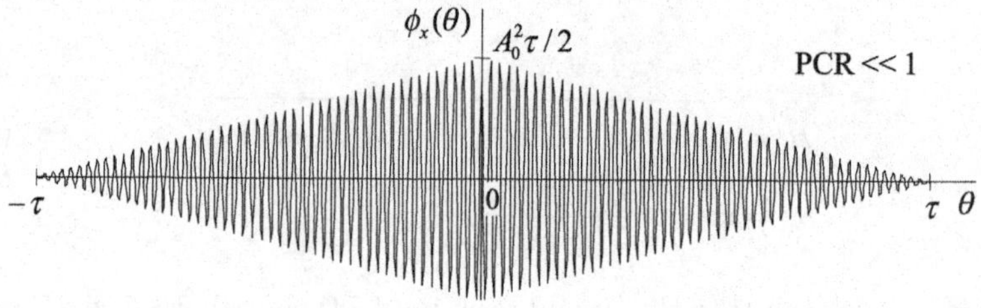

Fig. 4.17 Autocorrelation function of a rectangular LFM pulse with PCR $\ll 1$.

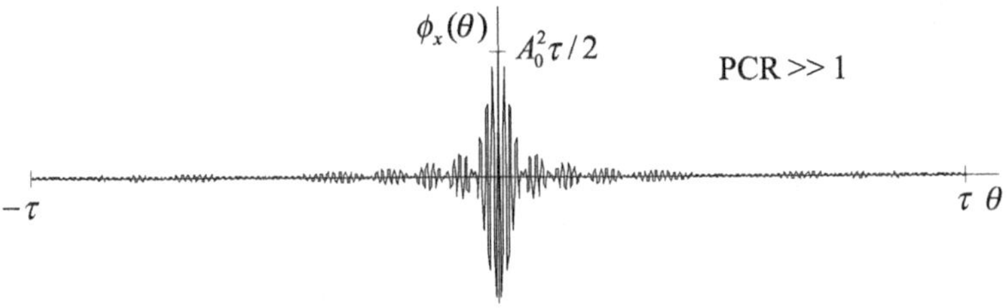

Fig. 4.18 Autocorrelation function of a rectangular LFM pulse with PCR $\gg 1$ and $\alpha(\tau - |\theta|)/2 \ll \omega_0$.

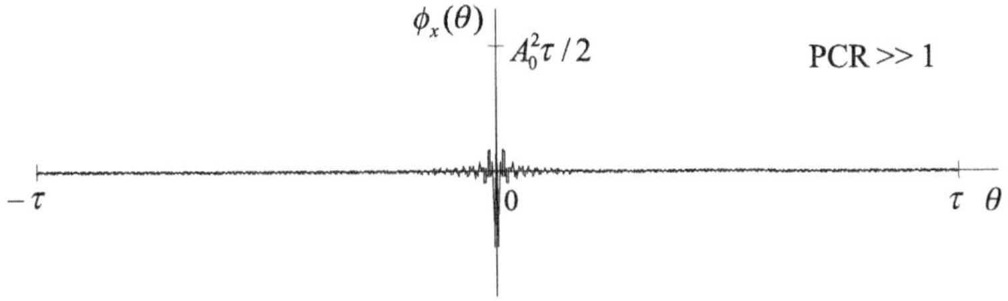

Fig. 4.19 Autocorrelation function of a rectangular LFM pulse with PCR $\gg 1$ and $\alpha(\tau - |\theta|)/2 \simeq \omega_0$.

becomes almost delta-shaped when the low bound of the ESD function (Fig. 4.16c) tends toward zero $\omega_0 - \Delta\omega/2 \to 0$; that is $\Delta\omega \simeq 2\omega_0$. We notice that this case is hard to be realized practically, since the pulse spectrum occupies a very wide range.

Two important generalizations follow from an analysis of (4.65) and (4.77) and observation of Fig. 4.16 and Figs. 4.17–4.19:

- The wider $G_x(\omega)$, the narrower $\phi_x(\theta)$.
- The width of the main lobe of $\phi_x(\theta)$ changes as a reciprocal of the frequency deviation Δf. Assuming that the width W_{xx} is defined by zeros of the main lobe of the sinc function and supposing that $\theta \ll \tau$, we have $W_{xx} = 2\pi/\alpha\tau$. Using (3.97) and substituting $\alpha = 2\pi\Delta f/\tau$ yields $W_{xx} = 1/\Delta f$.

4.6 Complex Phase-Coded Signals

Examining the autocorrelation function, one comes to the conclusion that the function becomes narrower if a signal undergoes modulation stretching its spectrum (see LFM signal). One can also observe that the autocorrelation function of a pulse-burst may satisfy the requirement of narrowness if to suppress somehow the side lobes in Fig. 4.14b, for example. An opportunity to do it is realized in the phase coding that we learned earlier in Chapter 3. In such signals, the RF pulse-burst is formed to keep the phase angle in each of a single pulse by the certain code. Later we show that the effect produced in the autocorrelation function becomes even more appreciable than that in the LFM pulse.

4.6.1 Essence of Phase Coding

Before formulating the rules of phase coding, it seems in order to consider three simple particular situations giving an idea of how the energy autocorrelation function is shaped and how it occurs to be narrow.

The first case is trivial (Fig. 4.20) corresponding to the above-learned rectangular RF pulse, which correlation function has a triangular envelope (4.29). It is important to notice that an initial signal phase (Fig. 4.20a and b) does not affect its correlation function. Here and in the following we symbolize the pulse envelope by "1" if the initial phase is zero and by "–1" if it is π.

Let us now combine the pulse with two same pulses, each of duration τ, and assign their phases by "1" and "–1" (Fig. 4.21a). The correlation function is formed as follows. When the time shift is $\tau \leqslant |\theta| \leqslant 2\tau$, the function is shaped by the interaction of two elementary pulses. Accordingly, its envelope rises linearly starting with $|\theta| = 2\tau$ from zero and then attaining a maximum E_1 (energy of a single pulse) at $|\theta| = \tau$. Because the phases are shifted on π, it is in order to trace the lower envelope. At the next stage of $0 \leqslant |\theta| \leqslant \tau$, the lower envelope rises linearly to attain the signal total energy $2E_1$ at $\theta = 0$. It is seen that the envelope passes through a zero point (Fig. 4.21c). The inverted signal (Fig. 4.21b) does not change the picture. So, we resume, $\phi_x(\theta)$ has acquired two side lobes attenuated by the factor of 2 for the main lobe.

We now complicate an example with three single pulses having the phases as in Fig. 4.22a. Reasoning similarly, we pass over the particular shifts, calculate the autocorrelation function, and arrive at Fig. 4.22c. The principle point is that the level of the side lobes is still E_1, whereas the peak value of the main lobe is three times larger; that is the total energy of the burst is $3E_1$.

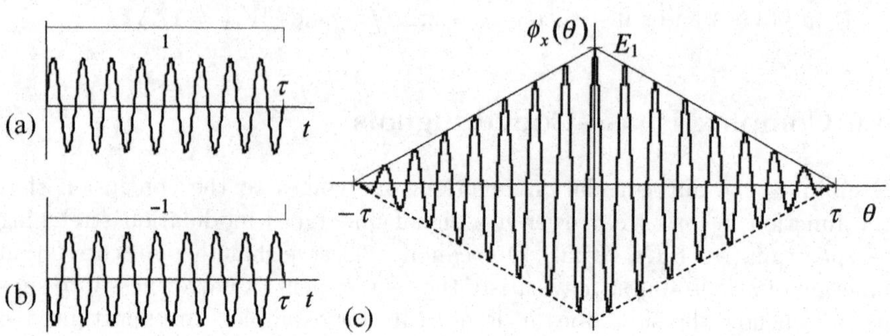

Fig. 4.20 A rectangular RF pulse: **(a)** in-phase, **(b)** inverted, and **(c)** energy autocorrelation function.

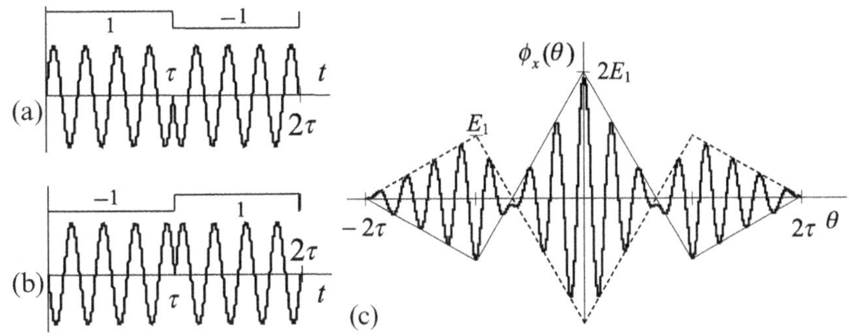

Fig. 4.21 A burst of two rectangular RF pulses: **(a)** in-phase, **(b)** inverted, and **(c)** autocorrelation function.

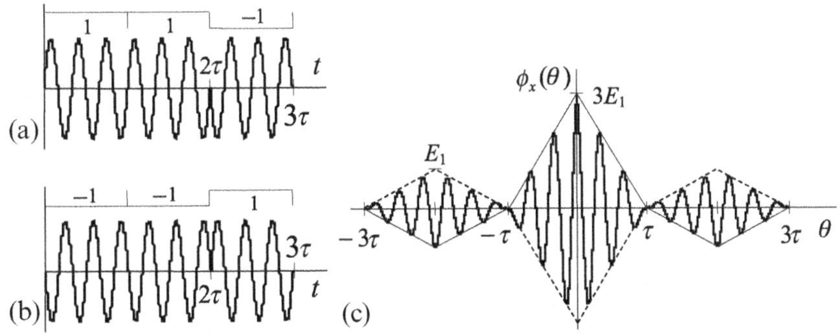

Fig. 4.22 A burst of three rectangular RF pulses: **(a)** in-phase, **(b)** inverted, and **(c)** autocorrelation function.

The problem with phase coding formulates hence as follows. The phase code for the pulse-burst must be chosen in a way to obtain the maximum value NE_1 of $\phi_x(\theta)$ at $\theta = 0$ and E_1 if $0 < |\theta| \leqslant N\tau$.

4.6.2 Autocorrelation of Phase-Coded Pulses

In spite of a certain difference between the correlation functions in the three above-considered cases, an important common feature may also be indicated. As it is seen, between the neighboring points multiple to τ, the envelope of $\phi_x(\theta)$ behaves linearly. This means that, describing $\phi_x(\theta)$, we may operate at discrete points $n\tau$ and then connect the results linearly.

The model of a phase-coded signal may be performed by a sequence of numbers

$$\{m_1, \quad m_2, \quad \ldots, \quad m_{N-1}, \quad m_N\},$$

in which every symbol m_i, $i \in [1, N]$, takes one of the allowed values either $+1$ or -1. We will also tacitly assume that if a signal is not determined in some time range then it becomes zero.

For example, let us perform a signal {1, 1, –1} (Fig. 4.22) as

$$\ldots 0\;0\;0\;0\;1\;1\;{-1}\;0\;0\;0\;0\ldots$$

and calculate its correlation function as the sum of the products of the multiplications of the shifted copies. For the sake of clarity, we show below the original signal and several such copies:

$$\underline{\ldots 0\;0\;0\;0\;1\;1\;-1\;0\;0\;0\;0\ldots}$$
$$\ldots 0\;0\;0\;0\;1\;1\;-1\;0\;0\;0\;0\ldots$$
$$\ldots 0\;0\;0\;0\;0\;1\;1\;-1\;0\;0\;0\ldots$$
$$\ldots 0\;0\;0\;0\;0\;0\;1\;1\;-1\;0\;0\ldots$$
$$\ldots 0\;0\;0\;0\;0\;0\;0\;1\;1\;-1\;0\ldots$$

Since the multiplication obeys the rule,

$$\{1\}\{1\} = 1, \quad \{-1\}\{-1\} = 1, \quad \text{and} \quad \{1\}\{-1\} = \{-1\}\{1\} = -1,$$

we arrive at

$$\begin{aligned} \text{for } \theta = 0 &: \{1\}\{1\} + \{1\}\{1\} + \{-1\}\{-1\} = 3, \\ \text{for } |\theta| = \tau &: \{1\}\{1\} + \{-1\}\{1\} = 0, \\ \text{for } |\theta| = 2\tau &: \{-1\}\{1\} = -1, \\ \text{for } |\theta| = 3\tau &: 0. \end{aligned}$$

The discrete correlation function may now be written as

$$\hat{\phi}_x(n) = \ldots\;0\;\;0\;-1\;\;0\;\;3\;\;0\;-1\;\;0\;\;0\;\ldots,$$

where the value "3" corresponds to $n = 0$, and we indicate that the result is consistent to that shown in Fig. 4.22c.

The procedure may now be generalized to the arbitrary code length. To calculate (4.28), it needs substituting the operation of integration by the operation of summation and instead of the continuous-time shift θ employ its discrete value $n\tau$. The function then reads

$$\hat{\phi}_x(n) = \sum_{i=-\infty}^{\infty} m_i m_{i-n}. \tag{4.79}$$

Obviously that this discrete-time function possesses many of the properties of the relevant continuous-time function $\phi_x(\theta)$. Indeed, the function is even,

$$\hat{\phi}_x(n) = \hat{\phi}_x(-n), \tag{4.80}$$

and, by $n = 0$, it determines the signal energy,

$$\hat{\phi}_x(0) = \sum_{i=-\infty}^{\infty} m_i^2 = E_x. \tag{4.81}$$

It also follows that such a "discrete-time" approach has obvious advantages against the continuous-time calculus, since the envelope of the autocorrelation function may easily be restored by lines connecting the neighboring discrete values.

4.6.3 Barker Phase Coding

As we have learned in Chapter 3, signals with Barker's phase coding have wide spectra. It is not, however, the prime appointment of the Barker codes. The codes were found to squeeze the correlation function of signals as it is demonstrated in Figs. 4.21 and 4.22. The effect of the Barker code of length N is reduced by the factor of N in the level of the side lobes.

Independently on the Barker code length N, the main lobe reaches the signal energy NE_1 at $\theta = 0$ and the envelope of the side lobes does not exceed E_1. The structures of the autocorrelation functions of the phase-coded signals associated with the known Barker codes are given in Table 4.3. Unfortunately, the procedure to derive these codes is not trivial and we do not know the ones exceeding $N13$.

Example 4.19. Given a rectangular pulse-burst of 13 pulses with $q = 1$. The phase in an elementary pulse is coded by the Barker code $N13$. The code structure and the pulse-burst are shown in Fig. 4.23*a* and *b*, respectively. The autocorrelation function is sketched in Fig. 4.23*c*. The relevant ESD is shown in Fig. 4.24 about the carrier ω_0. □

Two strong advantages of Barker, the phase-coded signal, may now be pointed out:

- By $N13$, the side lobes do not exceed the level of about 7.7%. LFM signals allow for no lower than about 21%.
- Phase-coded signals are representatives of impulse techniques. Therefore, they are produced easier than LFM signals (analog technique).

Table 4.3 Barker phase coding

Code length N	Code elements	Autocorrelation function
$N2$	01 or 00	2, −1
$N3$	001	3, 0, −1
$N4$a	0001	4, 1, 0, −1
$N4$b	0010	4, −1, 0, 1
$N5$	00010	5, 0, 1, 0, 1
$N7$	0001101	7, 0, −1, 0, −1, 0, −1
$N11$	00011101101	11, 0, −1, 0, −1, 0, −1, 0, −1, 0, −1
$N13$	0000011001010	13, 0, 1, 0, 1, 0, 1, 0, 1, 0, 1, 0, 1

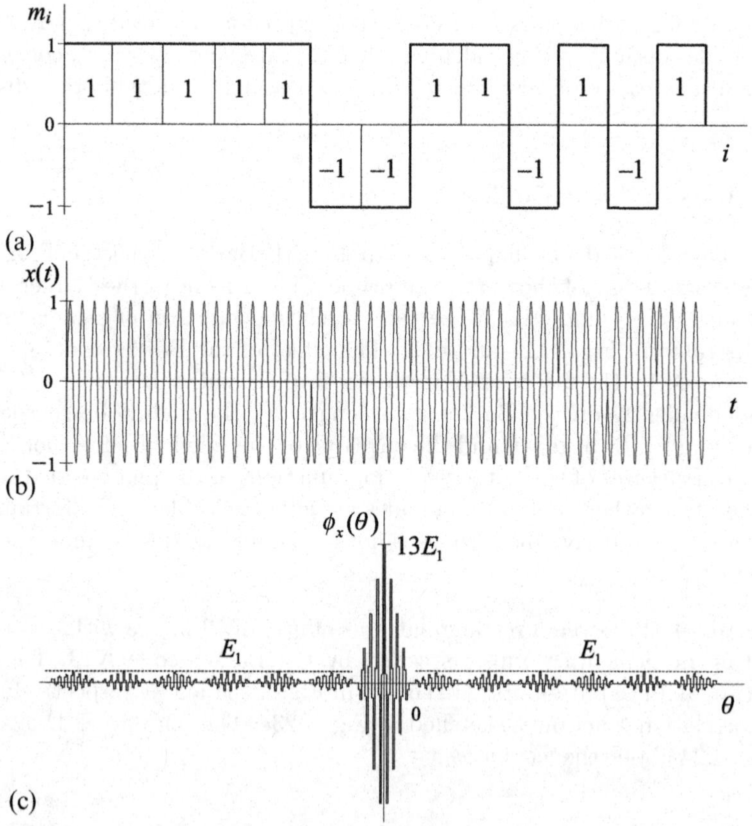

Fig. 4.23 Phase-coded rectangular pulse-burst (Barker code $N13$): **(a)** code structure, **(b)** signal, and **(c)** energy autocorrelation function.

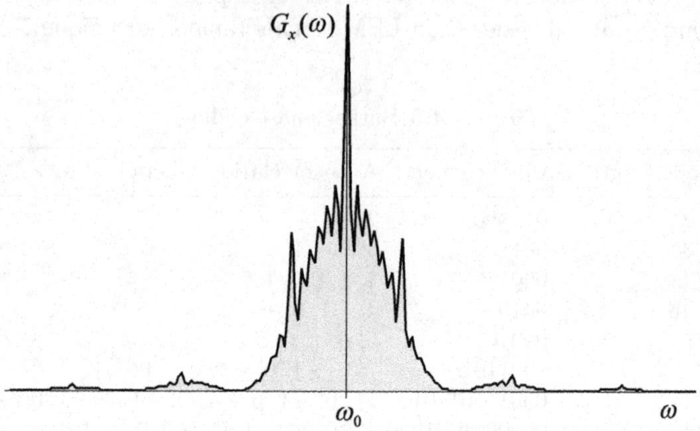

Fig. 4.24 ESD function of the phase-coded rectangular pulse-burst (Barker code $N13$).

4.7 Signal Cross-correlation

In practical applications, when any signal is shifted in time it is also changed, at least by the system channel or noise. Therefore, rigorously, when we speak about two shifted version of the same signal, we tacitly imply two different signals. In some cases, the difference may be negligible and the autocorrelation properties are studied. In some others, two signals differ by nature or the shifted version is too much "deformed." To evaluate correlation of different signals, the concept of "autocorrelation" can no longer be useful and the other concept of "cross-correlation" is employed. In fact, the autocorrelation is intended to evaluate the power of "similarity," whereas the cross-correlation gives an idea about the power of "diversity" of signals.

Like the autocorrelation, the cross-correlation is also associated with energy and power signals. In bottom, the cross-correlation function is the autocorrelation function, in which two different shifted signals $x(t)$ and $y(t-\theta)$ are used instead of two same shifted signals $x(t)$ and $x(t-\theta)$. So long as signals are either energy or power, the cross-correlation function may also be either energy or power. It cannot be related to different types of signals, e.g., to energy and power or power and energy.

4.7.1 Energy Cross-correlation of Signals

Analogous to (4.28), the energy cross-correlation function is also defined by the inner product

$$
\begin{aligned}
\phi_{xy}(\theta) = \langle x, y_\theta \rangle &= \lim_{T\to\infty} \int_{-T}^{T} x(t)y^*(t-\theta)\mathrm{d}t \\
&= \int_{-\infty}^{\infty} x(t)y^*(t-\theta)\mathrm{d}t. \qquad (4.82)
\end{aligned}
$$

Examining this relation, first of all we would like to know if the function is symmetric or it is not, to calculate only for positive or negative θ? Figure 4.25 demonstrates two different real waveforms $x(t)$ and $y(t)$ and the product of their multiplication. In the first case (a), $x(t)$ is time-fixed and $y(t)$ is delayed on θ. The cross-correlation is thus evaluated by the area (shaded) of the product $x(t)y(t-\theta)$. In the second case (b), we consider $y(t)$ and $x(t-\theta)$ and evaluate the cross-correlation by the area of the product $y(t)x(t-\theta)$.

Even a quick look at Fig. 4.25 allows to deduce that two integrals will produce different values, since the shaded areas are not equal. This means that the cross-correlation function is not even and thus

$$
\int_{-\infty}^{\infty} x(t)y^*(t-\theta)\mathrm{d}t \neq \int_{-\infty}^{\infty} y(t)x^*(t-\theta)\mathrm{d}t.
$$

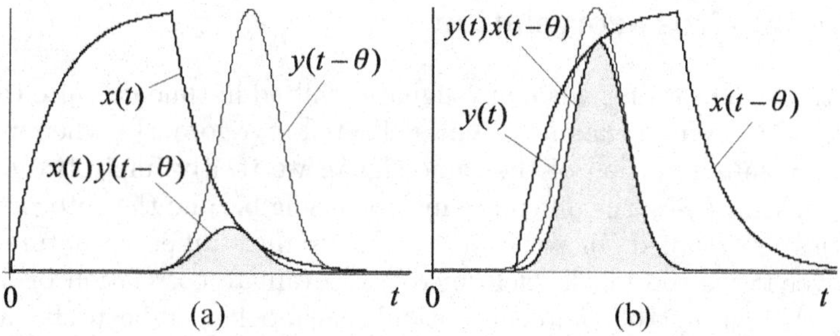

Fig. 4.25 Cross-correlation of two different waveforms: **(a)** $x(t)$ and $y(t-\theta)$ and **(b)** $y(t)$ and $x(t-\theta)$.

In fact, by chance of variables in (4.82), it can be seen that

$$\phi_{xy}(\theta) = \int_{-\infty}^{\infty} x(t)y^*(t-\theta)\mathrm{d}t = \int_{-\infty}^{\infty} y^*(t)x(t+\theta)\mathrm{d}t.$$

A simple geometrical interpretation of $\phi_{xy}(\theta)$ is given in Fig. 4.26 for two reverse truncated ramp pulses of duration τ. A geometrical explanation of the calculus of $\phi_{xy}(\theta)$ coincides with that given for the autocorrelation function in Fig. 4.9. Here we just point out two obvious features of $\phi_{xy}(\theta)$:

- The function $\phi_{xy}(\theta)$ is not symmetric and it is not even.
- The maximum of $\phi_{xy}(\theta)$ is not obligatorily equal to the joint energy of signals at $\theta = 0$.

In our example (Fig. 4.26), the joint energy of two signals has appeared to be between the maximum and zero of the cross-correlation function. It is in order to suppose that there may be some other waveforms allowing exactly for $\phi_{xy}(0) = 0$ and $\phi_{xy}(0) = \max$. It is also out of doubt that $\phi_{xy}(0) < 0$ if one of the signals is negative.

Example 4.20. Given real rectangular and ramp pulses of amplitude A and duration τ (Fig. 4.27a and b) described, respectively, by

$$x(t) = Au(t) - Au(t-\tau),$$

$$y(t) = \begin{cases} At/\tau, & 0 \leqslant t \leqslant \tau \\ 0, & \text{otherwise} \end{cases}.$$

The cross-correlation function is calculated in two special ranges. If $\theta \geqslant 0$, then

$$\phi_{xy}(\theta \geqslant 0) = \frac{A^2}{\tau}\int_{\theta}^{\tau}(t-\theta)\mathrm{d}t = \frac{A^2}{\tau}\left(1 - \frac{2\theta}{\tau} + \frac{\theta^2}{\tau^2}\right).$$

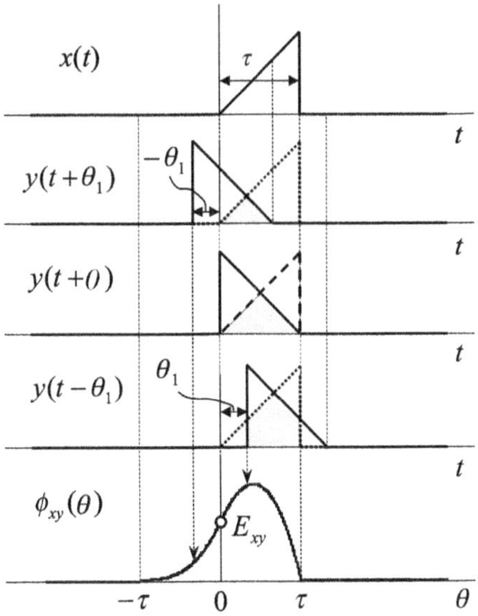

Fig. 4.26 Geometrical interpretation of the energy cross-correlation function.

If $\theta \leqslant 0$, then

$$\phi_{xy}(\theta \leqslant 0) = \frac{A^2}{\tau} \int\limits_0^{\tau-|\theta|} (t - |\theta|)\mathrm{d}t = \frac{A^2}{\tau}\left(1 - \frac{\theta^2}{\tau^2}\right).$$

By $\theta = 0$, both formulas produce the joint signal energy

$$\phi_{xy}(\theta = 0) = \frac{A^2}{\tau}.$$

Figure 4.27c illustrates the cross-correlation function derived. Unlike the case of Fig. 4.26, here the joint energy E_{xy} coincides with the maximum of $\phi_{xy}(\theta = 0)$. We notice that if one of the signals will appear to be negative, then the cross-correlation function and joint energy will be negative as well.

□

Example 4.21. Given two real pulses shown in Fig. 4.28a and b and described, respectively, by

$$x(t) = Au(t) - Au(t - \tau),$$
$$y(t) = Au(t) - 2Au\left(t - \frac{\tau}{2}\right) + Au\left(t - \tau\right).$$

The cross-correlation function is calculated by

$$\phi_{xy}(\theta) = \begin{cases} -\frac{A^2\tau}{2}\left(1 - \frac{2}{\tau}\left|\theta + \frac{\tau}{2}\right|\right), & -\tau \leqslant \theta < 0 \\ \frac{A^2\tau}{2}\left(1 - \frac{2}{\tau}\left|\theta - \frac{\tau}{2}\right|\right), & 0 \leqslant \theta \leqslant \tau \\ 0, & \text{otherwise} \end{cases}$$

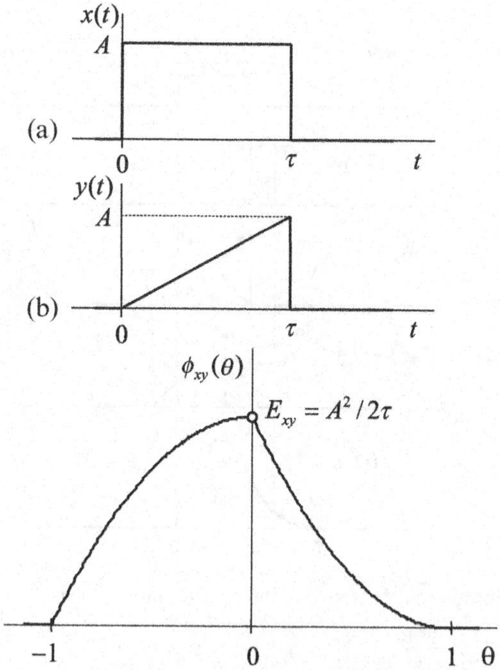

Fig. 4.27 Energy function of two signals: **(a)** rectangular pulse, **(b)** ramp pulse, and **(c)** cross-correlation.

and shown in Fig. 4.28c. As it is seen, $\phi_{xy}(\theta)$ crosses zero when $\theta = 0$. Moreover, the negative sign in one of the signals does not change its property to be zero at zero.

□

Example 4.21 turns us back to the definition of orthogonality. We recall that two signals, $x(t)$ and $y(t)$, are orthogonal if the inner product $\langle x, y\rangle$ is zero. Since, $\theta = 0$, the joint energy (inner product) of $x(t)$ and $y(t)$ in this example is zero, they are orthogonal. Calculating $\phi_{xy}(0)$ for any pair of orthogonal functions (periodic or short), we inevitably will produce zero in each of the cases. The class of orthogonal functions is basic for the Fourier transforms, wavelet transforms, and other useful transforms. In helping to recognize this class, the cross-correlation function calculates the power of diversity or "nonorthogonality" of two signals. Therefore, it is often used as an excellent indicator of the signals orthogonality to indicate "stability" of the orthogonality state of signals in systems with different time shifts.

4.7.2 Power Cross-correlation of Signals

In line with the energy cross-correlation function, the *power cross-correlation function* is also calculated by using two different shifted signals in the

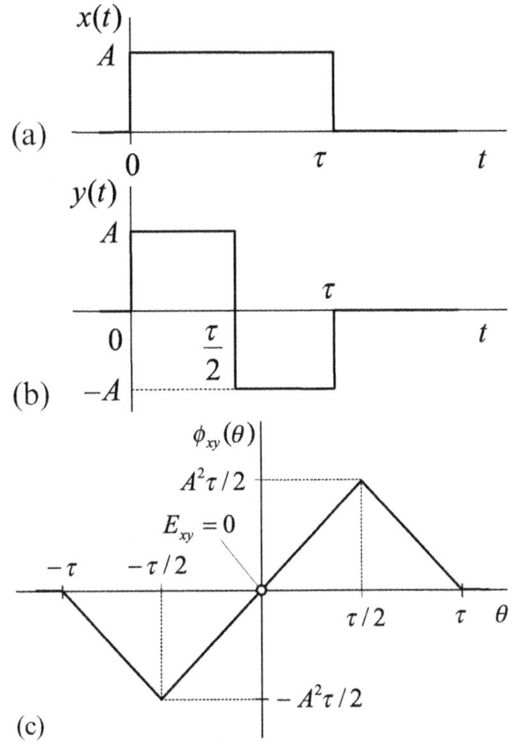

Fig. 4.28 Energy function of two signals: **(a)** rectangular pulse, **(b)** pulse-burst, and **(c)** cross-correlation.

integrand. Using (4.39), we may thus write

$$R_{xy}(\theta) = \lim_{T\to\infty} \frac{1}{2T} \int_{-T}^{T} x(t)y^*(t-\theta)\mathrm{d}t. \tag{4.83}$$

Of applied importance is that (4.83) possesses almost all the properties of the energy cross-correlation, as it may alternatively be calculated by

$$R_{xy}(\theta) = \lim_{T\to\infty} \frac{\phi_{xy}(\theta)}{2T}$$

and hence the principle difference could be found not between $R_{xy}(\theta)$ and $\phi_{xy}(\theta)$, but in the signals, nature (energy or power).

Example 4.22. Given two harmonic signals $x(t) = A\cos\omega_0 t$ and $y(t) = B\sin\omega_0 t$. The cross-correlation function may be calculated by (4.83) if to integrate over the period T by

$$R_{xy}(\theta) = \frac{1}{T} \int_{-T/2}^{T/2} x(t)y^*(t-\theta)\mathrm{d}t \tag{4.84}$$

that yields

$$
\begin{aligned}
R_{xy}(\theta) &= \frac{AB}{T} \int_{-T/2}^{T/2} \cos\omega_0 t \sin\omega_0 (t-\theta) \mathrm{d}t \\
&= \frac{AB}{2T} \left[\int_{-T/2}^{T/2} \sin\omega_0 \left(2t-\theta\right) \mathrm{d}t - \int_{-T/2}^{T/2} \sin\omega_0 \theta \mathrm{d}t \right] \\
&= -\frac{AB}{2} \sin\omega_0 \theta . \qquad (4.85)
\end{aligned}
$$

By $\theta = 0$, (4.85) produces zero and thus the signals are orthogonal. It is obviously not an unexpected conclusion if to recall that such a property of harmonic functions is a basis for the Fourier transforms. We also notice that the function (4.85) is periodic with the period T of signals.

□

4.7.3 Properties of Signals Cross-correlation

The most important and widely used properties of cross-correlation are the following:

- The Fourier transform of the cross-correlation function is called the *cross PSD function*. The *cross ESD function* and the *cross PSD function* are complex and given, respectively, by

$$
G_{xy}(j\omega) = \mathcal{F}\{\phi_{xy}(\theta)\} = \int_{-\infty}^{\infty} \phi_{xy}(\theta) \mathrm{e}^{-j\omega\theta} \mathrm{d}\theta , \qquad (4.86)
$$

$$
S_{xy}(j\omega) = \mathcal{F}\{R_{xy}(\theta)\} = \int_{-\infty}^{\infty} R_{xy}(\theta) \mathrm{e}^{-j\omega\theta} \mathrm{d}\theta . \qquad (4.87)
$$

- The functions $\phi_{xy}(\theta)$ and $R_{xy}(\theta)$ are provided by the inverse Fourier transform, respectively,

$$
\phi_{xy}(\theta) = \mathcal{F}^{-1}\{G_{xy}(\omega)\} = \frac{1}{2\pi} \int_{-\infty}^{\infty} G_{xy}(\omega) \mathrm{e}^{j\omega\theta} \mathrm{d}\omega , \qquad (4.88)
$$

$$
R_{xy}(\theta) = \mathcal{F}^{-1}\{S_{xy}(\omega)\} = \frac{1}{2\pi} \int_{-\infty}^{\infty} S_{xy}(\omega) \mathrm{e}^{j\omega\theta} \mathrm{d}\omega . \qquad (4.89)
$$

- The cross ESD and PSD functions are calculated through the spectral densities of signals, respectively, by

$$G_{xy}(j\omega) = X(j\omega)Y^*(j\omega)\,, \tag{4.90}$$

$$\begin{aligned} S_{xy}(j\omega) &= \lim_{T\to\infty} \frac{1}{2T} G_x(j\omega) \\ &= \lim_{T\to\infty} \frac{1}{2T} X(j\omega)Y^*(j\omega). \end{aligned} \tag{4.91}$$

- The cross PSD function is coupled with the amplitudes of the Fourier series by

$$S_{xy}(\omega) = 2\pi \sum_{k=-\infty}^{\infty} C_{xk}C_{yk}^*\delta(\omega - k\Omega). \tag{4.92}$$

- The function $\phi_{xy}(\theta)$ is coupled with the signals, spectral densities by

$$\phi_{xy}(\theta) = \frac{1}{2\pi}\langle X, Y_\theta\rangle = \frac{1}{2\pi}\int_{-\infty}^{\infty} X(j\omega)Y_\theta^*(j\omega)\mathrm{d}\omega\,. \tag{4.93}$$

- The joint energy of two signals is calculated by

$$E_{xy} = \phi_{xy}(0) = \frac{1}{2\pi}\int_{-\infty}^{\infty} G_{xy}(\omega)\mathrm{d}\omega = \int_{-\infty}^{\infty} x(t)y^*(t)\mathrm{d}t \tag{4.94}$$

  and may be either positive, negative, or zero.

- The joint power of two power signals is calculated by

$$R_{xy}(0) = \frac{1}{2\pi}\int_{-\infty}^{\infty} S_{xy}(\omega)\mathrm{d}\omega = \lim_{T\to\infty}\frac{1}{2T}\int_{-T}^{T} x(t)y^*(t)\mathrm{d}t \tag{4.95}$$

  and may be either positive, negative, or zero.

- The following properties hold for the cross-correlation functions,

$$\phi_{xy}(\theta) = \phi_{yx}(-\theta)\,, \tag{4.96}$$

$$R_{xy}(\theta) = R_{yx}(-\theta). \tag{4.97}$$

- The cross ESD and PSD functions are not even and not real-valued.

- The conjugate symmetry property of the Fourier transform holds,

$$G_{xy}(-j\omega) = G_{xy}^*(j\omega)\,, \tag{4.98}$$

$$S_{xy}(-j\omega) = S_{xy}^*(j\omega). \tag{4.99}$$

- The Cauchy–Bunyakovskii inequality claims that

$$|\phi_{xy}(\theta)| = |(x, y_\theta)| \leqslant \|x\| \cdot \|y_\theta\|$$
$$= \sqrt{\int_{-\infty}^{\infty} x^2(t)\mathrm{d}t} \sqrt{\int_{-\infty}^{\infty} y^2(t-\theta)\mathrm{d}t} = |E_{xy}| \,. \quad (4.100)$$

- If $x(t)$ and $y(t)$ are periodic with period T, then the cross-correlation function is also periodic; i.e., $\phi_{xy}(\theta) = \phi_{yx}(\theta \pm nT)$ and $R_{xy}(\theta) = R_{xy}(\theta \pm nT)$.

- If $x(t)$ and $y(t)$ are uncorrelated, then their cross-correlation function is zero everywhere; i.e., $\phi_{xy}(\theta) = 0$ and $R_{xy}(\theta) = 0$. The random signals $x(t)$ and $y(t)$ generated by different physical sources are usually assumed to be uncorrelated.

- If $x(t)$ and $y(t)$ are orthogonal, then their cross-correlation function is zero at $\theta = 0$; i.e., $\phi_{xy}(0) = 0$, and $R_{xy}(0) = 0$. This function is odd, $\phi_{xy}(\theta) = -\phi_{xy}(-\theta)$ and $R_{xy}(\theta) = -R_{xy}(-\theta)$.

- If $x(t)$ and $y(t)$ compose a signal $z(t) = x(t) + y(t)$, then

$$\phi_z(\theta) = \phi_x(\theta) + \phi_y(\theta) + \phi_{xy}(\theta) + \phi_{yx}(\theta)$$

and

$$G_z(j\omega) = G_x(j\omega) + G_y(j\omega) + G_{xy}(j\omega) + G_{yx}(j\omega).$$

Herewith, if $x(t)$ and $y(t)$ are uncorrelated, then

$$\phi_z(\theta) = \phi_x(\theta) + \phi_y(\theta)$$

and

$$G_z(j\omega) = G_x(j\omega) + G_y(j\omega).$$

The same is valid for power signals.

- If $x(t)$ is the input of a linear time-invariant system (LTI) with known impulse response $h(t)$, the output is produced by the convolution $y(t) = h(t) * x(t)$ and then

$$\phi_{xy}(\theta) = \int_{-\infty}^{\infty} x(t)y(t-\theta)\mathrm{d}t$$
$$= \int_{-\infty}^{\infty} x(t) \int_{-\infty}^{\infty} h(z)x(t-\theta-z)\mathrm{d}z\,\mathrm{d}t$$

$$= \int_{-\infty}^{\infty} h(z) \int_{-\infty}^{\infty} x(t)x(t-\theta-z)\mathrm{d}t\,\mathrm{d}z = \int_{-\infty}^{\infty} h(z)\phi_x(\theta+z)\mathrm{d}z$$

$$= -\int_{\infty}^{-\infty} h(-z)\phi_x(\theta-z)\mathrm{d}z = \int_{-\infty}^{\infty} h(-z)\phi_x(\theta-z)\mathrm{d}z$$

$$= h(-\theta) * \phi_x(\theta). \quad (4.101)$$

The following also holds,

$$\phi_{yx}(\theta) = h(\theta) * \phi_x(\theta)\,, \quad (4.102)$$

$$\phi_y(\theta) = h(\theta) * \phi_{xy}(\theta) = h(\theta) * h(-\theta) * \phi_x(\theta). \quad (4.103)$$

- If $H(j\omega)$ is the Fourier transform of $h(t)$ and $H^*(j\omega)$ is its complex conjugate, then

$$G_{xy}(j\omega) = H^*(j\omega)G_x(j\omega)\,, \quad (4.104)$$

$$G_{yx}(j\omega) = H(j\omega)G_x(j\omega)\,, \quad (4.105)$$

$$G_y(j\omega) = |H(j\omega)|^2\, G_x(j\omega). \quad (4.106)$$

We notice that the properties (4.101)–(4.106) are fundamental for LTI systems.

4.8 Width of the Autocorrelation Function

The most informative comparative measure of autocorrelation is the equivalent width of the autocorrelation function. It may also be specified in the frequency domain by the signal spectral density. Moreover, the definitions of signal widths given in Chapter 2 are applicable to autocorrelations.

4.8.1 Autocorrelation Width in the Time Domain

In the time domain, the measure of the autocorrelation width is given by

$$W_{xx} = \frac{1}{\phi_x(0)} \int_{-\infty}^{\infty} \phi_x(\theta)\mathrm{d}\theta = \frac{\int_{-\infty}^{\infty} x(t)\mathrm{d}t \int_{-\infty}^{\infty} x^*(t)\mathrm{d}t}{\int_{-\infty}^{\infty} |x(t)|^2\, \mathrm{d}t}$$

$$= \frac{1}{E_x} \int_{-\infty}^{\infty} x(t)\mathrm{d}t \int_{-\infty}^{\infty} x^*(t)\mathrm{d}t. \quad (4.107)$$

It may easily be deduced that the measure (4.107) is invariant to the signal shift, since any energy signal satisfies an equality

$$\int_{-\infty}^{\infty} x(t)\mathrm{d}t = \int_{-\infty}^{\infty} x(t-\theta)\mathrm{d}t.$$

Note that this conclusion does not hold true for the signal equivalent width, since, by (2.174), it depends on a signal shift regarding zero.

Example 4.23. Given a rectangular single pulse with the autocorrelation function (4.29). The autocorrelation width is determined by (4.107), to be

$$W_{xx} = \frac{1}{A^2\tau} \int_{-\tau/2}^{\tau/2} A\mathrm{d}t \int_{-\tau/2}^{\tau/2} A\mathrm{d}t = \tau. \tag{4.108}$$

□

Example 4.24. Given a Gaussian pulse $x(t) = A\mathrm{e}^{-(\alpha t)^2}$. Using an identity $\int_{-\infty}^{\infty} \mathrm{e}^{-px^2}\mathrm{d}x = \sqrt{\frac{\pi}{p}}$, its autocorrelation width is specified by (4.107), to be

$$W_{xx} = \frac{\int_{-\infty}^{\infty} \mathrm{e}^{-(\alpha t)^2}\mathrm{d}t \int_{-\infty}^{\infty} \mathrm{e}^{-(\alpha t)^2}\mathrm{d}t}{\int_{-\infty}^{\infty} \mathrm{e}^{-2(\alpha t)^2}\mathrm{d}t} = \frac{\sqrt{2\pi}}{\alpha}. \tag{4.109}$$

□

4.8.2 Measure by the Spectral Density

An alternative form of the autocorrelation width may be found in the frequency domain. It is enough to use the property (2.50), allowing for

$$X(j0) = \int_{-\infty}^{\infty} x(t)\mathrm{d}t \quad \text{and} \quad X^*(j0) = \int_{-\infty}^{\infty} x^*(t)\mathrm{d}t\,,$$

recall the Rayleigh theorem (2.62),

$$E_x = \frac{1}{\pi} \int_{0}^{\infty} |X(j\omega)|^2 \,\mathrm{d}\omega\,,$$

substitute these values to (4.107), and go to

$$W_{xx} = \frac{\pi X(0)X^*(0)}{\int_{0}^{\infty} |X(j\omega)|^2 \,\mathrm{d}\omega}. \tag{4.110}$$

We thus have one more option in calculating the autocorrelation width and, again, the choice depends on the signal waveform and its spectral density. Note that the class of functions for which (4.110) has a meaning is limited, since the spectral density is not allowed to be zero-valued at zero. Therefore, this measure is not appropriate for narrowband and high-pass signals.

Example 4.25. Given a rectangular single pulse of duration τ and amplitude A. Its spectral density is given by (2.80). Using an identity $\int_0^\infty \sin^2 ax/x^2 \mathrm{d}x = \pi a/2$, the autocorrelation width is calculated by (4.110), to be

$$W_{xx} = \frac{\pi X(0)X^*(0)}{\int\limits_0^\infty |X(j\omega)|^2 \,\mathrm{d}\omega} = \frac{\pi}{\int\limits_{-\infty}^\infty \frac{\sin^2 \omega\tau/2}{(\omega\tau/2)^2}\mathrm{d}\omega} = \tau .$$

As we see, this measure is identical to that (4.108) derived by the autocorrelation function.

□

Example 4.26. Given a Gaussian pulse $x(t) = A\mathrm{e}^{-(\alpha t)^2}$. By (4.110) and an identity $\int_{-\infty}^\infty \mathrm{e}^{-px^2}\mathrm{d}x = \sqrt{\pi/p}$, its autocorrelation width is evaluated by

$$W_{xx} = \frac{\int\limits_{-\infty}^\infty \mathrm{e}^{-(\alpha t)^2}\mathrm{d}t \int\limits_{-\infty}^\infty \mathrm{e}^{-(\alpha t)^2}\mathrm{d}t}{\int\limits_{-\infty}^\infty \mathrm{e}^{-2(\alpha t)^2}\mathrm{d}t} = \frac{\sqrt{2\pi}}{\alpha} \tag{4.111}$$

that is identical to (4.109).

□

4.8.3 Equivalent Width of ESD

The equivalent width of the ESD function is specified as a reciprocal of (4.110),

$$W_{XX} = \frac{\int\limits_0^\infty |X(j\omega)|^2 \,\mathrm{d}\omega}{\pi X(0)X^*(0)} . \tag{4.112}$$

The measure (4.112) needs the same care as for (4.110), since it just loses any meaning if the spectral density of a signal reaches zero at $\omega = 0$.

A comparison of (4.110) and (4.112) leads to the useful relation

$$W_{xx} = \frac{1}{W_{XX}} \tag{4.113}$$

that, in turn, satisfies the *uncertainty principle*

$$W_{xx}W_{XX} \geqslant 1. \tag{4.114}$$

We thus conclude that the widths W_{xx} and W_{XX} of the same signal cannot be independently specified; so the wider the autocorrelation function, the narrower the spectral density and vice versa.

4.9 Summary

We now know what the energy and power signals are and what correlation is. We are also aware of the most important characteristics of such signals. The following observations would certainly be useful in using this knowledge:

- Energy signals have finite energy and are characterized by the energy correlation function and ESD.
- Power signals have infinite energy and are characterized by the power correlation function and PSD.
- All finite energy signals vanish at infinity.
- If two signals are orthogonal, their joint spectral density is zero.
- The energy autocorrelation function is even and its value with zero shift is equal to the signal energy.
- The power autocorrelation function is even and its value with zero shift is equal to the signal average power.
- ESD and PSD functions are coupled with the energy and power correlation functions, respectively, by the Fourier transform.
- The wider ESD or PSD function, the narrower the relevant energy and power correlation function.
- An envelope of the ESD function of the LFM pulse is consistent with the pulse waveform.
- The autocorrelation function of the LFM pulse has a peak at zero shift and the side lobes at the level of about 21%.
- The side lobes of the autocorrelation function of the phase-coded signal (Barker code $N13$) are attenuated to the level of about 7.7% of the main lobe.
- The cross-correlation function is not symmetric and it is not even.
- The value of the cross-correlation function with zero shift may be either negative, positive, or zero.
- If the cross-correlation function with zero shift is zero, then the signals are orthogonal.
- The correlation width changes as a reciprocal of the spectral width.

4.10 Problems

4.1. Given two signals,

$$x(t) = at \quad \text{and} \quad y(t) = \begin{cases} bt, & -\tau/2 \leqslant t \leqslant \tau/2 \\ 0, & \text{otherwise} \end{cases}.$$

Determine, which signal is a power signal and which is an energy signal.

4.2 (Energy signals). Given the waveforms (Figs. 1.20 and 2.64). Using the Rayleigh theorem, determine the energy of the signals for your own number M.

4.3. Given the waveform (Figs. 1.20 and 2.64). Determine and plot its ESD function.

4.4. Using the cross ESD function of the waveform (Problem 4.3), determine its energy and compare the result with that achieved in Problem 4.2.

4.5. Given a sinc-shaped pulse $x(t) = A(\sin \alpha t)/\alpha t$. Determine its energy.

4.6. The ESD function is assumed to have a symmetric rectangular shape about zero. What can you say about the signal waveform?

4.7 (Power signals). Given a pulse-train of the waveform (Figs. 1.20 and 2.64) with period $T = (N+1)\tau$, where τ is the pulse duration and N is your own integer number. Determine the average power of this signal.

4.8. Solve Problem 4.7 using the coefficients of the Fourier series derived in Problem 2.13.

4.9 (Energy autocorrelation function). Given a symmetric triangular waveform of duration τ and amplitude A. Calculate its energy autocorrelation function and give a geometrical interpretation.

4.10. Given the waveform (Figs. 1.20 and 2.64). Calculate its energy autocorrelation function and give a geometrical interpretation.

4.11. Given the waveform (Figs. 1.20 and 2.64) filled with the carrier signal of a frequency f_0. Calculate the energy autocorrelation function of such an RF signal and give a geometrical interpretation.

4.12 (Power autocorrelation function). Determine the power autocorrelation function of the given signal:

1. $x(t) = 4\sin(\omega_0 t + \pi/8)$
2. $x(t) = 2\cos(\omega_0 t - \pi/3)$
3. $x(t) = 2\cos\omega_0 t + 3\sin(\omega_0 t + \pi/4)$
4. $x(t) = \sin(\omega_0 t - \pi/3) - 2\cos\omega_0 t$

4.13 (Power spectral density). Given a signal with known power autocorrelation function (Problem 4.12). Define the relevant PSD function.

4.14 (Energy cross-correlation function). Determine the energy cross-correlation function of the given rectangular pulse and truncated ramp waveform, respectively,

1. $x(t) = \begin{cases} A, & -\tau/2 \leqslant t \leqslant \tau/2 \\ 0, & \text{otherwise} \end{cases}$,
2. $y(t) = \begin{cases} At, & 0 \leqslant t \leqslant \tau \\ 0, & \text{otherwise} \end{cases}$.

4.15. Given a rectangular waveform (Problem 4.14.1). Determine the cross-autocorrelation function of this signal with the waveform given in Figs. 1.20 and 2.64.

4.16 (Energy cross-spectral density). Determine the cross ESD function of the signals given in Problem 4.14.

4.17. Using the derived energy cross-correlation function (Problem 4.15), determine the cross ESD function for the rectangular waveform and the waveform given in Figs. 1.20 and 2.64.

4.18 (Width). Using the autocorrelation function of the triangular waveform (Problem 4.9), calculate its autocorrelation width.

4.19. Determine the energy spectral density of the triangular waveform (Problem 4.9) and calculate its equivalent width.

4.20. Based on the solved Problems 4.18 and 4.19, examine the uncertainty principle (4.114).

5

Bandlimited Signals

5.1 Introduction

Our life teaches us that all real physical processes and thus their representatives, the modulating and modulated electronic signals, are bandlimited by nature; so, their spectral width W_{XX} cannot be infinite in practice. The frequency content of the simplest speech signals falls in the range of 300 Hz to 3 kHz, music contains spectral components from 20 Hz to 20 kHz, and television information is distributed over 0–5 MHz. When such signals modulate a carrier signal, which frequency ω_0 is typically much larger than the modulating signal spectral width, $W_{XX} \ll \omega_0$, the modulated signal becomes *narrowband*. Examples of *narrowband signals* may be found in a broad area of wireless applications, e.g., communications, radars, positioning, sensing, and remote control. Since energy of the narrowband signal is concentrated around ω_0, its major performance is the *envelope*, *phase*, or/and *instantaneous frequency*. The mathematical theory of this class of signals is developed both in presence and absence of noise and is based on applications of the Hilbert transform that is associated with the generalized complex model of a narrowband signal called the *analytic signal*.

An important property of bandlimited signals is that they are slowly changing in time. If such signals are presented by measured samples, then it is very often desired to interpolate the signals between the samples that is possible to do by using, for example, Lagrange or Newton[1] methods. On the other hand, any intention to provide digital processing of continuous signals requires *sampling*. With this aim, the *sampling theorem* is used allowing also an exact reverse reconstruction of signals. Finally, to establish a link between an *analog* part of any bandlimited system and its *digital* part, special devices are employed. An *analog-to-digital converter* (ADC) obtains a translation of a continuous-time signal to samples and then to a digital code. A *digital-to-analog converter* (DAC) solves an inverse problem: it first transforms a digital code to samples and then interpolates a signal between samples. We discuss all these problems in the following sections.

[1] Sir Isaac Newton, English physicist, mathematician, astronomer, philosopher, and alchemist, 25 December 1642–20 March 1727.

5.2 Signals with Bandlimited Spectrum

Describing signals, which bandwidth is limited, turns us back to the earlier learned material, where spectrums of signals were analyzed. To carry out our talk in a more or less generalized context, we consider below only the most common models of bandlimited signals, e.g., the ideal LP and band-pass signals and the narrowband signal.

5.2.1 Ideal Low-pass Signal

Let us assume that some signal $x(t)$ has a constant spectral density X_0 within the bandwidth $|\omega| \leqslant W$ and that the value of this density falls down to zero beyond W. Mathematically, we write a function

$$X(j\omega) = \begin{cases} X_0, & \text{if } |\omega| \leqslant W \\ 0, & \text{otherwise} \end{cases} \tag{5.1}$$

that, by the inverse Fourier transform, becomes a time signal

$$x(t) = \frac{X_0}{2\pi} \int_{-W}^{W} e^{j\omega t} d\omega = \frac{X_0 W}{\pi} \frac{\sin Wt}{Wt} . \tag{5.2}$$

Thanks to the rectangular shape of a spectral density (5.1), a signal (5.2) is said to be the *ideal LP signal.* Figures 5.1a and b illustrates (5.1) and (5.2), respectively. It follows that an ideal LP signal is noncausal, once it exists in the infinite time range from $-\infty$ to ∞. Noncausality follows straightforwardly from the duality property of the Fourier transforms, meaning that the signal is not physically realizable, and explains the term "ideal."

5.2.2 Ideal Band-pass Signal

Let us now use a signal $x(t)$ (5.2) to modulate a purely harmonic wave $z(t) = \cos \omega_0 t$ with a carrier frequency ω_0. The spectral density of this modulated signal $y(t) = x(t)z(t)$ is determined by

$$Y(j\omega) = \begin{cases} 0.5X_0, & \text{if } \omega_0 - W \leqslant \omega \leqslant \omega_0 + W \\ & \text{and } -\omega_0 - W \leqslant \omega \leqslant -\omega_0 + W, \\ 0, & \text{otherwise} \end{cases} \tag{5.3}$$

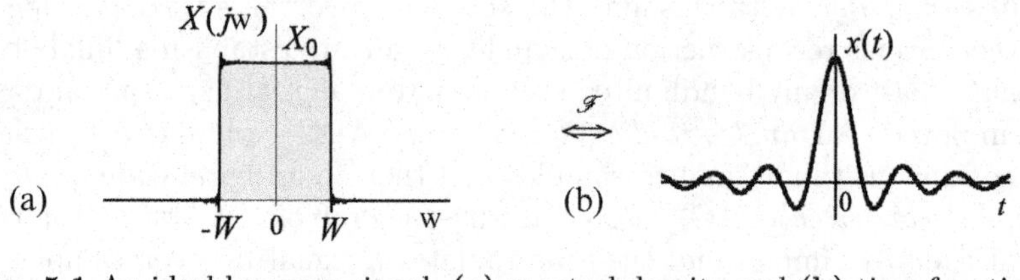

Fig. 5.1 An ideal low-pass signal: **(a)** spectral density and **(b)** time function.

and its inverse Fourier transform yields a real signal

$$y(t) = \mathrm{Re}\left(\frac{X_0}{2\pi}\int\limits_{\omega_0-W}^{\omega_0+W} \mathrm{e}^{j\omega t}\mathrm{d}\omega\right) = \frac{X_0 W}{\pi}\frac{\sin Wt}{Wt}\cos\omega_0 t\,. \tag{5.4}$$

Signal (5.4) has the same basic properties as (5.2). Therefore, it is called an *ideal band-pass signal.* Its spectral density and time presentation are given in Fig. 5.2. By the modulation process, an LP $x(t)$ is replaced at a carrier frequency ω_0 and is filled with the harmonic wave. Owing to this carrier, a signal $y(t)$ oscillates with the amplitude $x(t)$ that may take either positive or negative values. This inherent property of $x(t)$ imposes a certain inconvenience. To circumvent, the positive-valued envelope $|x(t)|$ is very often used. A concept of the *envelope* as well as the signal *phase* and *instantaneous frequency* is rigorously specified in the other model named *narrowband.*

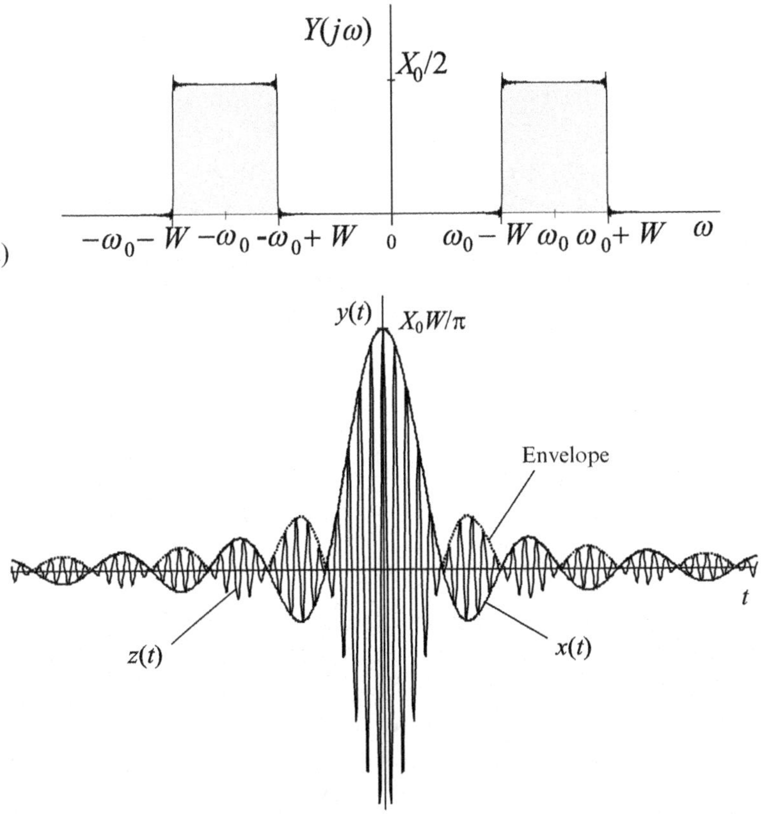

(a)

(b)

Fig. 5.2 An ideal passband signal: **(a)** double-sided spectral density and **(b)** time function.

5.2.3 Narrowband Signal

Most generally, allowing shift in the carrier signal for some time, the band-pass signal, like (5.4), may be performed as follows:

$$y(t) = A_c(t)\cos\omega_0 t - A_s(t)\sin\omega_0 t \tag{5.5}$$

$$= A(t)\cos[\omega_0 t + \psi(t)]\,, \tag{5.6}$$

where $A_c(t) = A(t)\cos\psi(t)$ is called the *in-phase* amplitude and $A_s(t) = A(t)\sin\psi(t)$ is said to be the *quadrature phase* amplitude. Both $A_c(t)$ and $A_s(t)$ may take either positive or negative values, $-\infty < A_c, A_s < \infty$, and it is claimed that they are slowly changing in time with the transforms localized in a narrow frequency band close to zero.

The real signal (5.5) with its generalized version (5.6) is called the *narrowband signal.* However, its more general complex presentation is also used. As in the Fourier series case, the Euler formula represents both harmonic functions in (5.5) by the exponential functions that performs (5.5) by

$$y(t) = \mathrm{Re}[\dot{A}(t)\ \mathrm{e}^{j\omega_0 t}], \tag{5.7}$$

where $\dot{A}(t)$ is defined by

$$\dot{A}(t) = A_c(t) + jA_s(t) = A(t)\mathrm{e}^{j\psi(t)}$$

and is called the *complex envelope.* In this relation, $A(t)$ has a meaning of a physical positive-valued *envelope* and $\psi(t)$ is an *informative phase.*

The spectral density of a signal (5.7) may be defined as follows:

$$Y(j\omega) = \int_{-\infty}^{\infty} \mathrm{Re}[\dot{A}(t)\mathrm{e}^{j\omega_0 t}]\mathrm{e}^{-j\omega t}\mathrm{d}t$$

$$= \frac{1}{2}\int_{-\infty}^{\infty} \dot{A}(t)\mathrm{e}^{-j(\omega-\omega_0)t}\mathrm{d}t + \frac{1}{2}\int_{-\infty}^{\infty} \dot{A}^*(t)\mathrm{e}^{-j(\omega+\omega_0)t}\mathrm{d}t$$

$$= \frac{1}{2}A(j\omega - j\omega_0) + \frac{1}{2}A^*(-j\omega - j\omega_0), \tag{5.8}$$

where $A(j\omega)$ is a spectral density of a complex envelope of a narrowband signal and $A^*(j\omega)$ is its complex conjugate. Relation (5.8) gives a very useful rule. In fact, the spectral density $Y(j\omega)$ of a narrowband signal determines the spectral density $A(j\omega)$ of its complex envelope and vice versa. The inverse Fourier transform of $A(j\omega)$ allows getting the quadrature amplitudes $A_c(t)$ and $A_s(t)$. The latter, in turn, specify the physical envelope, informative phase, and instantaneous frequency of a narrowband signal.

5.2.3.1 Envelope

The physical *envelope* $A(t)$ of a narrowband signal $y(t)$ is determined by the in-phase and quadrature phase amplitudes by

$$A(t) = \sqrt{A_c^2 + A_s^2}, \tag{5.9}$$

demonstrating three major properties for applications. First, it is a positive-valued function that may range from zero until infinity, $0 \leqslant A(t) \leqslant \infty$, but never be negative. Next, for different frequencies and phases, the envelope may take equal values as it is shown in Fig. 5.3a for two time instances when $A(t_1) = A(t_2)$. Finally, at any time instant, a signal value cannot exceed the value of its envelope, so that $|y(t)| \leqslant A(t)$.

5.2.3.2 Phase

The *total phase* $\Psi(t)$ of a signal (5.6) is given by

$$\Psi(t) = \omega_0 t + \psi(t), \tag{5.10}$$

where the informative slowly changing phase $\psi(t)$ ranges basically in infinite bounds, $|\psi(t)| \leqslant \infty$ (Fig. 5.3b). Such a behavior is associated, e.g., with a Brownian[2] motion in $\psi(t)$ produced by the white frequency noise in electronic systems. In other applications, $\psi(t)$ is calculated by

$$\psi(t) = \arctan \frac{A_s}{A_c} \tag{5.11}$$

to lie in the range from $-\pi/2$ to $\pi/2$. The most common imagination about a behavior of this phase in electronic systems is associated with its 2π physical periodicity induced by harmonic functions. In the latter case, $\psi(t)$ is said to be the modulo 2π phase being specified by

$$\psi(t) = \begin{cases} \arctan(A_s/A_c), & A_c \geqslant 0 \\ \arctan(A_s/A_c) \pm \pi, & A_c < 0, \quad \begin{cases} A_s \geqslant 0 \\ A_s < 0 \end{cases} \end{cases} \tag{5.12}$$

to range from $-\pi$ to π. Figure 5.3b demonstrates an example of the phase $\psi(t)$ and its mod 2π version.

5.2.3.3 Instantaneous Frequency

The instantaneous frequency $\omega_y(t)$ of a narrowband signal $y(t)$ is defined by the first time derivative of its total phase $\Psi(t)$ (5.10). The derivative is

[2] Ernest William Brown, English-born American scientist, 29 November 1866–23 July 1938.

invariant to the phase modulo (except for the points where the modulo phase jumps). Therefore, (5.11) defines the instantaneous frequency in an exhaustive way by

$$\omega_y(t) = \frac{\mathrm{d}}{\mathrm{d}t}\Psi(t) = \omega_0 + \frac{\mathrm{d}}{\mathrm{d}t}\psi(t) = \omega_0 + \frac{\mathrm{d}}{\mathrm{d}t}\arctan\frac{A_s}{A_c}$$

$$= \omega_0 + \frac{A_s' A_c - A_c' A_s}{A_c^2 + A_s^2}, \quad (5.13)$$

where $A_c' = \mathrm{d}A_c/\mathrm{d}t$ and $A_s' = \mathrm{d}A_s/dt$. An example of the instantaneous frequency corresponding to the phase (Fig. 5.3b) is shown in Fig. 5.3c.

One may deduce from the above definitions that $A(t)$, $\psi(t)$, and $\omega_y(t)$ are specified by the LP amplitudes, $A_c(t)$ and $A_s(t)$, and thus they are also LP functions. It may also be observed that $A_c(t)$, $A_s(t)$, $A(t)$, $\psi(t)$, and $\omega_y(t)$ are informative, whereas $\cos\omega_0 t$ and $\sin\omega_0 t$ in (5.5) are just two quadrature auxiliary harmonic functions intended at removing information to the carrier frequency.

Example 5.1. Given a narrowband signal $y(t)$, which spectral density is asymmetric about the frequency ω_1. In the positive frequency domain, the spectral density is described by (Fig. 5.4a)

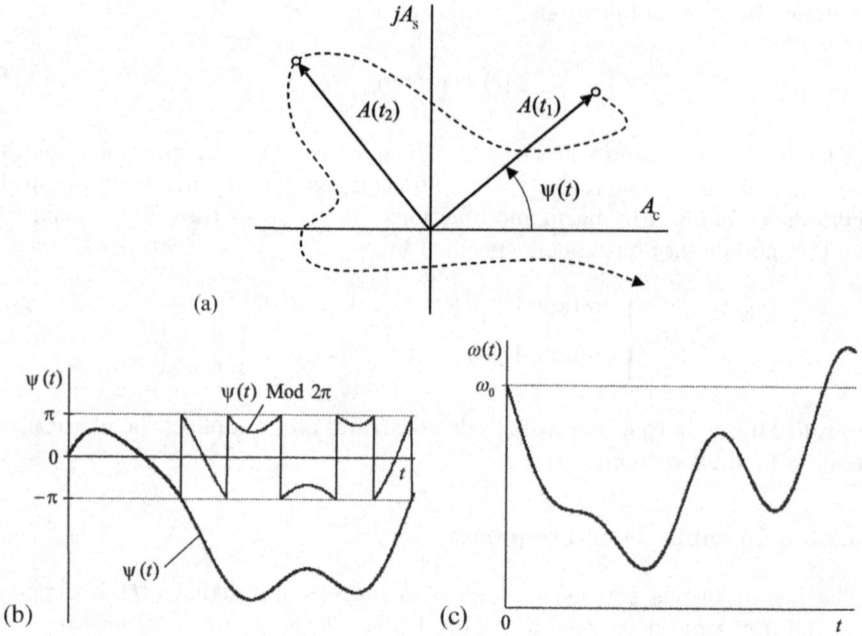

Fig. 5.3 Representation of a narrowband signal: **(a)** complex vector, **(b)** phase, and **(c)** instantaneous frequency.

$$Y(j\omega) = \begin{cases} \frac{1}{2}Y_0 e^{-a(\omega-\omega_1)}, & \text{if } \omega \geqslant \omega_1 \\ 0, & \text{otherwise} \end{cases}, \quad a > 0.$$

By (5.8), the spectral density of the complex envelope of this signal is provided to be (Fig. 5.4b)

$$A(j\omega) = \begin{cases} Y_0 e^{-a\omega}, & \text{if } \omega \geqslant 0 \\ 0, & \text{otherwise}. \end{cases}$$

The inverse Fourier transform applied to $A(j\omega)$ produces the complex envelope,

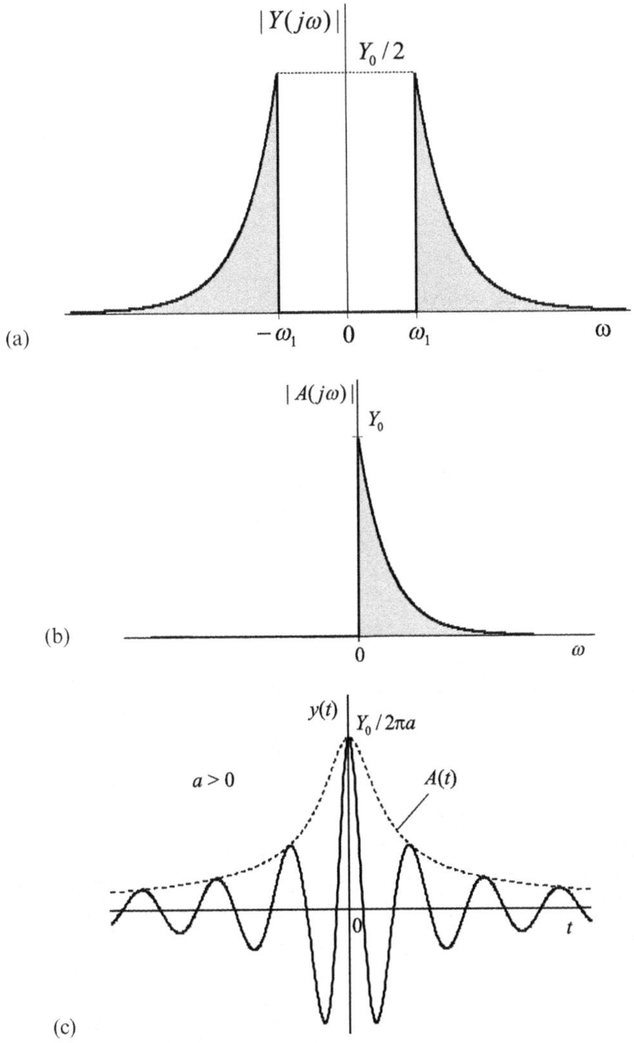

Fig. 5.4 A real signal with an asymmetric spectral density: **(a)** double-side spectral density; **(b)** spectral density of the complex envelope; and **(c)** time presentation.

$$\dot{A}(t) = \frac{Y_0}{2\pi}\int\limits_0^{\infty} e^{(-a+jt)\omega} d\omega = \frac{Y_0}{2\pi(a-jt)}$$

$$= \frac{aY_0}{2\pi(a^2+t^2)} + j\frac{tY_0}{2\pi(a^2+t^2)} = A_c(t) + jA_s(t).$$

By (5.9), the physical envelope of this signal is performed as

$$A(t) = \frac{Y_0}{2\pi\sqrt{a^2+t^2}}.$$

The informative phase is calculated, by (5.11), to be

$$\psi(t) = \arctan\frac{t}{a},$$

and the instantaneous frequency is provided, by (5.13), to possess the form of

$$\omega_y(t) = \omega_1 + \frac{d}{dt}\psi(t) = \omega_1 + \frac{a}{a^2+t^2}$$

Finally, the real signal $y(t)$ becomes

$$y(t) = \frac{Y_0}{2\pi\sqrt{a^2+t^2}}\cos\left(\omega_1 t + \arctan\frac{t}{a}\right).$$

It follows, that the signal is an RF pulse symmetric about zero (Fig. 5.4c). Its maximum $Y_0/2\pi a$ is placed at zero and its envelope approaches zero asymptotically with $|t| \to \infty$. The instantaneous frequency has the value of $\omega_y(t=0) = \omega_1 + 1/a$ at zero time point. It then tends toward ω_1 when $|t| \to \infty$.

□

5.2.3.4 Classical Application of the Narrowband Signal

An application of the narrowband model (5.5) in the coherent receiver, as it is shown in Fig. 5.5, is already classical. Here, an impulse signal with a carrier signal $\cos\omega_0 t$ is transmitted toward some object and its reflected version (delayed and attenuated) is performed by a narrowband model (5.5). In this model, the amplitudes $A_c(t)$ and $A_s(t)$ bear information about the object and therefore need to be detected. At the receiver, the signal is multiplied with $\cos\omega_0 t$ in the in-phase channel and with $\sin\omega_0 t$ in the quadrature one yielding two signals, respectively,

$$y_I(t) = \frac{1}{2}[A_c(t) + A_c\cos 2\omega_0 t - A_s \sin 2\omega_0 t],$$

$$y_Q(t) = -\frac{1}{2}[A_s(t) - A_c\sin 2\omega_0 t - A_c\cos 2\omega_0 t].$$

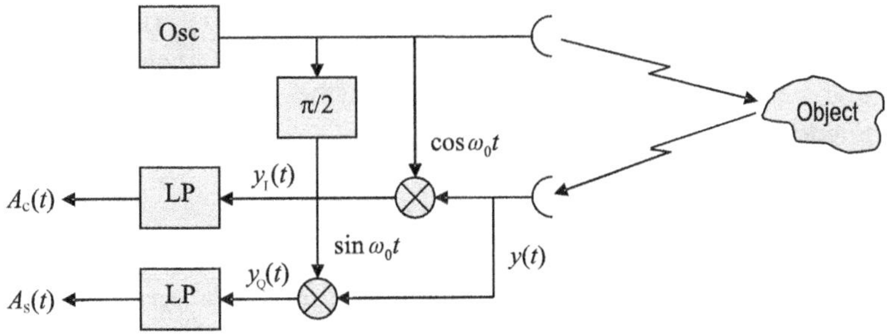

Fig. 5.5 Application of the narrowband signal in a wireless detection of an object.

The amplitudes $A_c(t)$ and $A_s(t)$ are then filtered with the LP filters having gain factors 2 and the envelope, phase, and/or instantaneous frequency are calculated by (5.9)–(5.13).

Observing the narrowband model in more detail, we arrive at an important conclusion. Although $A(t)$, $\psi(t)$, and $\omega_y(t)$ describe the narrowband signal in an exhaustive manner, its application in the theory of signals and systems has the same inconvenience as an application of the Fourier transform has with an orthogonal set of harmonic (cosine and sine) functions. More generally, a real narrowband model (5.5) needs to be accompanied with its complex conjugate version that is provided by the Hilbert transforms. The model then becomes complex promising all benefits of a generalized analysis.

5.3 Hilbert Transform

An important rule introduced by Hilbert and therefore called the *Hilbert transform* allows finding an imaginary version $\hat{y}(t)$ of a signal $y(t)$ and thereby define the *analytic signal.* An application of the Hilbert transform in communication systems was first showed and developed in 1946 by Gabor in his theory of communication. One of its results was what Gabor called the "complex signal" that is now widely cited as the analytic signal.

5.3.1 Concept of the Analytic Signal

To comprehend the essence of the Hilbert transform, it is useful first to assume an arbitrary narrowband signal $y(t)$, which transform $Y(j\omega)$ is known. The inverse Fourier transform applied to $Y(j\omega)$ gives

$$\begin{aligned} y(t) &= \frac{1}{2\pi}\int_{-\infty}^{\infty} Y(j\omega)e^{j\omega t}d\omega \\ &= \frac{1}{2\pi}\int_{-\infty}^{0} Y(j\omega)e^{j\omega t}d\omega + \frac{1}{2\pi}\int_{0}^{\infty} Y(j\omega)e^{j\omega t}d\omega. \end{aligned} \tag{5.14}$$

The one-sided spectral density of a real signal $y(t)$, that is the doubled second term in (5.14), is said to be related to the *analytic signal* $y_a(t)$ associated with $y(t)$. An instantaneous value of the analytic signal is therefore determined by

$$y_a(t) = \frac{1}{\pi} \int_0^\infty Y(j\omega) e^{j\omega t} d\omega. \tag{5.15}$$

Changing the sign of a variable and interchanging the integration bounds in the first term of (5.14) produces a complex conjugate of the analytic signal

$$y_a^*(t) = \frac{1}{\pi} \int_0^\infty Y(-j\omega) e^{-j\omega t} d\omega. \tag{5.16}$$

A real signal $y(t)$ may now be represented as follows

$$y(t) = \frac{1}{2} \left[y_a(t) + y_a^*(t) \right]. \tag{5.17}$$

It is seen that the real component of the analytic signal is equal to $y(t)$ and that its imaginary part is a complex conjugate of $y(t)$, respectively,

$$y(t) = \mathrm{Re}\, y_a(t), \tag{5.18}$$

$$\hat{y}(t) = \mathrm{Im}\, y_a(t). \tag{5.19}$$

An analytic signal may therefore be performed as a complex vector

$$y_a(t) = y(t) + j\hat{y}(t) \tag{5.20}$$

that plays a fundamental role in the theory of narrowband signals. This vector is sometimes called the *preenvelope* of a real signal and demonstrates two important properties:

- It is a complex signal created by taking an arbitrary real signal and then adding in quadrature its complex conjugate.
- Its spectral density exists only in the positive frequency domain.

Example 5.2. An ideal LP signal is given with the rectangular spectral density (5.1). An analytic signal associated with (5.1) is calculated by (5.15) to be

$$x_a(t) = \frac{X_0}{\pi} \int_0^W e^{j\omega t} d\omega = \frac{X_0}{j\pi t} (e^{j\omega_0 t} - 1)$$

that gives a real and imaginary part, respectively,

$$x(t) = \frac{X_0 W}{\pi} \frac{\sin Wt}{Wt},$$

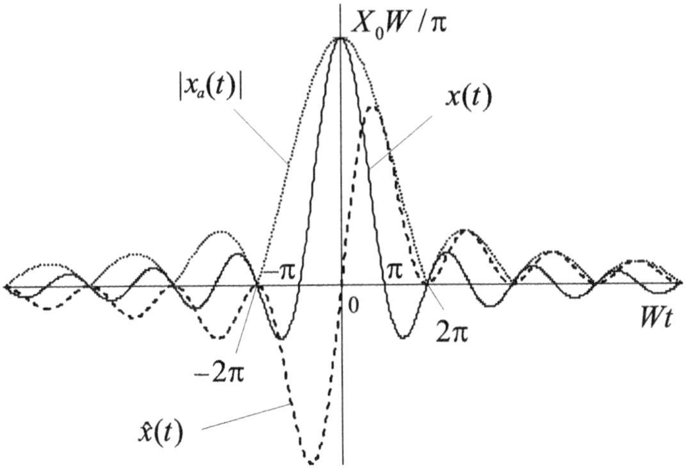

Fig. 5.6 Ideal analytic low-pass signal: $|x_a(t)|$ is the envelope, $x(t)$ and $\hat{x}(t)$ are real and imaginary components, respectively.

$$\hat{x}(t) = \frac{X_0 W}{\pi} \frac{\sin^2 Wt/2}{Wt/2} .$$

Figure 5.6 illustrates the components of this analytic signal along with its envelope $|x_a(t)|$.

□

Example 5.3. The one-sided spectral density of a LP signal is given by

$$X_{\mathrm{a}}(j\omega) = \begin{cases} X_0 \mathrm{e}^{-a\omega}, & \text{if} \omega \geqslant 0 \\ 0, & \text{if} \omega < 0 \end{cases} .$$

The analytic signal, its real part, and its imaginary part are calculated, respectively, by

$$x_{\mathrm{a}}(t) = \frac{X_0}{\pi} \int_0^{\infty} \mathrm{e}^{(-a+jt)w} \mathrm{d}\omega = \frac{X_0}{\pi(a-jt)},$$

$$x(t) = \frac{aX_0}{\pi(a^2+t^2)},$$

$$\hat{x}(t) = \frac{tX_0}{\pi(a^2+t^2)}.$$

The real asymmetric one-sided spectral density of this signal is shown in Fig. 5.7a. Figure 5.7b exhibits the real and imaginary parts of this analytic signal along with its envelope.

□

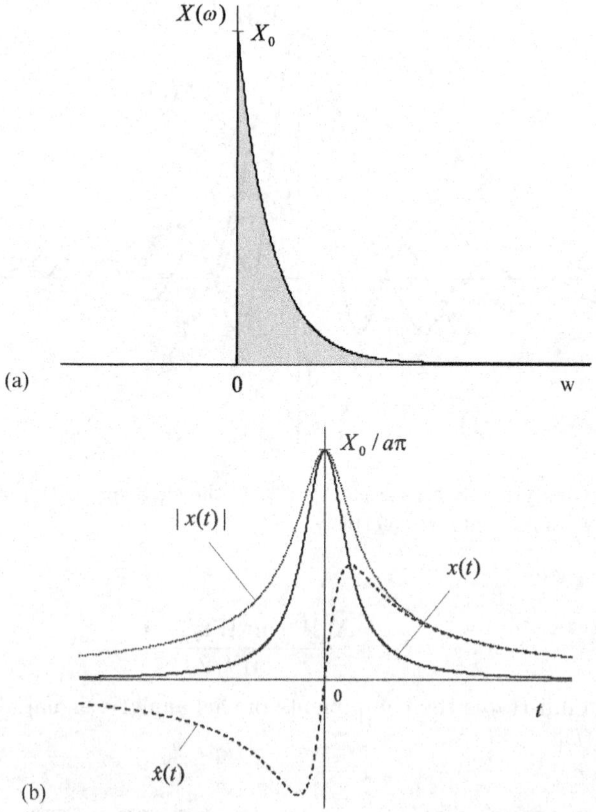

Fig. 5.7 An analytic signal: **(a)** one-sided spectral density and **(b)** envelope $|x_a(t)|$, real part $x(t)$, and imaginary part $\hat{x}(t)$.

5.3.2 Hilbert Transform

Let us now examine the spectral density of the analytic signal $y_{\mathrm{a}}(t)$,

$$Y_{\mathrm{a}}(j\omega) = \int_{-\infty}^{\infty} y_{\mathrm{a}}(t)\, \mathrm{e}^{-j\omega t} \mathrm{d}t. \tag{5.21}$$

The second of the above-listed properties claims that (5.21) exists only for positive frequencies and thus (5.21) may be rewritten as

$$Y_{\mathrm{a}}(j\omega) = \begin{cases} 2Y(j\omega), & \text{if } \omega \geqslant 0 \\ 0, & \text{if } \omega < 0 \end{cases}, \tag{5.22}$$

where $Y(j\omega)$ corresponds to the relevant narrowband signal $y(t)$. On the other hand, if to define the spectral density of a conjugate signal $\hat{y}(t)$ by $\hat{Y}(j\omega)$, then the following obvious relation

$$Y_{\mathrm{a}}(j\omega) = Y(j\omega) + j\hat{Y}(j\omega) \tag{5.23}$$

will claim that (5.22) holds true when the spectral components in (5.23) are coupled in the following manner:

$$\hat{Y}(j\omega) = -j\,\mathrm{sgn}(\omega)Y(j\omega) = \begin{cases} -jY(j\omega), & \text{if } \omega \geqslant 0 \\ jY(j\omega), & \text{if } \omega < 0 \end{cases}, \tag{5.24}$$

where $\mathrm{sgn}(t)$ is a signum function,

$$\mathrm{sgn}(t) = \begin{cases} 1, & \text{if } t > 0 \\ 0, & \text{if } t = 0 \\ -1, & \text{if } t < 0 \end{cases}. \tag{5.25}$$

A resume follows immediately: the spectral density $\hat{Y}(j\omega)$ of the conjugate signal $\hat{y}(t)$ is a product of the multiplication of the spectral density $Y(j\omega)$ of a real signal $y(t)$ and an auxiliary function $Q(j\omega) = -j\,\mathrm{sgn}(\omega)$. This means, by the properties of the Fourier transforms, that the conjugate signal $\hat{y}(t)$ is defined by the convolution of $y(t)$ and the inverse Fourier transform $q(t)$ of $Q(j\omega)$. The function $q(t)$ is therefore unique and its instantaneous value is determined, using the integral

$$\int_0^\infty \sin bx\,\mathrm{d}x = \lim_{\alpha\to 1}\int_0^\infty x^{\alpha-1}\sin bx\,\mathrm{d}x = \frac{\Gamma(\alpha)}{b^\alpha}\sin\frac{\alpha\pi}{2}\bigg|_{\alpha=1} = \frac{1}{b},$$

to be

$$\begin{aligned} q(t) &= \frac{1}{2\pi}\int_{-\infty}^{\infty}(-j\,\mathrm{sgn})e^{j\omega t}\mathrm{d}\omega = \frac{j}{2\pi}\int_{-\infty}^{0}e^{j\omega t}\mathrm{d}\omega - \frac{j}{2\pi}\int_0^\infty e^{j\omega t}\mathrm{d}\omega \\ &= \frac{j}{2\pi}\int_0^\infty\left(e^{-j\omega t} - e^{j\omega t}\right)\mathrm{d}\omega = \frac{1}{\pi}\int_0^\infty \sin\omega t\mathrm{d}\omega = \frac{1}{\pi t}. \end{aligned} \tag{5.26}$$

The conjugate signal is thus defined by

$$\hat{y}(t) = y(t) * \frac{1}{\pi t} = \frac{1}{\pi}\int_{-\infty}^{\infty}\frac{y(\theta)}{t-\theta}\mathrm{d}\theta. \tag{5.27}$$

Relation (5.24) easily produces $Y(j\omega) = j\,\mathrm{sgn}(\omega)\hat{Y}(j\omega)$ to mean that the inverse calculus differs from (5.27) only by the sign changed, therefore

$$y(t) = -\hat{y}(t) * \frac{1}{\pi t} = \frac{1}{\pi}\int_{-\infty}^{\infty}\frac{\hat{y}(\theta)}{\theta - t}\mathrm{d}\theta. \tag{5.28}$$

Both (5.27) and (5.28) are known as the *direct Hilbert transform* $\hat{y}(t) = \mathcal{H}y(t)$ and the *inverse Hilbert transform* $y(t) = \mathcal{H}^{-1}\hat{y}(t)$, respectively.

Example 5.4. Given a sinc signal

$$x(t) = A\frac{\sin Wt}{Wt}.$$

By the integral identity

$$\int_{-\infty}^{\infty} \frac{\sin x}{x(x-y)}\mathrm{d}x = \frac{\pi}{y}(\cos y - 1),$$

its Hilbert transform is performed in the straightforward way as

$$\hat{x}(t) = \frac{A}{W\pi}\int_{-\infty}^{\infty} \frac{\sin W\theta}{\theta(t-\theta)}\mathrm{d}\theta = \frac{A}{Wt}(1-\cos Wt) = A\frac{\sin^2 Wt/2}{Wt/2}.$$

Inherently, by $A = X_0 W/\pi$, this result becomes that obtained in Example 5.2 for the signal with a rectangular one-side spectral density.

□

The trouble with the Hilbert transform may arise in calculating (5.27) and (5.28), once integrals do not exist at $t = \theta$ in the Cauchy sense. If it occurs, one needs to investigate the limits, for the integrand $f(t) = y(t)/(t-\theta)$,

$$\lim_{\epsilon \to 0} \int_{-\infty}^{\theta-\epsilon} f(t)\mathrm{d}t \quad \text{and} \quad \lim_{\epsilon \to 0} \int_{\theta+\epsilon}^{\infty} f(t)\mathrm{d}t.$$

If both still exist, then the direct Hilbert transform (5.27) may be calculated by the improper integral

$$\hat{y}(t) = \frac{1}{\pi}\lim_{\epsilon \to 0}\left[\int_{-\infty}^{\theta-\epsilon} \frac{y(\theta)}{t-\theta}\mathrm{d}\theta + \int_{\theta+\epsilon}^{\infty} \frac{y(\theta)}{t-\theta}\mathrm{d}\theta\right] \tag{5.29}$$

and so may the inverse Hilbert transform (5.28).

Example 5.5. Given a rectangular impulse signal defined by

$$x(t) = \begin{cases} A, & \text{if } \tau/2 \leqslant t \leqslant \tau/2 \\ 0, & \text{otherwise} \end{cases}.$$

Its Hilbert transform is calculated, using (5.29), by

$$\hat{x}(t) = \frac{A}{\pi}\lim_{\epsilon \to 0}\left(\int_{-\tau/2}^{-\epsilon} \frac{\mathrm{d}\theta}{t-\theta} + \int_{\epsilon}^{\tau/2} \frac{\mathrm{d}\theta}{t-\theta}\right).$$

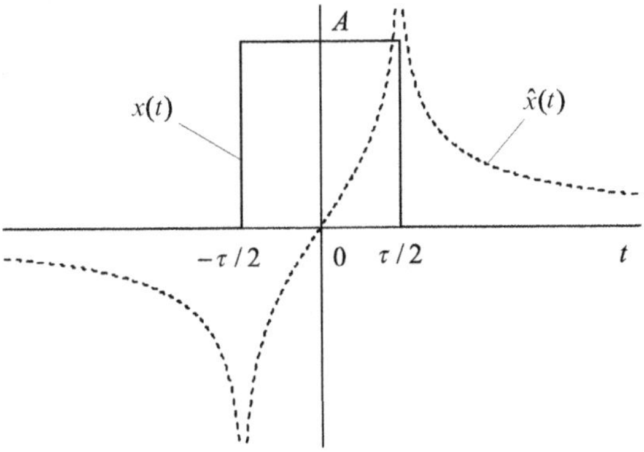

Fig. 5.8 A real rectangular pulse $x(t)$ and its Hilbert transform $\hat{x}(t)$.

By changing a variable to $\eta = t - \theta$, we go to

$$\hat{x}(t) = -\frac{A}{\pi} \lim_{\epsilon \to 0} \left(\int_{t+\tau/2}^{t+\epsilon} \frac{\mathrm{d}\eta}{\eta} + \int_{t-\epsilon}^{t-\tau/2} \frac{\mathrm{d}\eta}{\eta} \right) = \frac{A}{\pi} \ln \left| \frac{t+\tau/2}{t-\tau/2} \right| .$$

Figure 5.8 illustrates this pulse and its Hilbert transform.

□

5.3.3 Properties of the Hilbert transform

Like any other generalized technique, the Hilbert transform demonstrates many interesting properties that may be useful to reduce the mathematical burden and to arrive at the final result with minimum manipulations. The most widely used properties are listed below.

5.3.3.1 Filtering

The transform is equivalent to the filter called *Hilbert transformer*, which does not change the amplitudes of the spectral components, but alters their phases by $\pi/2$ for all negative frequencies and by $-\pi/2$ for all positive frequencies. This property states that

- The Hilbert transform of a constant is zero
- The Hilbert transform of a real function is a real function
- The Hilbert transform of an even function is an odd function and vice versa
- The Hilbert transform of a cosine function is a sine function, namely

$$\mathcal{H} \cos \omega_0 t = \sin \omega_0 t$$

Example 5.6. A rectangular pulse-train with the period-to-pulse duration ratio, $q = 2$, is approximated with the Fourier series

$$x(t) = \sum_{n=0}^{N-1} \varepsilon_n \frac{\sin(5n/\pi)}{5n/\pi} \cos nt\,,$$

where N is integer, $\varepsilon_0 = 1$, and $\varepsilon_{n>0} = 2$. Without the integral transformation, its Hilbert transform may be written directly by substituting $\cos nt$ with $\sin nt$ in the series,

$$\hat{x}(t) = \sum_{n=0}^{N-1} \varepsilon_n \frac{\sin(5n/\pi)}{5n/\pi} \sin nt\,.$$

The approximating function $x(t)$ and its Hilbert transform $\hat{x}(t)$ are shown in Fig. 5.9 for $N = 4$ and $N = 21$, respectively. It follows that, by large N, both approximations approach rigorous functions shown in Fig. 5.8. It is also seen that (1) $\hat{x}(t)$ is zero at $t = 0$, whereas $x(t)$ is constant here (has zero time derivative); (2) both $x(t)$ and $\hat{x}(t)$ are real functions; and (3) $x(t)$ is even, whereas $\hat{x}(t)$ is odd.

□

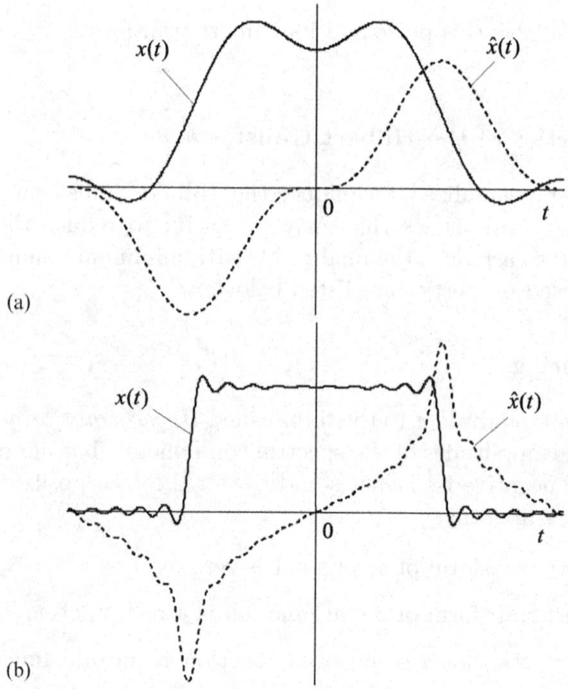

Fig. 5.9 Approximation of the rectangular pulse-train $x(t)$ and its Hilbert transform $\hat{x}(t)$ by the Fourier series: (**a**) $N = 4$ and (**b**) $N = 21$.

5.3.3.2 Causality

For a causal signal $y(t)$, the imaginary part of its Fourier transform is completely determined by a knowledge about its real part and vice versa. If the causal function $y(t)$ contains no singularities at the origin, then $\mathcal{F}y(t) = Y(j\omega) = Y_{\mathrm{r}}(\omega) + jY_{\mathrm{i}}(\omega)$ is its Fourier transform, in which $Y_{\mathrm{r}}(\omega)$ and $Y_{\mathrm{i}}(\omega)$ are coupled by the Hilbert transforms:

$$Y_{\mathrm{i}}(\omega) = \frac{1}{\pi}\int_{-\infty}^{\infty} \frac{Y_{\mathrm{r}}(\nu)}{\omega - \nu}\mathrm{d}\nu\,, \tag{5.30}$$

$$Y_{\mathrm{r}}(\omega) = -\frac{1}{\pi}\int_{-\infty}^{\infty} \frac{Y_{\mathrm{i}}(\nu)}{\omega - \nu}\mathrm{d}\nu\,, \tag{5.31}$$

Example 5.7. Consider a causal rectangular pulse

$$x(t) = \begin{cases} A, & \text{if } 0 \leqslant t \leqslant \tau \\ 0, & \text{otherwise} \end{cases} \tag{5.32}$$

Its Fourier transform is

$$\begin{aligned} X(j\omega) &= A\tau\left(\frac{\sin\omega\tau/2}{\omega\tau/2}\right)\mathrm{e}^{-j\omega\tau/2} = A\tau\left(\frac{\sin\omega\tau}{\omega\tau}\right) - jA\tau\left(\frac{\sin^2\omega\tau/2}{\omega\tau/2}\right) \\ &= X_{\mathrm{r}}(\omega) + jX_{\mathrm{i}}(\omega)\,. \end{aligned} \tag{5.33}$$

By applying (5.30) to $X_{\mathrm{r}}(\omega) = A\tau\,(\sin\omega\tau/\omega\tau)$, an imaginary part of the spectral density becomes $X_{\mathrm{i}}(\omega) = -A\tau\left(\sin^2\omega\tau/2/\omega\tau/2\right)$ that is identical to that in (5.33). Inversely, applying (5.31) to the imaginary part of (5.33) leads to its real component.

□

5.3.3.3 Linearity

If $y(t)$ is a linear combination of the weighted narrowband functions $\mathrm{a}y_1(t)$ and $\mathrm{b}y_2(t)$, where a and b are constants, then its Hilbert transform is

$$\begin{aligned} &\mathcal{H}[\mathrm{a}y_1(t) + \mathrm{b}y_2(t)] \\ &= \frac{1}{\pi}\int_{-\infty}^{\infty} \frac{\mathrm{a}y_1(\theta) + \mathrm{b}y_2(\theta)}{t-\theta}\mathrm{d}\theta = \frac{\mathrm{a}}{\pi}\int_{-\infty}^{\infty} \frac{y_1(\theta)}{t-\theta}\mathrm{d}\theta + \frac{\mathrm{b}}{\pi}\int_{-\infty}^{\infty} \frac{y_2(\theta)}{t-\theta}\mathrm{d}\theta \\ &= \mathrm{a}\mathcal{H}y_1(t) + \mathrm{b}\mathcal{H}y_2(t) = \mathrm{a}\hat{y}_1(t) + \mathrm{b}\hat{y}_2(t)\,. \end{aligned} \tag{5.34}$$

Example 5.8. The property of linearity was used in Example 5.6, where the Hilbert transform of the Fourier series is calculated as the series of the Hilbert transforms of the terms.

□

5.3.3.4 Time shifting

For all real a, the following time-shift theorem gives

$$\begin{aligned}
&\mathcal{H}y(t-\mathrm{a}) \\
&= \frac{1}{\pi}\int_{-\infty}^{\infty} \frac{y(\theta-\mathrm{a})}{t-\theta}\mathrm{d}\theta = \frac{1}{\pi}\int_{-\infty}^{\infty} \frac{y(\theta-\mathrm{a})}{(t-\mathrm{a})-(\theta-\mathrm{a})}\mathrm{d}(\theta-\mathrm{a}) \\
&= \frac{1}{\pi}\int_{-\infty}^{\infty} \frac{y(\eta)}{(t-\mathrm{a})-\eta}\mathrm{d}\eta \\
&= \hat{y}(t-\mathrm{a})
\end{aligned}$$

Example 5.9. A real carrier signal is shifted in time on τ to be $y(t) = A\cos\omega_0(t-\tau)$. By the filtering and time-shifting properties, its Hilbert transform easily becomes $\hat{y}(t) = A\sin\omega_0(t-\tau)$.

□

5.3.3.5 Time Scaling

For all real a > 0, the following mappings holds true:

$$\begin{aligned}
&\mathcal{H}y(\mathrm{a}t) \\
&= \frac{1}{\pi}\int_{-\infty}^{\infty} \frac{y(\mathrm{a}\theta)}{t-\theta}\mathrm{d}\theta = \frac{1}{\pi}\int_{-\infty}^{\infty} \frac{y(\mathrm{a}\theta)}{\mathrm{a}t-\mathrm{a}\theta}\mathrm{d}(a\theta) = \frac{1}{\pi}\int_{-\infty}^{\infty} \frac{y(\eta)}{\mathrm{a}t-\eta}\mathrm{d}(\eta) \\
&= \hat{y}(\mathrm{a}t) \\
&\mathcal{H}y(-\mathrm{a}t) \\
&= \frac{1}{\pi}\int_{-\infty}^{\infty} \frac{y(-\mathrm{a}\theta)}{t-\theta}\mathrm{d}\theta = -\frac{1}{\pi}\int_{-\infty}^{\infty} \frac{y(-\mathrm{a}\theta)}{-\mathrm{a}t-(-a\theta)}\mathrm{d}(-\mathrm{a}\theta) \\
&= -\frac{1}{\pi}\int_{-\infty}^{\infty} \frac{y(\eta)}{-\mathrm{a}t-\eta}\mathrm{d}(\eta) \\
&= -\hat{y}(-\mathrm{a}t)
\end{aligned}$$

Example 5.10. A real harmonic signal is scaled in time by ω_0 to be $A\cos\omega_0 t$. By the filtering and scaling properties, its Hilbert transform becomes $A\sin\omega_0 t$. The same signal is scaled by $-\omega_0$ to be $A\cos(-\omega_0 t) = A\cos\omega_0 t$. By the scaling property, its Hilbert transform becomes $-A\sin(-\omega_0 t) = A\sin\omega_0 t$ that is also stated by the filtering property.

□

5.3.3.6 Multiple Transform

If to denote the multiple Hilbert transform of $y(t)$ by $\mathcal{H}^n y(t) = \underbrace{\mathcal{H}...\mathcal{H}}_{n} y(t)$, then the following identities are valid:

- $\mathcal{H}^2 y(t) = -y(t)$
- $\mathcal{H}^3 y(t) = \mathcal{H}^{-1} y(t)$
- $\mathcal{H}^4 y(t) = y(t)$
- $\mathcal{H}^n y(t) \overset{\mathcal{F}}{\Leftrightarrow} [-j\,\mathrm{sgn}(\omega)]^n Y(j\omega)$

Example 5.11. The multiple transform property establishes the following link between the cosine signal and its multiple Hilbert transform

$$\cos\omega_0 t \overset{\mathcal{H}}{\to} \sin\omega_0 t \overset{\mathcal{H}}{\to} -\cos\omega_0 t \overset{\mathcal{H}}{\to} -\sin\omega_0 t \overset{\mathcal{H}}{\to} \cos\omega_0 t$$

□

5.3.3.7 Hilbert Transform of the Time Derivatives

The Hilbert transform of the time derivative $y'(t)$ of $y(t)$,

$$\mathcal{H}y'(t) = \frac{1}{\pi}\int_{-\infty}^{\infty} \frac{y'(\theta)}{t-\theta}\mathrm{d}\theta$$

is equivalent to the time derivative of the Hilbert transform of $y(t)$. To prove, it is enough to use a new variable $\theta = t - \eta$ and, by simple manipulations, go to

$$\mathcal{H}y'(t) = \frac{1}{\pi}\int_{-\infty}^{\infty} \frac{y'(t-\eta)}{\eta}\mathrm{d}\eta = \frac{\mathrm{d}}{\mathrm{d}t}\left(\frac{1}{\pi}\int_{-\infty}^{\infty} \frac{y(t-\eta)}{\eta}\mathrm{d}\eta\right)$$
$$= \frac{\mathrm{d}}{\mathrm{d}t}\left(\frac{1}{\pi}\int_{-\infty}^{\infty} \frac{y(\theta)}{t-\theta}\mathrm{d}\theta\right) = \frac{\mathrm{d}}{\mathrm{d}t}\mathcal{H}y(t) = \hat{y}'(t)\,. \tag{5.35}$$

The Hilbert transform of (5.35) then produces

$$y'(t) = -\mathcal{H}\frac{\mathrm{d}}{\mathrm{d}t}\mathcal{H}y(t) = -\mathcal{H}\hat{y}'(t)\,. \tag{5.36}$$

Example 5.12. Given a delta-shaped signal $y(t) = \delta(t)$. Its Hilbert transform, by the sifting property of a delta function, becomes

$$\mathcal{H}\delta(t) = \frac{1}{\pi}\int_{-\infty}^{\infty}\frac{\delta(\theta)}{t-\theta}\mathrm{d}\theta = \frac{1}{\pi t}\,. \tag{5.37}$$

By (5.36) and (5.37), the time derivative of the delta function $\delta(t)$ is readily defined to be

$$\delta'(t) = \mathcal{H}\frac{1}{\pi t^2}\,. \tag{5.38}$$

The Hilbert transforms of the time derivatives of $\delta(t)$ are then performed by

$$\mathcal{H}\delta'(t) = -\frac{1}{\pi t^2}\,, \quad \mathcal{H}\delta''(t) = \frac{2}{\pi t^3}\,, \quad \ldots . \tag{5.39}$$

□

5.3.3.8 Hilbert Transform of the Definitive Integrals

Since the definitive integral of $y(t)$ is a constant (time-invariant), then its Hilbert transform, by the filtering property, is zero.

5.3.3.9 Orthogonality

A real signal $y(t)$ and its Hilbert transform $\hat{y}(t)$ are orthogonal. Indeed, by the Parseval theorem, we obtain

$$\int_{-\infty}^{\infty} y(t)\hat{y}(t)\mathrm{d}t = \frac{1}{2\pi}\int_{-\infty}^{\infty} Y(j\omega)[-j\,\mathrm{sgn}(\omega)Y(j\omega)]^*\mathrm{d}\omega$$

$$= \frac{j}{2\pi}\int_{-\infty}^{\infty} \mathrm{sgn}(\omega)|Y(j\omega)|^2\mathrm{d}\omega\,.$$

In the latter relation, the squared magnitude of the signal spectral density is always even and the signum function is odd. Thus, the integral is zero

$$\int_{-\infty}^{\infty} y(t)\hat{y}(t)\mathrm{d}t = 0\,. \tag{5.40}$$

That is only possible if the nonzero integrant is combined with the orthogonal functions.

Example 5.13. Consider a signal $x(t)$ and its Hilbert transform $\hat{x}(t)$ examined in Example 5.6. Integrating the product of these functions yields

$$\sum_{n=0}^{N-1}\sum_{m=0}^{N-1} \varepsilon_n \varepsilon_m \frac{\sin(5n/\pi)}{5n/\pi}\frac{\sin(5m/\pi)}{5m/\pi}\int_{-\infty}^{\infty} \cos nt \, \sin mt \, \mathrm{d}t \,,$$

where the integral is zero and thus the signals are orthogonal.

□

5.3.3.10 Hilbert Transform of the Convolution

The Hilbert transform of the convolution of two signals $y_1(t)$ and $y_2(t)$ is the convolution of one with the Hilbert transform of the other.

$$\mathcal{H}[y_1(t) \times y_2(t)] = \frac{1}{\pi}\int_{-\infty}^{\infty}\int_{-\infty}^{\infty} \frac{y_1(\tau)y_2(\theta-\tau)}{t-\theta}\mathrm{d}\tau \, \mathrm{d}\theta$$

$$= \int_{-\infty}^{\infty} y_1(\tau)\left(\frac{1}{\pi}\int_{-\infty}^{\infty}\frac{y_2(\theta-\tau)}{t-\theta}\mathrm{d}\theta\right)\mathrm{d}\tau = \int_{-\infty}^{\infty} y_1(\tau)\mathcal{H}y_2(t-\tau)\mathrm{d}\tau$$

$$= y_1(t) * \mathcal{H}y_2(t) = y_1(t) * \hat{y}_2(t) \,. \tag{5.41}$$

It also follows that

$$\int_{-\infty}^{\infty} y_1(t)\hat{y}_2(t)\mathrm{d}t = \int_{-\infty}^{\infty} \hat{y}_1(t)y_2(t)\mathrm{d}t \,, \tag{5.42}$$

$$\int_{-\infty}^{\infty} y_1(t)y_2(t)\mathrm{d}t = \int_{-\infty}^{\infty} \hat{y}_1(t)\hat{y}_2(t)\mathrm{d}t \,, \tag{5.43}$$

$$\int_{-\infty}^{\infty} y^2(t)\mathrm{d}t = \int_{-\infty}^{\infty} \hat{y}^2(t)\mathrm{d}t \,. \tag{5.44}$$

Example 5.14. Consider a pair of the orthogonal signals, $z(t) = \cos\omega t$, $\hat{z}(t) = \sin\omega t$, $x(t) = A(\sin Wt/Wt)$, and $\hat{x}(t) = A(\sin^2 Wt/2/Wt/2)$. The difference of two integrals in (5.42) produces zero,

$$A\int\limits_{-\infty}^{\infty}\left(\frac{\sin^2 Wt/2}{Wt/2}\cos\omega t-\frac{\sin Wt}{Wt}\sin\omega t\right)\mathrm{d}t$$

$$=A\int\limits_{-\infty}^{\infty}\frac{\sin Wt/2}{Wt/2}\left(\cos\omega t\sin\frac{Wt}{2}-\sin\omega t\cos\frac{Wt}{2}\right)\mathrm{d}t$$

$$=A\int\limits_{-\infty}^{\infty}\frac{\sin Wt/2}{Wt/2}\sin\left(\omega t-\frac{Wt}{2}\right)\mathrm{d}t$$

$$=A\int\limits_{-\infty}^{\infty}\frac{\cos(\omega-W)t}{Wt}\mathrm{d}t-A\int\limits_{-\infty}^{\infty}\frac{\cos\omega t}{Wt}\mathrm{d}t=0\,,$$

and thus the left-hand side and the right-hand side of (5.42) may be used equivalently. In the same manner, (5.43) and (5.44) may be verified.

□

5.3.3.11 Energy

A signal $y(t)$ and its Hilbert transform $\hat{y}(t)$ have equal energies. Indeed, by the Rayleigh theorem, we may write

$$E_y=\int\limits_{-\infty}^{\infty}|y(t)|^2\mathrm{d}t=\frac{1}{2\pi}\int\limits_{-\infty}^{\infty}|Y(j\omega)|^2\mathrm{d}\omega\,, \tag{5.45}$$

$$E_{\hat{y}}=\int\limits_{-\infty}^{\infty}|\hat{y}(t)|^2\mathrm{d}t=\frac{1}{2\pi}\int\limits_{-\infty}^{\infty}|-j\,\mathrm{sgn}\,(\omega)Y(j\omega)|^2\mathrm{d}\omega\,. \tag{5.46}$$

Since $|-j\,\mathrm{sgn}\,(\omega)|^2=1$, except for $\omega=0$, then

$$E_y=E_{\hat{y}}\,. \tag{5.47}$$

The Rayleigh theorem also leads to the conclusion that a signal $y(t)$ and its Hilbert transform $\hat{y}(t)$ have equal energy spectrums,

$$|Y(j\omega)|^2=|\hat{Y}(j\omega)|^2\,. \tag{5.48}$$

Example 5.15. Given an LP signal $x(t)$ and its Hilbert transform $\hat{x}(t)$ considered in Example 5.2. The Fourier transform of $x(t)$ has a rectangular shape and (5.24) easily produces the Fourier transform of $\hat{x}(t)$. Accordingly, we write

$$x(t) = \frac{X_0 W}{\pi} \frac{\sin Wt}{Wt} \quad \stackrel{\mathcal{F}}{\Leftrightarrow} \quad X(j\omega) = \begin{cases} X_0, & \text{if } |\omega| \leqslant W \\ 0, & \text{otherwise} \end{cases},$$

$$\hat{x}(t) = \frac{X_0 W}{\pi} \frac{\sin^2 Wt/2}{Wt/2} \quad \stackrel{\mathcal{F}}{\Leftrightarrow} \quad \hat{X}(j\omega) = \begin{cases} -jX_0, & \text{if } 0 \leqslant \omega \leqslant W \\ jX_0, & \text{if } -W \leqslant \omega < 0 \\ 0, & \text{otherwise} \end{cases}.$$

By (5.45) and (5.46), the signal energy is calculated by the spectral densities to be

$$\frac{1}{2\pi} \int_{-\infty}^{\infty} |X(j\omega)|^2 \mathrm{d}\omega = \frac{1}{2\pi} \int_{-\infty}^{\infty} |\hat{X}(j\omega)|^2 \mathrm{d}\omega = \frac{X_0^2 W}{\pi}.$$

The same result may be found to use the integral identities $\int_{-\infty}^{\infty} (\sin^2 bt/t^2)\, \mathrm{d}t = \pi b$ and $\int_{-\infty}^{\infty} (\sin^4 bt/t^2)\, \mathrm{d}t = \pi b/2$, and calculate the signal energy by $x(t)$ and $\hat{x}(t)$ employing (5.45) and (5.46), respectively.

□

5.3.3.12 Autocorrelation

The signal $y(t)$ and its Hilbert transform $\hat{y}(t)$ have equal autocorrelation functions.

$$\int_{-\infty}^{\infty} y(t)y(t-\theta)\mathrm{d}t = \int_{-\infty}^{\infty} \hat{y}(t)\hat{y}(t-\theta)\mathrm{d}t. \tag{5.49}$$

Example 5.16. Consider a sinc-shaped pulse and its Hilbert transform, respectively,

$$x(t) = \frac{\sin t}{t} \quad \text{and} \quad \hat{x}(t) = \frac{\sin^2(t/2)}{t/2}.$$

The autocorrelation function of the pulse is defined by

$$\begin{aligned} \phi_x(\theta) &= \int_{-\infty}^{\infty} \frac{\sin t}{t} \frac{\sin(t-\theta)}{t-\theta} \mathrm{d}t \\ &= \frac{\cos\theta}{2} \int_{-\infty}^{\infty} \frac{\mathrm{d}t}{t(t-\theta)} - \frac{1}{2} \int_{-\infty}^{\infty} \frac{\cos\,(2t-\theta)}{t(t-\theta)} \mathrm{d}t. \end{aligned}$$

The first integral produces zero,

$$\int \frac{\mathrm{d}t}{t(t-\theta)} = -\frac{1}{\theta} \ln \left| \frac{t}{t-\theta} \right| \Bigg|_{-\infty}^{\infty} = 0.$$

The second integral, by changing a variable, $z = 2t - \theta$, and using an identity $\int\limits_0^{\infty} \frac{\cos(ax)}{b^2 - x^2} dx = \frac{\pi}{2b} \sin{(ab)}$, is transformed as follows:

$$\int\limits_{-\infty}^{\infty} \frac{\cos{(2t-\theta)}}{t(t-\theta)} dt = 2 \int\limits_{-\infty}^{\infty} \frac{\cos z}{z^2 - \theta^2} dz = -\frac{2\pi}{\theta} \sin\theta \, .$$

Thus, the function is

$$\phi_x(\theta) = \frac{\pi}{\theta} \sin\theta \, .$$

The autocorrelation function of $\hat{x}(t)$, by using an identity $\sin^2 x = 0.5\,(1 - \cos 2x)$, is transformed to

$$\begin{aligned}
\phi_{\hat{x}}(\theta) &= \int\limits_{-\infty}^{\infty} \frac{\sin^2(t/2)}{t/2} \frac{\sin^2[(t-\theta)/2]}{(t-\theta)/2} dt \\
&= \int\limits_{-\infty}^{\infty} \frac{(1-\cos t)[1-\cos(t-\theta)]}{t(t-\theta)} dt \\
&= \int\limits_{-\infty}^{\infty} \frac{\mathrm{d}t}{t(t-\theta)} - \int\limits_{-\infty}^{\infty} \frac{\cos{(t-\theta)}}{t(t-\theta)} dt \\
&\quad - \int\limits_{-\infty}^{\infty} \frac{\cos t}{t(t-\theta)} dt + \int\limits_{-\infty}^{\infty} \frac{\cos t \cos(t-\theta)}{t(t-\theta)} dt \, .
\end{aligned}$$

The first integral is zero. It may be shown by simple manipulations with variables that the second and third integrals compensate. Finally, the last integral produces

$$\phi_{\hat{x}}(\theta) = -\frac{\pi}{\theta} \sin\theta \, .$$

Since $\phi_{\hat{x}}(\theta) = \phi_{\hat{x}}(-\theta)$, the autocorrelation function $\phi_{\hat{x}}(\theta)$ becomes identical to $\phi_x(\theta)$.

□

We have already considered many useful properties of the Hilbert transform. Some others are coupled with its application to the analytic signals that we will learn in further.

5.3.4 Applications of the Hilbert Transform in Systems

The above-considered examples have virtually shown already how the Hilbert transform may be used in solving the system problems. A principal point here is that the Hilbert transform completes the in-phase signal component with the quadrature-phase component. Thus we have two informative signals instead of only one and the accuracy of system operation may be increased. This is efficiently used in modulation and demodulation.

5.3.4.1 Amplitude Detection

A common and very efficient technique for envelope detection is based on the Hilbert Transform $\hat{y}(t)$ of an original HF signal $y(t)$. In such a detector, the absolute values of $y(t)$ and $\hat{y}(t)$ are summed and the peak hold operation provides tracking the maximum of one of these signals when the other one has zero (Fig. 5.10). The Hilbert transformer provides a 90° phase shift that is typically implemented as a finite impulse response (FIR) filter with a linear phase. Therefore, the original signal must be delayed to match the group delay of the Hilbert transform. The peak hold function is often implemented as a one-pole infinite impulse response (IIR) filter.

The effect of employing $y(t)$ and its Hilbert transform $\hat{y}(t)$ instead of a direct detection of a signal waveform by $y(t)$ is neatly demonstrated in Fig. 5.11 for the Gaussian-waveform RF pulse

$$y(t) = \mathrm{e}^{-\alpha(t-t_1)^2} \cos \omega_0 t \,,$$

where $\alpha \ll \omega_0$ and t_1 correspond to the peak value of the Gaussian waveform. The Hilbert transformer provides a $\pi/2$ version of $y(t)$,

$$\hat{y}(t) \cong \mathrm{e}^{-\alpha(t-t_1)^2} \sin \omega_0 t \,.$$

The absolute value $|y(t)|$ of an original signal (a) is formed as in (c) having an envelope $A(t)$. If to exploit two signals (b), then the maximums of their absolute values produce twice larger number of the points (d) resulting in a higher accuracy of the envelope detection by the peak hold. In Fig. 5.11d, the envelope is shown for the analytic signal. It is important that the sum of $y(t)$ and $\hat{y}(t)$ fits this envelope only at discrete points when one of these signals attains a maximum and the other becomes zero.

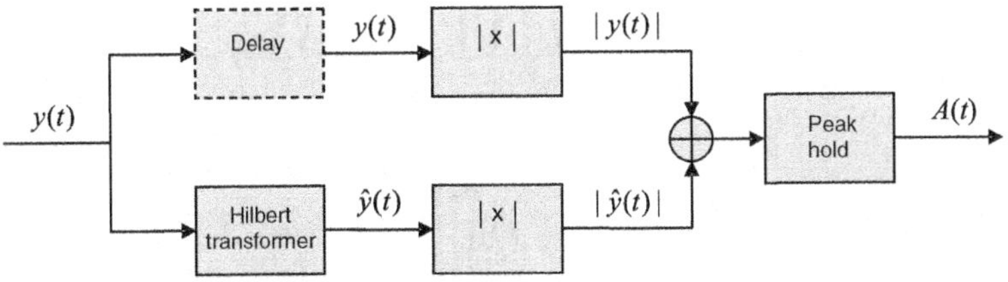

Fig. 5.10 Detecting the envelope $A(t)$ of a narrowband signal $y(t)$ employing the Hilbert transformer.

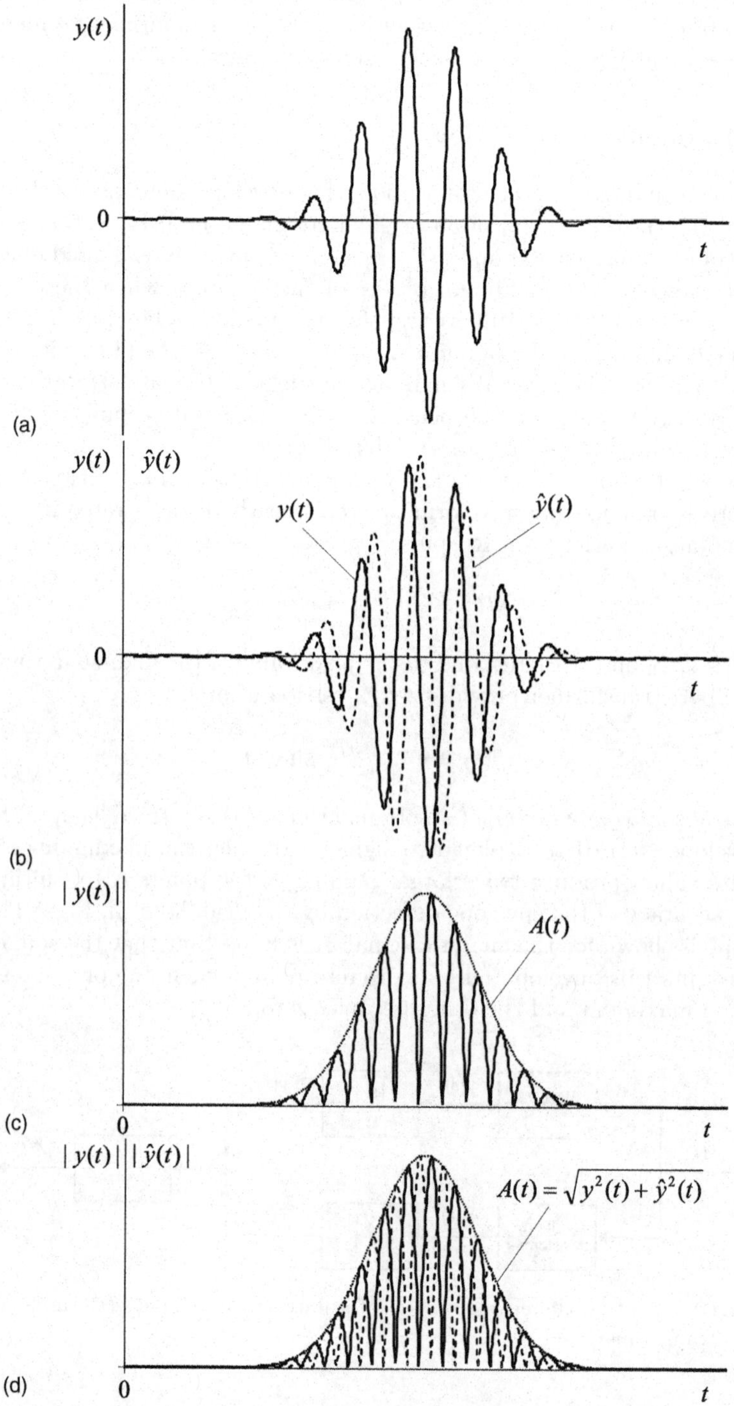

Fig. 5.11 Signals in a system (Fig. 5.10): **(a)** Gaussian RF pulse $y(t)$; **(b)** $y(t)$ and its Hilbert transform $\hat{y}(t)$; **(c)** absolute signal $|y(t)|$ and its envelope $A(t)$, and **(d)** $|y(t)|$ and $|\hat{y}(t)|$; and their envelope (root-mean-square sum) $A(t) = \sqrt{y^2(t) + \hat{y}^2(t)}$.

5.3.4.2 Hilbert Modulator

The other efficient application of the Hilbert transform is to create a modulated signal with SSB. In Fig. 5.12, we show a typical structure of a modulator. To simplify its presentation, we will think that the modulating signal $x(t)$ is a purely harmonic wave with unit amplitude and modulating frequency Ω, so that $x(t) = \cos \Omega t$ and $\hat{x}(t) = \sin \Omega t$. As in the case of Fig. 5.10, some time delay of the original signal is necessary to compensate the group delay induced by FIR filter of the Hilbert transformer. The carrier signal $\cos \omega_0 t$ is transformed here to $\sin \omega_0 t$ by a $\pi/2$ lag phase shifter. Such a Hilbert transformer is usually called a *phase-shift filter*, a *$\pi/2$-shifter*, or a *quadrature filter*.

Modulation of the original and shifted carrier signals produces the real modulated signal and its Hilbert transform, respectively,

$$y(t) = \cos \Omega t \cos \omega_0 t\,,$$
$$\hat{y}(t) = \sin \Omega t \sin \omega_0 t\,.$$

These signals may now be either summed to obtain an SSB signal with an LSB,

$$s(t) = \cos \Omega t \cos \omega_0 t + \sin \Omega t \sin \omega_0 t = \cos(\omega_0 - \Omega)t\,,$$

or subtracted, yielding an SSB signal with an USB,

$$s(t) = \cos \Omega t \cos \omega_0 t - \sin \Omega t \sin \omega_0 t = \cos(\omega_0 + \Omega)t\,.$$

In theory, the Hilbert modulator promises no other sideband, thus an infinite suppression and zero transition band. This, however, may be a bit broken by the phase errors in the structure.

5.3.4.3 Quadrature Demodulator

We have already shown in Fig. 5.5 how the Hilbert transform of a reference signal may be used in a wireless coherent detection of an object. Most generally, this principle results in the *quadrature demodulator* shown in Fig. 5.13.

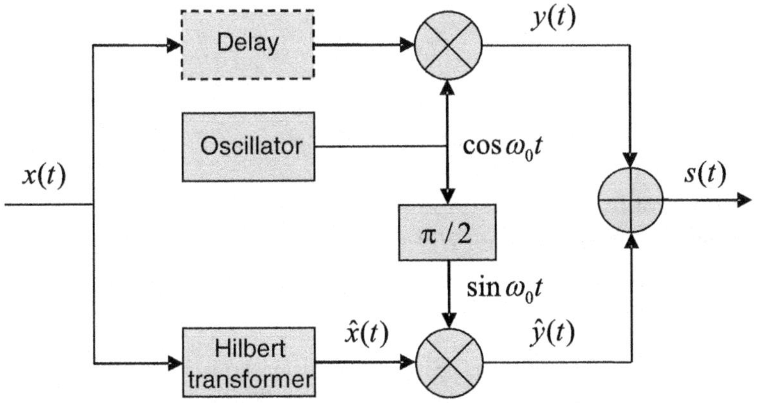

Fig. 5.12 Creating an SSB signal with the Hilbert transformer.

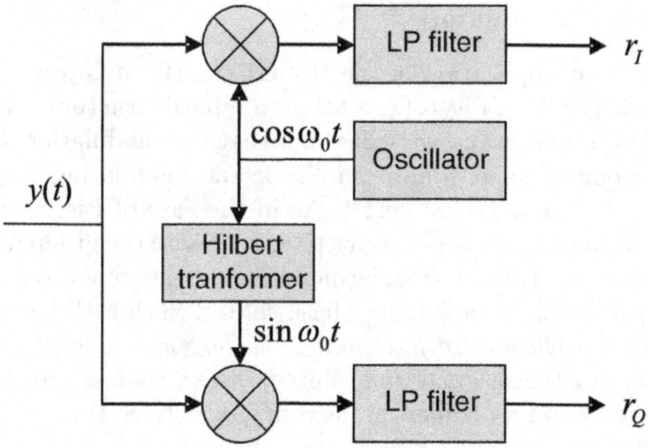

Fig. 5.13 Quadrature demodulation with the Hilbert transform of a reference signal.

Here, an input signal with the carrier frequency ω_0,

$$y(t) = \cos(\omega_0 t + \varphi)$$

goes to two channels. In the first channel, it is multiplied with the in-phase carrier signal $z(t) = \cos\omega_0 t$ and the product is

$$y(t)z(t) = \frac{1}{2}\cos\varphi + \frac{1}{2}\cos(2\omega_0 t + \varphi).$$

In the second channel, $y(t)$ is multiplied with the Hilbert transform of a carrier signal, $\hat{z}(t) = \sin\omega_0 t$, that produces

$$y(t)\hat{z}(t) = -\frac{1}{2}\sin\varphi + \frac{1}{2}\sin(2\omega_0 t + \varphi).$$

Two LP filters with the gain factors 2 and -2 select the in-phase and quadrature phase amplitudes, respectively,

$$r_\mathrm{I} = \cos\varphi,$$
$$r_\mathrm{Q} = \sin\varphi,$$

and the demodulation is complete. The Hilbert transformer is designed here as a $\pi/2$-*shifter.*

The quadrature demodulator is used as a building block in many systems for more elaborate modulation and demodulation schemes.

5.3.4.4 Costas loop

In binary phase shift keying (BPSK), the carrier signal is modulated by shifting its phase by 0° or 180° at a specified rate to be

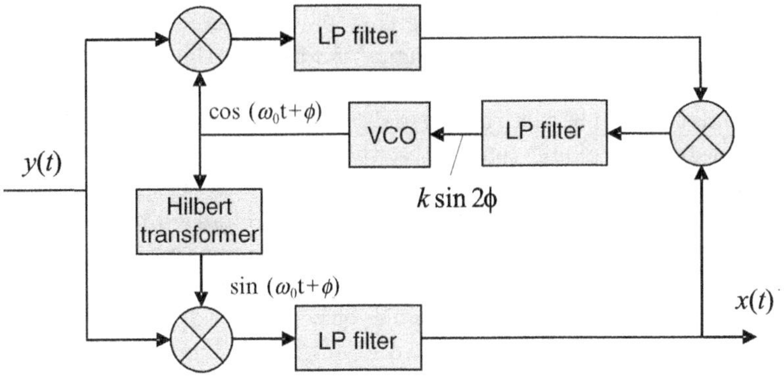

Fig. 5.14 Costas loop for quadrature demodulation of BPSK.

$$y(t) = \cos(\omega_0 t \pm \pi/2)\,.$$

To demodulate this signal, the quadrature demodulator may be used to supply the coherent reference signal for product detection in what is called the Costas phase-locked loop (PLL). A Costas loop for BPSK demodulation is shown in Fig. 5.14. It is assumed that the voltage-controlled oscillator (VCO) is locked to the input suppressed carrier frequency with a constant phase error ϕ and that the Hilbert transformer ($\pi/2$ shifter) produces its quadrature version. The VCO signals (in-phase and quadrature phase) modulate the input signal. The products then go through LP filters, which bandwidths are predetermined by the modulation data rate. Accordingly, two signals appear at the outputs of the LP filters in the in-phase and quadrature phase channels, respectively,

$$\frac{1}{2}\cos\left(\pm\frac{\pi}{2} - \phi\right),$$

$$-\frac{1}{2}\sin\left(\pm\frac{\pi}{2} - \phi\right).$$

These signals are multiplied together and the product goes to an LP filter, which cutoff frequency lies near zero so that the filter acts as an integrator to produce the necessary control voltage $k\sin 2\phi$. The output of the quadrature channel then yields a signal that is proportional to the phase modulation law,

$$x(t) = \mp\frac{1}{2}\cos\phi\,.$$

With $\phi = 0$, the output of a Costas loop generates simply ± 0.5. In communications, this loop is also adopted to demodulate QPSK signals.

5.4 Analytic Signal

Nowadays, the Hilbert transform has found a wide application in the theory of signals and systems as well as in the systems design owing to its splendid property to couple the real and imaginary parts (1) of an analytic signal; (2) of the spectral density of a causal signal; and (3) of the transfer function of any system. Moreover, the Hilbert transform is used to generate complex-valued analytic signals from real signals. Since a signal is analytic when its components are harmonic conjugates, then the Hilbert transform generates the components, which are harmonic conjugates to the original signal.

Any real systems operate with real values. Therefore a direct application of analytic (complex) signals in systems is impossible and an SP part is included whenever necessary to provide mathematical manipulations (Fig. 5.15).

A system may produce, for example, either only the in-phase component (I) or both the in-phase and quadrature phase (Q) components of a real signal. The SP part will then create complex analytic signals, operate with them, and produce some result that may be final. The SP part may also transfer a real component of the result produced to the same system (Fig. 5.15a) or to several systems (Fig. 5.15b). An example is in Fig. 5.13, where the SP block may be used in order to calculate the signal envelope, phase or instantaneous frequency that cannot be accomplished in a simple way physically. In modern systems various configurations of electronic parts and computer-aided SP parts are used aimed at achieving a highest system efficiency.

Let us now come back to the beginning of this chapter and recall that the analytic signal is nothing more than a complex vector

$$y_a(t) = y(t) + j\hat{y}(t)\,, \tag{5.50}$$

in which $y(t)$ is an arbitrary real signal and $\hat{y}(t)$ is its Hilbert transform. If a signal is narrowband, then this pair of functions becomes

$$y(t) = A_c(t)\cos\omega_0 t - A_s(t)\sin\omega_0 t\,, \tag{5.51}$$

$$\hat{y}(t) = A_c(t)\sin\omega_0 t + A_s(t)\cos\omega_0 t\,. \tag{5.52}$$

As any other complex vector, any analytic signal is characterized with its envelope, phase, and instantaneous frequency.

5.4.1 Envelope

The envelope of an analytic signal is defined by

$$A(t) = \sqrt{y^2(t) + \hat{y}^2(t)}\,. \tag{5.53}$$

If to involve (5.51) and (5.52) and provide elementary manipulations, then (5.53) will be converted to the formula $A(t) = \sqrt{A_c^2(t) + A_s^2(t)}$ defined by

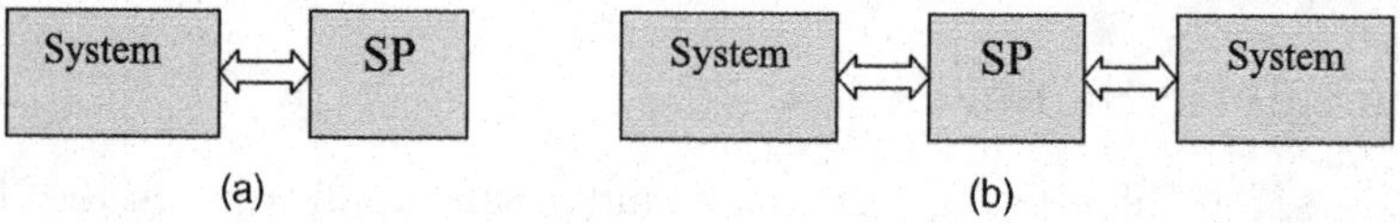

Fig. 5.15 Signal processing of analytic signals in systems: **(a)** single system and **(b)** multiple system.

(5.9) by the amplitudes of the narrowband signal, A_c and A_s. This means that the envelopes of a real narrowband signal and its analytic signal are equal and thus have the same properties.

Example 5.17. Given an oscillatory system of the second order with a high-quality factor. Such a system is specified with the impulse response

$$h(t) = A\mathrm{e}^{-\sigma t} \sin \omega_0 t \,,$$

where a damping factor α is much smaller than the carrier frequency ω_0; $\sigma \ll \omega_0$. For such a relation, the oscillations amplitude is almost constant during the period of repetition $T = 2\pi/\omega_0$. Therefore, the Hilbert transform of $h(t)$ is approximately performed by

$$\hat{h}(t) \cong -A\mathrm{e}^{-\sigma t} \cos \omega_0 t$$

and the envelope is calculated as

$$\sqrt{h^2(t) + \hat{h}^2(t)} = A\mathrm{e}^{-\sigma t} \,.$$

Figure 5.16 neatly shows that the accuracy of the envelope detection is higher if to use $h(t)$ and $\hat{h}(t)$, rather than if to detect only by $h(t)$.

□

5.4.2 Phase

By definition, the total phase $\Psi(t)$ of any real signal $y(t)$ is equal to the argument of its analytic signal $y_a(t)$,

$$\Psi(t) = \arg y_a(t) = \arctan \frac{\hat{y}(t)}{y(t)} \,. \tag{5.54}$$

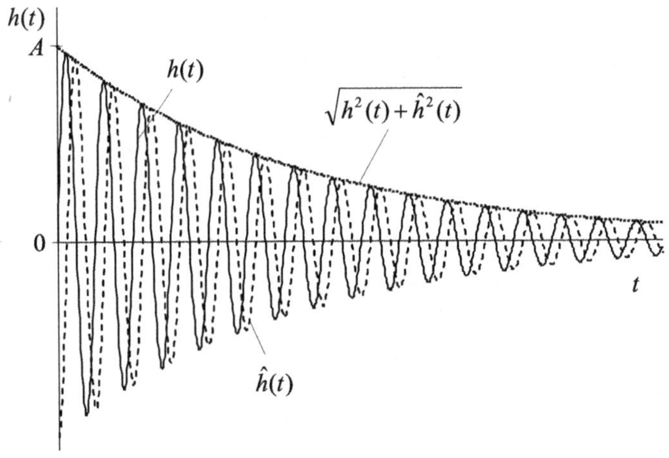

Fig. 5.16 Impulse response $h(t)$ of an oscillatory system of the second order, its Hilbert transform $\hat{h}(t)$, and the envelope $\sqrt{h^2(t) + \hat{h}^2(t)}$.

If a signal is narrowband with the carrier frequency ω_0, then its slowly changing informative phase $\psi(t)$ is calculated by

$$\psi(t) = \Psi(t) - \omega_0 t. \tag{5.55}$$

As it may be observed, transition between the phase $\psi(t)$ defined by the amplitudes A_c and A_s of a narrowband signal, (5.11) and (5.12), and that defined by (5.55) is not so straightforward as between (5.53) and (5.9).

5.4.3 Instantaneous Frequency

The instantaneous frequency $\omega_y(t)$ of any analytic signal is defined by the first time derivative of its total phase $\Psi(t)$ by

$$\omega_y(t) = \frac{\mathrm{d}}{\mathrm{d}t}\Psi(t) = \frac{\mathrm{d}}{\mathrm{d}t}\arctan\frac{\hat{y}(t)}{y(t)} = \frac{\hat{y}'(t)y(t) - y'(t)\hat{y}(t)}{y^2(t) + \hat{y}^2(t)}\,. \tag{5.56}$$

Again we notice that, for narrowband analytic signals, the carrier frequency ω_0 cannot be extracted in (5.56) instantly and extra manipulations are required in each of the particular cases.

Example 5.18. Given a carrier signal $z(t) = \cos\omega_0 t$ and its Hilbert transform $\hat{z}(t) = \sin\omega_0 t$. This signal does not bear any information, because of its envelope is unity, $A(t) = \sqrt{\cos^2\omega_0 t + \sin^2\omega_0 t} = 1$; the total phase is linear, $\Psi(t) = \omega_0 t$; the informative phase is zero, $\psi(t) = \Psi(t) - \omega_0 t = 0$; and the instantaneous frequency is constant, $\omega_y(t) = \omega_0$.

□

Example 5.19. An ideal band-pass signal has a symmetric one-side spectral density

$$Y(j\omega) = \begin{cases} X_0, & \text{if } \omega_0 - W \leqslant \omega \leqslant \omega_0 + W \\ 0, & \text{otherwise} \end{cases}.$$

An analytic signal corresponding to this spectral density is calculated by

$$\begin{aligned} y_a(t) &= \frac{X_0}{\pi}\int\limits_{\omega_0 - W}^{\omega_0 + W} \mathrm{e}^{j\omega t}\,\mathrm{d}\omega = \frac{X_0}{j\pi t}\left[\mathrm{e}^{j(\omega_0 + W)t} - \mathrm{e}^{j(\omega_0 - W)t}\right] \\ &= \frac{X_0}{\pi t}\{[\sin(\omega_0 + W)t - \sin(\omega_0 - W)t] \\ &\quad - j[\cos(\omega_0 + W)t - \cos(\omega_0 - W)t]\} \end{aligned}$$

with its real and imaginary (Hilbert transform) parts, respectively,

$$\begin{aligned} y(t) &= \frac{2X_0 W}{\pi}\frac{\sin Wt}{Wt}\cos\omega_0 t\,, \\ \hat{y}(t) &= \frac{2X_0 W}{\pi}\frac{\sin Wt}{Wt}\sin\omega_0 t\,. \end{aligned}$$

By (5.51), the envelope is calculated to be

$$A(t) = \frac{2X_0 W}{\pi}\left|\frac{\sin Wt}{Wt}\right|$$

that is consistent with (5.2). The total phase is linear, $\Psi(t) = \omega_0 t$; the informative phase is zero, $\psi(t) = \Psi(y) - \omega_0 t = 0$; and the instantaneous frequency is constant, $\omega_y(t) = \omega_0$.

□

Example 5.20. An analytic signal with an unknown carrier frequency ω_0 is given with an asymmetric one-side spectral density

$$Y(j\omega) = \begin{cases} X_0 e^{-a(\omega-\omega_1)}, & \text{if } \omega \geqslant \omega_1 \\ 0, & \text{otherwise}. \end{cases}$$

Its time function is provided by (5.15) as

$$\begin{aligned} y_a(t) &= \frac{X_0}{\pi}\int_{\omega_1}^{\infty} e^{-a(\omega-\omega_1)+j\omega t}\,d\omega = \frac{X_0}{\pi(a-jt)}e^{-(a-jt)\omega_1} \\ &= \frac{X_0}{\pi(a^2+t^2)}[(a\cos\omega_1 t - t\sin\omega_1 t) + j(t\cos\omega_1 t + a\sin\omega_1 t)] \end{aligned}$$

and thus the real signal and its Hilbert transform are, respectively,

$$y(t) = \frac{X_0}{\pi(a^2+t^2)}(a\cos\omega_1 t - t\sin\omega_1 t)\,,$$

$$\hat{y}(t) = \frac{X_0}{\pi(a^2+t^2)}(t\cos\omega_1 t + a\sin\omega_1 t)\,,$$

where the boundary frequency ω_1 has appeared to be a carrier frequency ω_0, by definition.

The envelope of this signal is performed by

$$A(t) = \frac{X_0}{\pi\sqrt{a^2+t^2}}$$

having a maximum $X_0/a\pi$ at $t = 0$ (Fig. 5.17a). Its total phase calculates

$$\Psi(t) = \arctan\frac{t\cos\omega_1 t + a\sin\omega_1 t}{a\cos\omega_1 t - t\sin\omega_1 t}$$

and thus the informative phase is not zero, $\psi(t) = \Psi(t) - \omega_1 t \neq 0$. The instantaneous frequency is calculated by

$$\omega_y(t) = \frac{d}{dt}\Psi(t) = \omega_1 + \frac{a}{a^2+t^2}$$

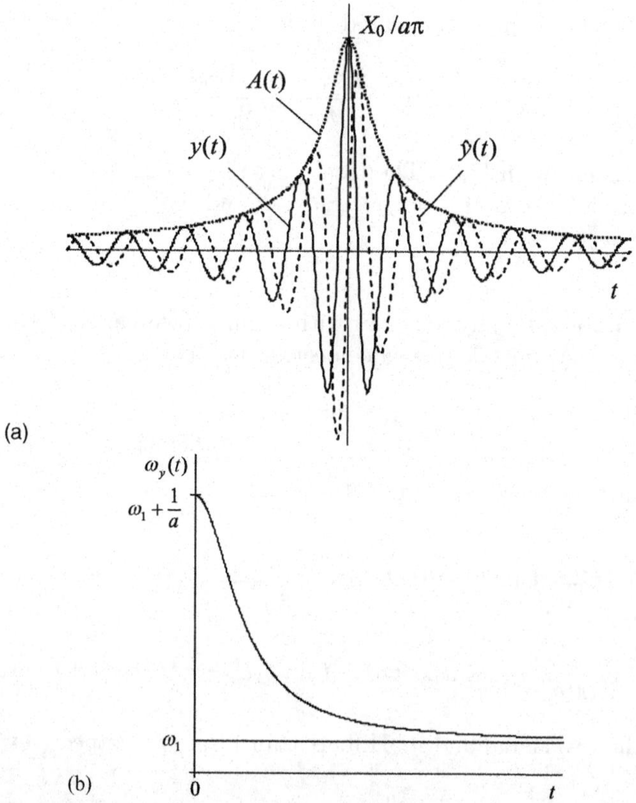

Fig. 5.17 An analytic signal with an asymmetric spectral density (Example 5.3.16): **(a)** analytic signal $y(t)$, its Hilbert transform $\hat{y}(t)$, and envelope $A(t)$ and **(b)** instantaneous frequency $\omega_y(t)$.

and, hence, is time-varying (Fig. 5.17b). At zero point $t = 0$, the frequency $\omega_y(t)$ starts with $\omega = \omega_1 + 1/a$ and then asymptotically approaches the carrier frequency $\omega_0 = \omega_1$ with $t \to \infty$. Now note that the results achieved in this example for a signal with a one-side spectral density are consistent with those obtained in Example 5.1 by the complex envelope of a signal with the double-sided spectral density.

□

5.4.4 Hilbert Transform of Analytic Signals

An analysis of signals generated by real systems may require not only single analytic signals but also their assemblages and Hilbert transforms. A direct application of the Hilbert transform to such signal by the weighted integration (5.15) usually entails difficulties and is embarrassing. In finding a shortest way, some other properties of the Hilbert transform may be used.

First of all, let us show that the Hilbert transform of the analytic signal $y_{\mathrm{a}}(t) = y(t) + j\hat{y}(t)$ is calculated in a very simple way; i.e.,

$$\mathcal{H}y_{\mathrm{a}}(t) \triangleq \hat{y}_{\mathrm{a}}(t) = \mathcal{H}[y(t) + j\hat{y}(t)] = \hat{y}(t) - jy(t) = -jy_{\mathrm{a}}(t)\,. \tag{5.57}$$

Now, the following important products may be found for two analytic signals, $y_{\mathrm{a}1}(t)$ and $y_{\mathrm{a}1}(t)$, and their Hilbert transforms:

- The product of $\hat{y}_{\mathrm{a}1}(t)y_{\mathrm{a}2}(t)$ is identical with the product of $y_{\mathrm{a}1}(t)\hat{y}_{\mathrm{a}2}(t)$:

$$\hat{y}_{\mathrm{a}1}(t)y_{\mathrm{a}2}(t) = -jy_{\mathrm{a}1}(t)y_{\mathrm{a}2}(t) = y_{\mathrm{a}1}(t)[-jy_{\mathrm{a}2}(t)] = y_{\mathrm{a}1}(t)\hat{y}_{\mathrm{a}2}(t) \tag{5.58}$$

- The product of $y_{\mathrm{a}1}(t)y_{\mathrm{a}2}(t)$ is identical with the product of $j\hat{y}_{\mathrm{a}1}(t)y_{\mathrm{a}2}(t) = jy_{\mathrm{a}1}(t)\hat{y}_{\mathrm{a}2}(t)$:

$$y_{\mathrm{a}1}(t)y_{\mathrm{a}2}(t) = j\hat{y}_{\mathrm{a}1}(t)y_{\mathrm{a}2}(t) = jy_{\mathrm{a}1}(t)\hat{y}_{\mathrm{a}2}(t)\,. \tag{5.59}$$

- The Hilbert transform of the real and imaginary products of two analytic signals produces the following identities:

$$\mathcal{H}\mathrm{Re}[y_{\mathrm{a}1}(t)y_{\mathrm{a}2}(t)] = \mathrm{Im}[y_{\mathrm{a}1}(t)y_{\mathrm{a}2}(t)]\,, \tag{5.60}$$

$$\mathcal{H}\mathrm{Im}[y_{\mathrm{a}1}(t)y_{\mathrm{a}2}(t)] = -\mathrm{Re}[y_{\mathrm{a}1}(t)y_{\mathrm{a}2}(t)]\,. \tag{5.61}$$

- The Hilbert transform of the product of $y_{\mathrm{a}1}(t)y_{\mathrm{a}2}(t)$ is identical to the product of $-jy_{\mathrm{a}1}(t)y_{\mathrm{a}2}(t)$:

$$\begin{aligned}\mathcal{H}y_{\mathrm{a}1}(t)y_{\mathrm{a}2}(t) &= \mathcal{H}[y_1(t) + j\hat{y}_1(t)][y_2(t) + j\hat{y}_2(t)]\\ &= \mathcal{H}\{y_1(t)y_2(t) - \hat{y}_1(t)\hat{y}_2(t) + j[y_1(t)\hat{y}_2(t) + \hat{y}_1(t)y_2(t)]\}\\ &= y_1(t)\hat{y}_2(t) + \hat{y}_1(t)y_2(t) - j[y_1(t)y_2(t) - \hat{y}_1(t)\hat{y}_2(t)]\\ &= -j[y_1(t)y_2(t) + jy_1(t)\hat{y}_2(t) + j\hat{y}_1(t)y_2(t) - \hat{y}_1(t)\hat{y}_2(t)]\\ &= -j[y_1(t) + j\hat{y}_1(t)][y_2(t) + j\hat{y}_2(t)] = -jy_{\mathrm{a}1}(t)y_{\mathrm{a}2}(t)\end{aligned} \tag{5.62}$$

that yields the following useful identities:

$$\mathcal{H}[y_{\mathrm{a}1}(t)y_{\mathrm{a}2}(t)] = \hat{y}_{\mathrm{a}1}(t)y_{\mathrm{a}2}(t) = y_{\mathrm{a}1}(t)\hat{y}_{\mathrm{a}2}(t) = -jy_{\mathrm{a}1}(t)y_{\mathrm{a}2}(t) \tag{5.63}$$

$$\mathcal{H}y_{\mathrm{a}}^{n}(t) = \hat{y}_{\mathrm{a}}(t)y_{\mathrm{a}}^{n-1}(t) = -jy_{\mathrm{a}}(t)y_{\mathrm{a}}^{n-1}(t) = -jy_{\mathrm{a}}^{n}(t) \tag{5.64}$$

Example 5.21. Given two analytic carrier signals

$$z_{\mathrm{a}1}(t) = \cos\omega_1 t + j\sin\omega_1 t\,,$$

$$z_{\mathrm{a}2}(t) = \cos\omega_2 t + j\sin\omega_2 t\,.$$

By (5.57), the Hilbert transforms of these signals are produced, respectively,

$$\hat{z}_{a1}(t) = \sin\omega_1 t - j\cos\omega_1 t\,,$$
$$\hat{z}_{a2}(t) = \sin\omega_2 t - j\cos\omega_2 t\,.$$

Employing (5.58), the following product is calculated to be

$$\hat{z}_{a1}(t)z_{a2}(t) = z_{a1}(t)\hat{z}_{a2}(t) = \sin\,(\omega_1+\omega_2)t - j\cos\,(\omega_1+\omega_2)t\,.$$

Each of the relations in (5.59), produces

$$z_{a1}(t)z_{a2}(t) = \cos\,(\omega_1+\omega_2)t + j\sin\,(\omega_1+\omega_2)t\,.$$

It follows, by (5.60) and (5.61), that

$$\mathcal{H}\mathrm{Re}\,[y_{a1}(t)y_{a2}(t)] = \mathrm{Im}\,[y_{a1}(t)y_{a2}(t)] = \sin\,(\omega_1+\omega_2)t\,,$$
$$\mathcal{H}\mathrm{Im}\,[y_{a1}(t)y_{a2}(t)] = -\mathrm{Re}\,[y_{a1}(t)y_{a2}(t)] = -\cos\,(\omega_1+\omega_2)t\,,$$

and thus the Hilbert transform of the product $z_{a1}(t)z_{a2}(t)$ is

$$\mathcal{H}[z_{a1}(t)z_{a2}(t)] = \sin\,(\omega_1+\omega_2)t - j\cos(\omega_1+\omega_2)t$$

that is also calculated by (5.63).

□

5.5 Interpolation

An *interpolation* problem arises when a continuous-time bandlimited signal needs to be reconstructed correctly from a set of its discrete-time measured samples. The system thus must be able to interpolate the signal between samples. Alternatively, in the SP block (Fig. 5.15), transition is provided from samples to the interpolated function, which actually becomes not continuous but rather discrete with a much smaller sample time. Such a transition employs polynomials, which order, from the standpoint of practical usefulness, should not be large. Therefore, interpolation is usually associated with slowly change in time bandlimited signals.

Numerical analysis offers several methods. The Lagrange form is the classical technique of finding an order polynomial that passes through given points. An interpolation polynomial in the Newton form is also called Newton's divided differences interpolation polynomial because its coefficients are calculated using divided differences. The technique known as cubic splines fits a third-order polynomial through two points so as to achieve a certain slope at one of the points. There are also used Bezier's[3] splines, which interpolate a set of points using smooth curves and do not necessarily pass through the points. The Bernstein[4] polynomial that is a linear combination of Bernstein basis polynomials is the other opportunity for interpolation.

[3] Pierre Etienne Bezier, French engineer, 1 September 1910–25 November 1999.
[4] Sergei Bernstein, Ukrainian mathematician, 5 March 1880–26 October 1968.

5.5.1 Lagrange Form

The Lagrange form of an interpolation polynomial is a well-known, classical technique for interpolation. The formula published by Lagrange in 1795 was earlier found by Waring[5] in 1779 and soon after rediscovered by Euler in 1783. Therefore, it is also called Waring–Lagrange interpolation or, rarely, Waring–Euler–Lagrange interpolation.

The Lagrange interpolation is performed as follows. Let a bandlimited signal $x(t)$ be given at $m+1$ discrete-time points as $x(t_n)$, $n = 0, 1, \ldots, m$. Then the following unique order m polynomial interpolates a signal between these points by

$$L_m(t) = \sum_{i=0}^{m} x(t_i) l_{mi}(t)\,, \tag{5.65}$$

where

$$\begin{aligned} l_{mi}(t) &= \prod_{k=0,k\neq i}^{m} \frac{t-t_k}{t_i-t_k} \\ &= \frac{(t-t_0)(t-t_1)\ldots(t-t_{i-1})(t-t_{i+1})\ldots(t-t_m)}{(t_i-t_0)(t_i-t_1)\ldots(t_i-t_{i-1})(t_i-t_{i+1})\ldots(t_i-t_m)} \end{aligned} \tag{5.66}$$

is an elementary nth order polynomial satisfying the condition of

$$l_{mi}(t_n) = \begin{cases} 1 & \text{if } n = i \\ 0 & \text{if } n \neq i \end{cases}. \tag{5.67}$$

A simple observation shows that the numerator of the right-hand side of (5.66) has zeros at all the points except the kth and that the denominator here is a constant playing a role of a normalizing coefficient to satisfy (5.67). To find the coefficients of the interpolating polynomial (5.65), it needs solving the equations system of order polynomials that is usually provided by the Vandermonde[6] matrix.

Example 5.22. The envelope of a narrowband signal at the system output was measured after switched on at discrete-time points t_n, $n = 0, 1, \ldots, 5$ with a constant time-step of 1 s to possess the values of, in volts, $x(t_0) = 0$, $x(t_1) = 2.256$, $x(t_2) = 3.494$, $x(t_3) = 4.174$, $x(t_4) = 4.546$, and $x(t_5) = 4.751$. The Lagrange interpolation (5.65) applied to these samples step-by-step produces the polynomials

$$\begin{aligned} L_1(t) &= 2.256t\,, \\ L_2(t) &= 2.7648t - 0.509t^2\,, \\ L_3(t) &= 2.918t - 0.7386t^2 + 0.0765t^3\,, \\ L_4(t) &= 2.9698t - 0.8335t^2 + 0.12834t^3 - 8.6337 \times 10^{-3}t^4\,, \\ L_5(t) &= 2.9885t - 0.87248t^2 + 0.1556t^3 - 0.01642t^4 + 7.7909 \times 10^{-4}t^5\,. \end{aligned}$$

[5] Edward Waring, English mathematician, 1736–15 August 1798.

[6] Alexandre-Théophile Vandermonde, French mathematician, 28 February 1735–1 January 1796.

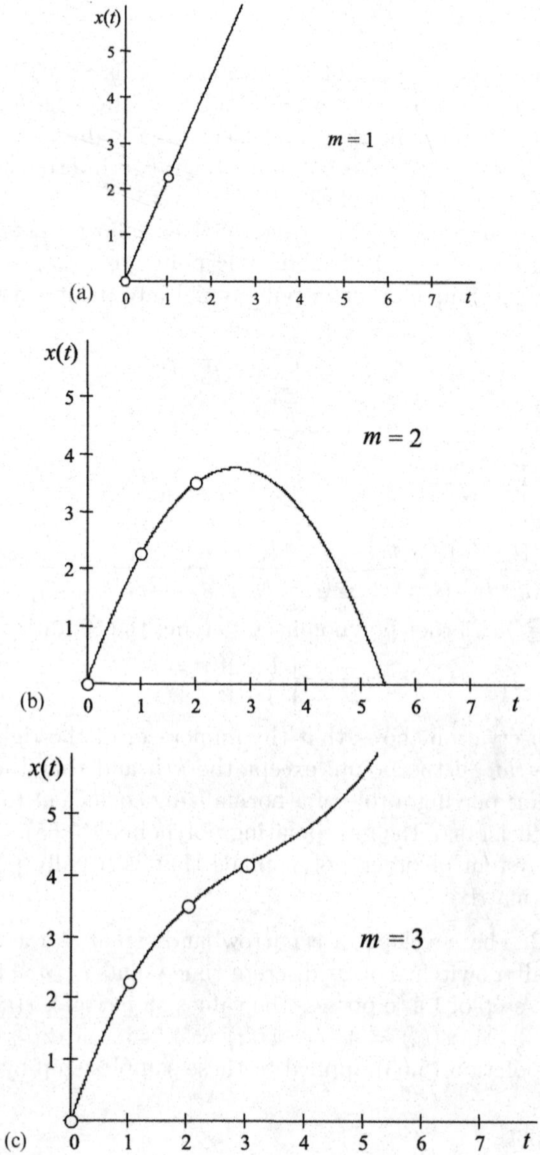

Fig. 5.18 Interpolation of the measured samples by Lagrange polynomials: **(a)** $m = 1$, **(b)** $m = 2$, **(c)** $m = 3$.

Figure 5.18 demonstrates that the method gives exact interpolations in all of the cases including large n when polynomials become long having not enough applied features. Figure 5.18f shows that the approximating function (dashed)

$$x(t) = 5(1 - e^{-0.6t})$$

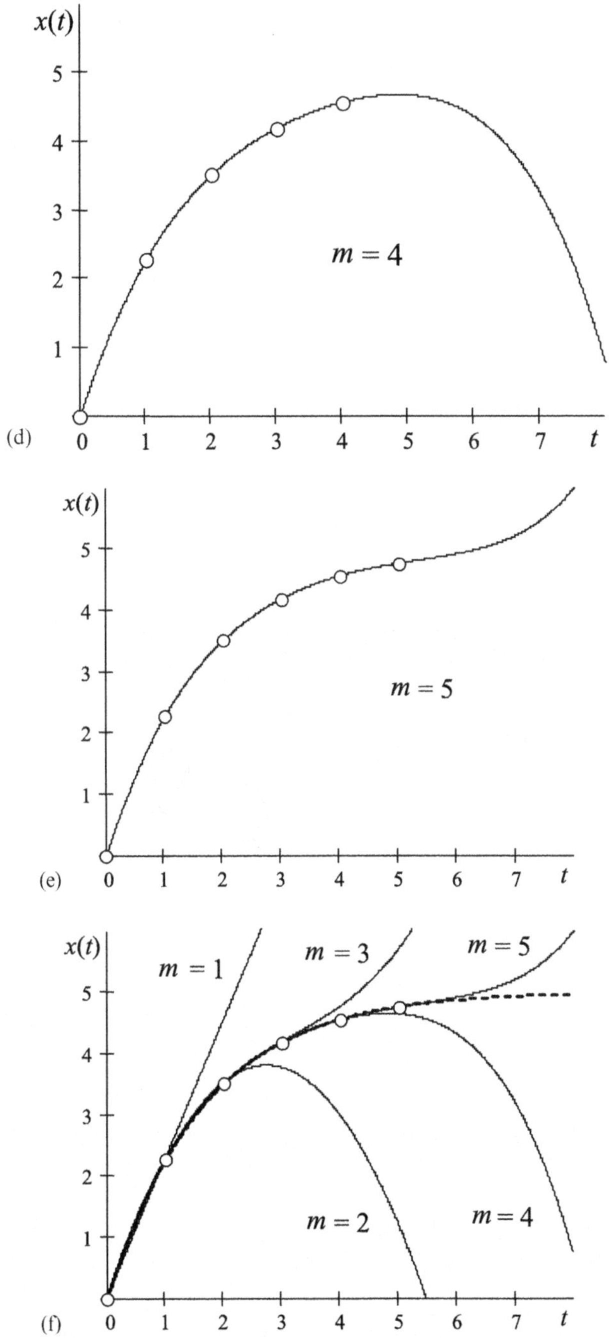

Fig. 5.18 *Continued* **(d)** $m = 4$, **(e)** $m = 5$, and **(f)** is a generalization; dashed line corresponds to $5(1 - e^{-0.6t})$.

fits samples claiming that it is a transient in the envelope associated with the LP filter of the first order.

□

The Lagrange form is convenient and therefore very often used in theoretical studies. But from the practical point of view, its usefulness is not so doubtless, since when we construct the $(m+1)$th order polynomial $L_{m+1}(t)$ we fully lose an information about the lower-order polynomial $L_m(t)$.

5.5.2 Newton Method

The other form of the interpolation polynomial is called the Newton polynomial or Newton's divided differences interpolation polynomial because the coefficients of the polynomial are calculated using divided differences. Again, we deal here with a bandlimited signal $x(t)$ that is given at $m+1$ discrete-time points as $x(t_n)$, $n = 0, 1, \ldots, m$.

The values

$$x(t_n, t_{n+1}) = \frac{x(t_{n+1}) - x(t_n)}{t_{n+1} - t_n}$$

are called the divided differences of the first order. The divided differences of the second order are defined by

$$x(t_n, t_{n+1}, t_{n+2}) = \frac{x(t_{n+1}, t_{n+2}) - x(t_n, t_{n+1})}{t_{n+2} - t_n}$$

and then those of an arbitrary $k \geqslant 2$ order are performed as

$$x(t_n, t_{n+1}, \ldots, t_{n+k}) = \frac{x(t_{n+1}, \ldots, t_{n+k}) - x(t_n, \ldots, t_{n+k-1})}{t_{n+k} - t_n}. \tag{5.68}$$

Utilizing (5.68), the Newton interpolation polynomial is written as follows

$$\begin{aligned} P_m(t) = x(t_0) + x(t_0, t_1)(t - t_0) + x(t_0, t_1, t_2)(t - t_0)(t - t_1) + \ldots \\ + x(t_0, \ldots, t_m)(t - t_0) \ldots (t - t_{m-1}). \end{aligned} \tag{5.69}$$

The value $\varepsilon_m = |x(t) - P_m(t)|$ is said to be the interpolation error or the rest interpolation term that is the same as in the Lagrange form. Since a bandlimited signal is typically changed slowly with its smoothed enough function, the approximate relation $x(t) - P_m(t) \approx P_{m+1}(t) - P_m(t)$ is usually used to estimate the interpolation error by

$$\varepsilon_m \cong |P_{m+1}(t) - P_m(t)|. \tag{5.70}$$

It may be shown that Newton's method leads to the same result as that of Lagrange (one may recalculate Example 5.22 employing the Newton approach). There is, however, one important difference between these two forms. It follows from the definition of the divided differences that new data points can be added to the data set to create a new interpolation polynomial without recalculating the old coefficients. Hence, when a data point changes one usually do not have to recalculate all coefficients. Furthermore, if a time-step is constant, the calculation of the divided differences becomes significantly easier. Therefore, the Newton form of the interpolation polynomial is usually preferred over the Lagrange form for practical purposes. Overall, when added a new sample, the Lagrange polynomial has to be recalculated fully, whereas the Newton form requires only an addition calculated for this new term.

5.6 Sampling

Let us now assume that we have a continuous-time bandlimited signal $y(t)$ and would like to present it with samples to make an analysis digitally, put to the digital SP block, or transmit at a distance. We may provide more samples so that an assemblage of samples will look almost as a continuous function. In such a case, the SP block or transmission may work slow and we will be in need to reduce the number of samples per time unit. The question is, then, how much samples should we save to represent $y(t)$ without a loss of information about its shape in the following interpolation. The sampling theory answers on this question.

A historical overview of the problem turns us back to the work of Whittaker[7] of 1915, in which he studied the Newton interpolation formula for an infinite number of equidistantly placed samples on both sides of a given point. In this work, he showed that, under certain conditions, the resulting interpolation converges to what he called the "cardinal function," which consists of a linear combination of shifted functions of the form $\sin(x)/x$. He has also shown that the Fourier transform of this interpolant has no periodic constituents of period less than $2W$, where W is the spectral band of samples. It then has become known that closely related variants of the cardinal function were considered earlier by some other authors without, however, connection with the bandlimited nature of signals.

The next important step ahead was made by Nyquist[8] in 1928, when he pointed out the importance of having a sampling interval equal to the reciprocal value of twice the highest frequency W of the signal to be transmitted. Referring to the works of Whittaker and Nyquist, Shannon[9] presented and proved in 1948–1949 the sampling theorem that has appeared to be fundamental for signals quantization, digital communications, and the information theory. The theorem was earlier formulated and proved by Kotelnikov[10] in

[7] Edmund Taylor Whittaker, English mathematician, 24 October 1873–24 March 1956.

[8] Harry Nyquist, Swedish-born engineer, 7 February 1889–4 April 1976.

[9] Claude Elwood Shannon, American mathematician-engineer, 30 April 1916–24 February 2001.

[10] Vladimir Kotelnikov, Russian scientist, 6 September 1908–1 February 2005.

1933 and by several other authors. In further, the theorem forced to think about the interpolation function as a linear combination of some basis functions that has found an application in many applied theories.

5.6.1 Sampling Theorem

We will now consider a continuous-time bandlimited signal $x(t)$ with a known spectral density $X(j\omega)$ coupled by the inverse Fourier transform as $x(t) = \mathcal{F}^{-1}X(j\omega)$. Since the term "bandlimited" presumes the spectral density to be zero beyond the range of $-2\pi W \leqslant \omega \leqslant 2\pi W$, where W is a boundary frequency, the inverse Fourier transform of this signal may be written as

$$x(t) = \frac{1}{2\pi}\int\limits_{-2\pi W}^{2\pi W} X(j\omega)\mathrm{e}^{j\omega t}\mathrm{d}\omega \tag{5.71}$$

and a signal $x(t)$ may be considered to be LP (but not obligatory).

We now would like to sample a signal equidistantly with interval $T = 1/2W$. Substituting a continuous time t with a discrete time $t_n = nT$, where n is an arbitrary integer, we rewrite (5.71) for samples $x_{\mathrm{s}}(nT) \triangleq x(t_n)$ as

$$x_{\mathrm{s}}(nT) = \frac{1}{2\pi}\int\limits_{-\pi/T}^{\pi/T} X(j\omega)\mathrm{e}^{j\omega nT}\mathrm{d}\omega$$

that, by a new angular variable in radians $\omega_{\mathrm{s}} = \omega T$, becomes

$$x_{\mathrm{s}}(n) = \frac{1}{2\pi T}\int\limits_{-\pi}^{\pi} X\left(j\frac{\omega_{\mathrm{s}}}{T}\right)\mathrm{e}^{jn\omega_{\mathrm{s}}}\mathrm{d}\omega_{\mathrm{s}}\,. \tag{5.72}$$

The inverse Fourier transform (5.72) claims that the spectral density of $x_{\mathrm{s}}(n)$ is determined in the range $|\omega_{\mathrm{s}}| \leqslant \pi$ by

$$X_{\mathrm{s}}(j\omega_{\mathrm{s}})\big|_{|\omega_{\mathrm{s}}|\leqslant\pi} = \frac{1}{T}X\left(j\frac{\omega_{\mathrm{s}}}{T}\right) \tag{5.73}$$

to mean that the spectral density of a sampled signal is just a scaled version of its origin in the observed frequency band. We now recall that a periodic signal has a discrete spectrum and thus, by the duality property of the Fourier transform, a sampled (discrete) signal has a periodic spectral density. Accounting for a 2π periodicity, we then arrive at the complete periodic spectral density of a sampled signal

$$X_{\mathrm{s}}(j\omega_{\mathrm{s}}) = \frac{1}{T}\sum_{k=-\infty}^{\infty} X\left(j\frac{\omega_{\mathrm{s}} - 2\pi k}{T}\right)\,. \tag{5.74}$$

To interpolate $x(t)$ between samples $x_s(nT)$, we employ (5.71), substitute $X(j\omega)$ by its discrete version taken from (5.73), and first write

$$x(t) = \frac{T}{2\pi} \int_{-\pi/T}^{\pi/T} X_s(j\omega T) e^{j\omega t} d\omega . \tag{5.75}$$

In analogous to (2.14), the spectral density $X_s(j\omega T)$ may be performed, by the duality property of the Fourier transform, by the Fourier series

$$X_s(j\omega T) = \sum_{n=-\infty}^{\infty} x(nT) e^{-j\omega nT}$$

and then the interpolated signal becomes

$$\begin{aligned} x(t) &= \frac{T}{2\pi} \int_{-\pi/T}^{\pi/T} \sum_{n=-\infty}^{\infty} x(nT) e^{-j\omega nT} e^{j\omega t} d\omega \\ &= \frac{T}{2\pi} \sum_{n=-\infty}^{\infty} x(nT) \int_{-\pi/T}^{\pi/T} e^{j\omega(t-nT)} d\omega \\ &= \sum_{n=-\infty}^{\infty} x(nT) \frac{\sin \pi(t/T - n)}{\pi(t/T - n)} \\ &= \sum_{n=-\infty}^{\infty} x\left(\frac{n}{2W}\right) \frac{\sin \pi(2Wt - n)}{\pi(2Wt - n)} . \end{aligned} \tag{5.76}$$

This exact reconstruction of a bandlimited signal by its samples proves the *sampling theorem* that, in the formulation of Shannon, reads as follows:

> ***Sampling theorem:*** If a function $x(t)$ contains no frequencies higher than W, Hz, it is completely determined by giving its ordinates at a series of points spaced $T = 1/2W$ s apart.

□

The sampling theorem is also known as the *Shannon sampling theorem* or the *Nyquist–Shannon sampling theorem*, or the *Kotelnikov theorem*, or the *Whittaker–Nyquist–Kotelnikov–Shannon sampling theorem*. The doubled frequency bound $2W$ is called the *Nyquist sample rate* and W is known as the *Nyquist frequency*.

Example 5.23. The Nyquist sample rate of a signal $x(t)$ is f_s. What is the Nyquist rate for a signal $y(t) = x(\alpha t)$? By the scaling property of the Fourier transform we have

$$Y(j\omega) = \frac{1}{\alpha} X\left(j\frac{\omega}{\alpha}\right) .$$

Therefore, the Nyquist rate for $x(\alpha t)$ is αf_s.

□

5.6.1.1 Orthogonality

It follows that the interpolation (5.76) is provided as a sum of the orthogonal $\operatorname{sinc} x$ functions that is illustrated in Fig. 5.19. The orthogonality is easily explained by the observation that each of these functions has only one nonzero value in the discrete time that lies at $x(nT)$. It means that when one of the functions attains its maximum at $x(nT)$ then all other functions cross zero at this point. In this regard, the interpolation by $\operatorname{sinc} x$ functions is akin to the Lagrange interpolation. Indeed, if to extend (5.65) over an infinite time then an interpolation formula may be generalized to

$$x(t) = \sum_{n=-\infty}^{\infty} x(nT)h(t-nT)\,, \tag{5.77}$$

where $h(t-nT) = \begin{cases} 1 & \text{if } t = nT \\ 0 & \text{if } t \neq nT \end{cases}$ demonstrates the same property as the Lagrange function (5.67) and the $\operatorname{sinc} x$ function in (5.76).

5.6.1.2 Nyquist Sample Rate

Sampling of continuous signals is usually provided in terms of the *sampling frequency* f_s or *sampling rate* that defines a number of samples per second taken from a continuous signal. A reciprocal of the sampling frequency is

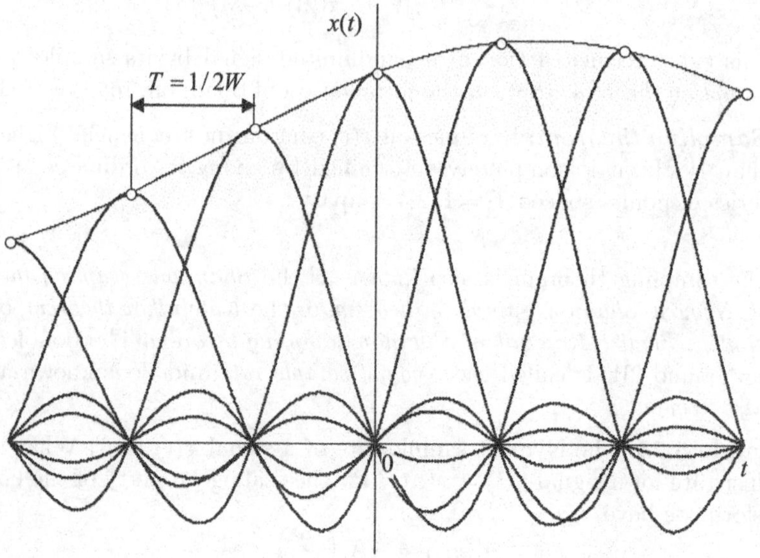

Fig. 5.19 Sampling and reconstruction of a continuous-time low-pass signal $x(t)$ with the Nyquist rate, $T = 1/2W$, over a finite interval of time.

the *sampling period* $T_s = 1/f_s$ or *sampling time*, which is the time between samples. Both f_s and T_s are claimed to be constants and thus the sampling theorem does not provide any rule for the cases when the samples are taken at a nonperiodic rate. It follows from Shannon's definition of sampling that the sampling frequency f_s must be equal to the Nyquist sample rate, i.e., $f_s = 2W$. Otherwise, the following quite feasible situations may be observed:

- If $f_s \cong 2W$, sampling satisfies the Shannon theorem and the signal is said to be *critically sampled.*
- If $f_s > 2W$ the signal is *oversampled* with a redundant number of points that, in applications, requires more memory and operation time, provided no loss of information about the signal waveform.
- If $f_s < 2W$, a signal is *undersampled* to mean that the number of points is not enough to avoid losing information about a signal waveform while recovering.

Example 5.24. Consider the Gaussian waveform $x(t) = Ae^{-(\alpha t)^2}$, for which spectral density is also Gaussian, $X(\omega) = \left(A\sqrt{\pi}/\alpha\right) e^{-\omega^2/4\alpha^2}$. Its Nyquist frequency W is conventionally determined as in Fig. 5.20a, where the sampling frequency is taken to be $f_s = 2W$ and sampling is thus critical. In Fig. 5.22b, the sampling frequency is set to be lower than the Nyquist rate. Accordingly, the waveform became undersampled causing the aliasing distortions. In Fig. 5.19c, the sampling frequency is higher than the Nyquist rate. Here no error occurs. The number of samples is just redundant.

□

The sampling theorem claims that a function $x(t)$ must contain no frequencies higher than W. However, the Fourier transform of many signal approaches zero with frequency asymptotically having no sharp bound, just as in the case of the Gaussian waveform (Fig. 5.20). Therefore, in applications, the Nyquist frequency W is conventionally determined for some reasonable attenuation level A_0. If to set A to be higher than its "true" value, $A > A_0$, sampling will be accompanied with aliasing. If $A < A_0$, the value of the sampling frequency will not be reasonable and a number of samples will be redundant (c).

5.6.1.3 Aliasing

Figure 5.20b neatly demonstrates that, by $f_s < 2W$, periodically repeated versions of a signal spectral density may overlap each other. The effect is called *aliasing* and, as we have shown above, is caused either by a small sampling frequency or by the Nyquist frequency that is lower than its "true" value. The phenomenon is also called *folding* and, in regard, the Nyquist frequency is sometimes mentioned as the *folding frequency.* Certainly, the aliasing effect needs to be avoided, once the frequency components of an original spectral density are distorted in the vicinity of an overlapped area by its image. This

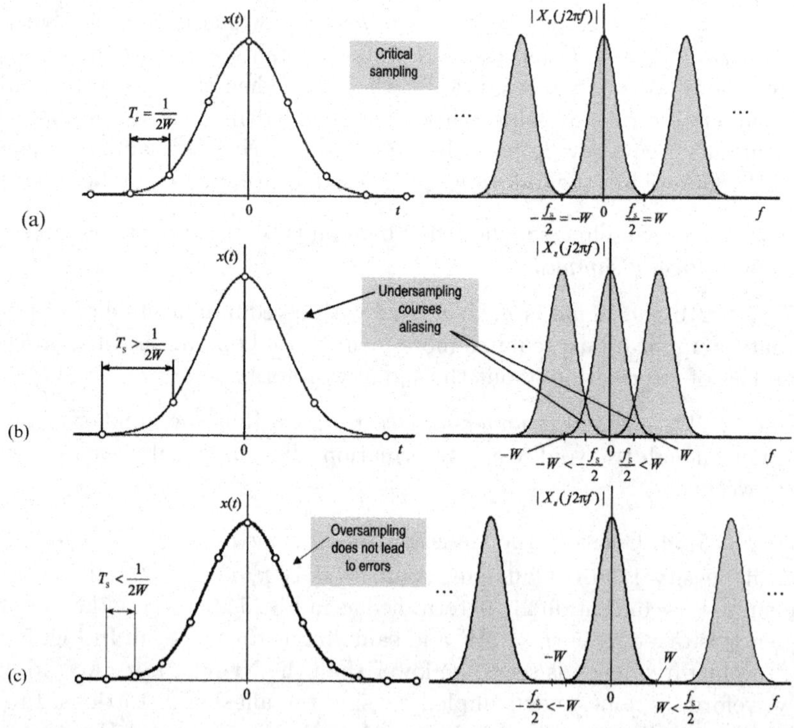

Fig. 5.20 Sampling of a Gaussian waveform: **(a)** critical sampling, **(b)** undersampling, and **(c)** oversampling.

distortion is called *aliasing distortion.* No one filter is able to restore the original signal exactly from its samples in presence of the aliasing distortions.

Example 5.25. A bandlimited signal $x(t)$ has a spectral density $X(j\omega)$ that does not equal to zero only in the range of $100\,\text{Hz} < f = \omega/2\pi < 3000$ Hz. What should be a sampling frequency to avoid aliasing distortions? The maximum frequency in this signal is $W = 3000$ Hz and thus the minimum sampling frequency must be $f_s = 2W = 6000$ Hz.

□

5.6.1.4 Sampling of Band-pass Signals

In its classical presentation, the sampling theorem considers an upper bound of the signal spectra and then offers a sampling frequency that is double the value of this bound. The approach is exact for any bandlimited signal. However, for substantially band-pass signals, the sampling frequency may significantly be reduced that is demonstrated in Fig. 5.21. Here we consider a spectral performance of a band-pass signal $y(t)$ that is centered about the carrier f_0 within a bandwidth BW.

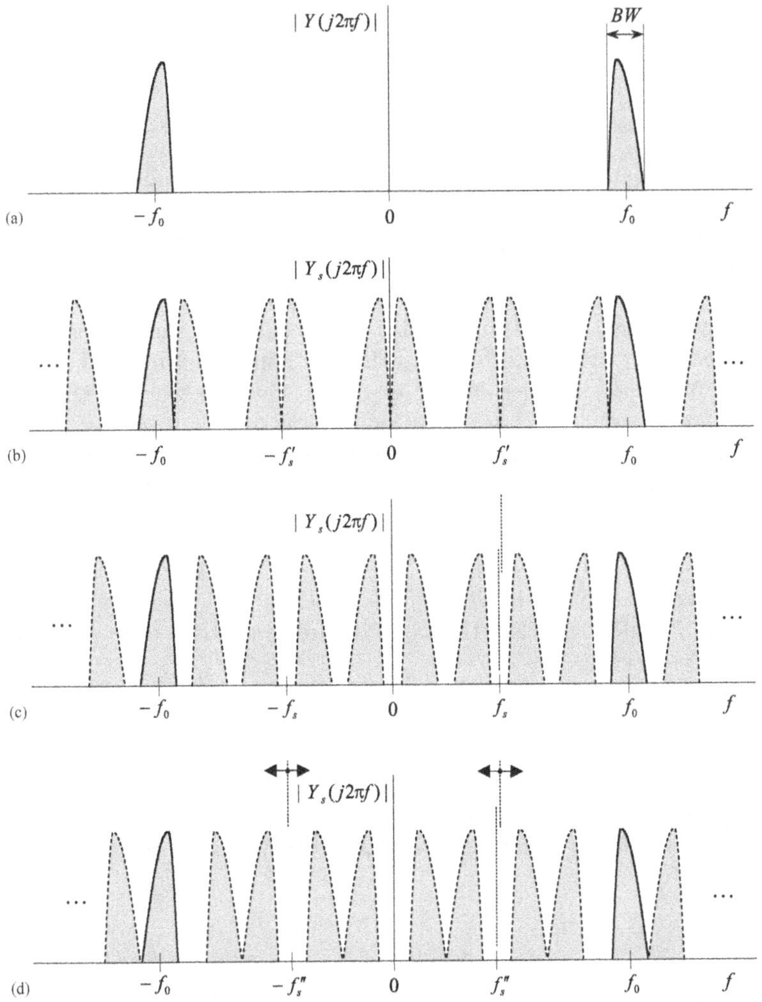

Fig. 5.21 Spectral density of a substantially band-pass sampled signal: **(a)** continuous-time; **(b)** sampled with a frequency f_s'; **(c)** sampled with $f_s'' \leqslant f_s \leqslant f_s'$; and (**d**) sampled with f_s''.

It may easily be found out that in the vicinity $-f_0 < f < f_0$ (Fig. 5.21a) there may be placed not more than k replications of a spectral density without aliasing (in Fig. 5.21b, $k = 4$). The placement cannot typically be done uniquely and we usually have two limiting cases to consider. In the first case (Fig. 5.21b), replications have a cross point at zero by a sampling frequency

$$f_s' = \frac{1}{k}(2f_0 - BW)\,.$$

We can also replace replications as in Fig. 5.21d that is achieved by reducing the sampling frequency to its second limiting value

$$f_s'' = \frac{1}{k+1}(2f_0 + BW)\,.$$

So, for the band-pass signal, a lowest sampling frequency lies in the range of

$$\frac{1}{k+1}(2f_0 + BW) \leqslant f_s \leqslant \frac{1}{k}(2f_0 - BW)\,. \tag{5.78}$$

As well as the Nyquist rate, the sampling frequency selected in the range (5.78) protects against aliasing distortions. In contrast, the approach has two practically important advantages. On the first hand, the substantially reduced sampling frequency produces larger spacing T_s between samples that requires a lower memory size. On the other hand, as it follows from Fig. 5.21b–d, a signal may be reconstructed from low frequencies rather than from the carrier frequency range that is preferable.

Example 5.26. A band-pass signal $y(t)$ has a nonzero Fourier transform in the range of $f_1 = 20$ kHz $\leqslant |f| \leqslant f_2 = W = 24$ kHz. Its central frequency is $f_0 = 22$ kHz and the bandwidth is BW $= 4$ kHz.

By the Shannon theorem, the sampling frequency is calculated to be $f_s = 2f_2 = 2W = 48$ kHz.

An integer number of replications k may be calculated by

$$k = 2\,\mathrm{int}\frac{f_0 - BW/2}{2BW}\,.$$

For the given signal, k calculates $k = 2\,\mathrm{int}\,(20/8) = 4$. Then, by (5.78), the range for the reduced sampling frequency is defined by 9.6 MHz $\leqslant f_s \leqslant 10$ MHz.

□

Example 5.27. The spectral density of a band-pass signal $y(t)$ has a bandwidth BW $= 5$ MHz about a carrier frequency of $f_0 = 20$ MHz.

By the Shannon theorem, the sampling frequency is calculated to be $f_s = 2W = 2(f_0 + BW/2) = 45$ MHz. The number of replications is $k = 2\ \mathrm{int}(17.5/10) = 2$. By (5.78), the range for the reduced sampling frequency calculates 15 MHz $\leqslant f_s \leqslant 17.5$ MHz.

□

5.6.1.5 Conversion

Sampling is usually associated with conversion of the continuous-time signal to its discrete-time samples performed by digital codes and vice versa. The devices are called ADC and DAC, respectively. More generally, we may say that both ADC and DAC are the direct and reverse linkers between the continuous-time real physical world and the discrete-time digital computer-based world, respectively.

5.6.2 Analog-to-digital Conversion

In applications, several common ways of implementing electronic ADCs are used:

- A *direct conversion* ADC for each decoded voltage range has a comparator that feeds a logic circuit to generate a code.
- A *successive-approximation* ADC uses a comparator to reject ranges of voltages, eventually settling on a final voltage range.
- A *delta-encoded* ADC has an up–down counter that feeds a DAC. Here both the input signal and the DAC output signal go to a comparator, where output adjusts the counter for a minimum difference between two signals.
- A *ramp-compare* ADC produces a sawtooth signal that ramps up, then quickly falls to zero. When the ramp starts, a timer starts counting. When the ramp voltage matches the input, a comparator fires, and the timer's value is recorded.
- A *pipeline* ADC uses two or more steps of subranging. First, a coarse conversion is done. In a second step, the difference to the input signal is determined and eliminated with a DAC.

In modelling ADC structures, three principle components are usually recognized (Fig. 5.22). The first component provides a conversion of the continuous signal to its samples and therefore is called the *sampler* or the *continuous-to-discrete converter.* The samples have a continuous range of possible amplitudes; therefore, the second component of the ADC is the *quantizer* intended to map the continuous amplitudes into a discrete set of amplitudes. The last component is the *encoder* that takes the digital signal and produces a sequence of binary codes and which is typically performed by the *pulse-code modulator* (PCM).

Figure 5.23 illustrates the process of analog-to-digital conversion assuming periodic sampling with period T_s. The value of T_s is usually set to be less than the reciprocal of the Nyquist rate, $T_s < 1/2W$, to avoid aliasing distortions. First, the continuous-time narrowband signal $x(t)$ (Fig. 5.23a) is multiplied by a *unit pulse-train*,

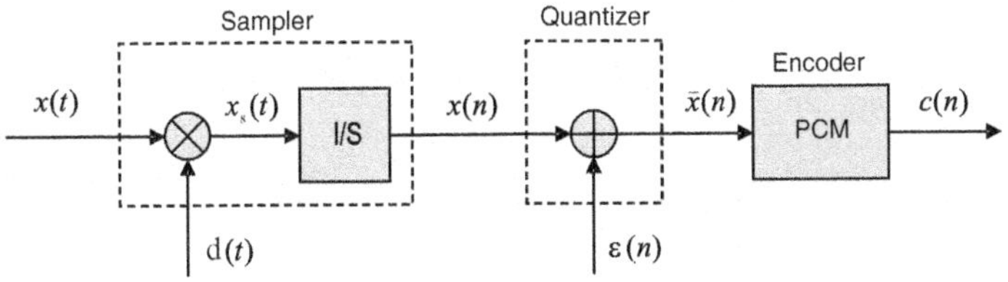

Fig. 5.22 Analog-to-digital converter.

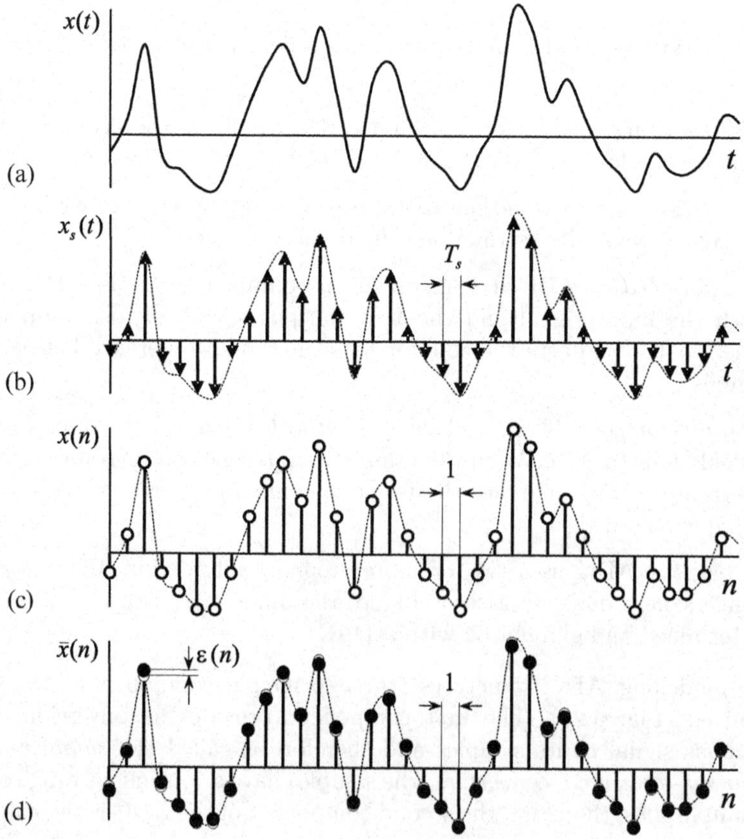

Fig. 5.23 Analog-to-digital conversion: **(a)** continuous-time signal; **(b)** sampled signal; **(c)** discrete-time samples; and **(d)** quantized discrete-time signal.

$$\mathrm{d}(t) = \sum_{n=-\infty}^{\infty} \delta(t - nT_\mathrm{s}) . \tag{5.79}$$

The product of the multiplication represents a *sampled signal* $x_\mathrm{s}(t)$ (Fig. 5.23b). Since the function $\mathrm{d}(t)$ does not equal zero only at discrete points nT_s, the product may be performed by

$$x_\mathrm{s}(t) = x(t)\mathrm{d}(t) = \sum_{n=-\infty}^{\infty} x(nT_\mathrm{s})\delta(t - nT_\mathrm{s}) . \tag{5.80}$$

In the frequency domain, the sampling process (5.80) is described by (5.72)–(5.74) to notice again that aliasing may take place if a sampling frequency $f_\mathrm{s} = 1/T_\mathrm{s}$ is not set to be higher than the Nyquist rate $2W$.

A sampled signal $x_\mathrm{s}(t)$ spaced with T_s is then converted into a discrete-time signal $x(n)$ (Fig. 5.23c), which values are indexed by the integer variable n, so that

$$x(n) = x(nT_s)\,.$$

This operation of *impulse-to-sample* (I/S) conversion may utilize a *zero-order hold* that holds a signal $x_s(t)$ between samples for T_s seconds producing a staircase function $\bar{x}_s(t)$. The I/S converter then has a constant input and converts $x_s(t)$ through $\bar{x}_s(t)$ to $x(n)$.

In the quantizer, the discrete-time signal $x(n)$ possesses the nearest of a finite number of possible values allowed by the discrete scale of amplitudes with usually a constant step Δ. The signal then becomes discrete both in the magnitude and time (Fig. 5.23d) owing to the operation of quantization,

$$\hat{x}(n) = \text{Quant}\left[x(n)\right].$$

Figure 5.24 shows a performance of the uniform quantizer. As it follows, the quantization process inevitably induces an error $\varepsilon(n)$ that is a difference between the discrete-time value $x(n)$ and the quantized value $\bar{x}(n)$. The error is bounded by $-\Delta/2 < \varepsilon(n) < \Delta/2$ and distributed uniformly, since the difference between $x(n)$ and $\hat{x}(n)$ may take any value within these bounds with equal probability.

Example 5.28. The continuous-time signal is measured in the range of -5 to 5 v. Resolution r of the ADC in this range is 12 bits, this is $2^r = 2^{12} = 4096$ quantization levels. Then the ADC voltage resolution is $(5 + 5)/4096 = 0.002441\ V$ or $2.441\ mV$.

□

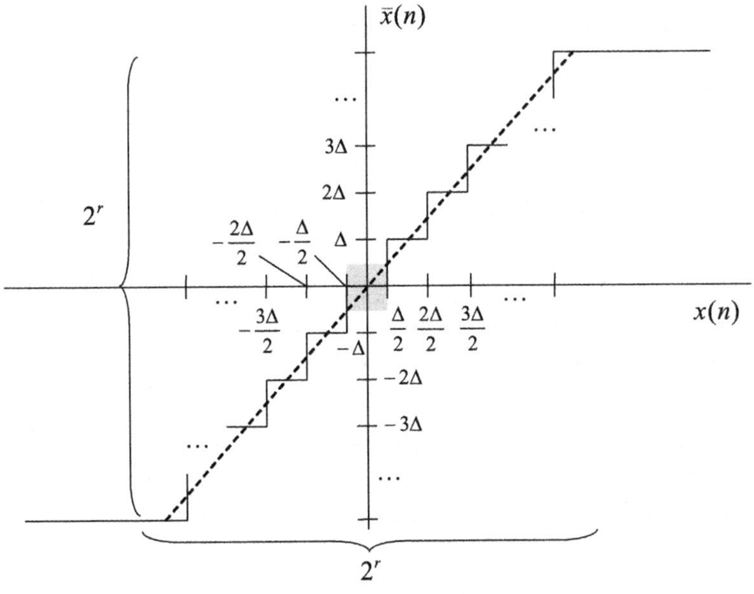

Fig. 5.24 Uniform quantizing: shadowed area corresponds to the quantization error.

In practice, resolution of ADC is limited by the noise intensity in signaling. With large noise, it is impossible to convert a signal accurately beyond a certain number of bits of resolution. This means that the output of ADC will not be accurate, since its lower bits are simply measuring noise. For example, if noise in the ADC has an intensity of about 3 mV, then the above-calculated resolution of 2.441 mV cannot be achieved and, practically, will be around double the resolution of the ADC.

5.6.3 Digital-to-analog Conversion

The reverse process of translating the discrete-time digital codes into a continuous-time electrical signals is called the *digital-to-analog conversion* . Accordingly, a DAC is said to be a device for converting a digital (usually binary) code to an analog signal (current, voltage or charges).

The following most common types of DACs are exploited:

- The *pulse width modulator* is the simplest DAC type. Here, a stable current (electricity) or voltage is switched into an LP analog filter with a duration determined by the digital input code.
- The *delta–sigma* DAC or the *oversampling* DAC represents a pulse density conversion technique that allows for the use of a lower resolution DAC internally (1-bit DAC is often chosen). The DAC is driven with a pulse-density-modulated signal, created through negative feedback that acts as a high-pass filter for the noise, thus pushing this noise out of the passband.
- The *binary-weighted* DAC contains one resistor or current source for each bit of the DAC. The resistors are connected to a summing point to produce the correct output value.
- The *R2R Ladder* DAC (a binary-weighted DAC) creates each value with a repeating structure of 2 resistor values, R and R times two.
- The *segmented* DAC contains an equal resistor or current source for each possible value of DAC output.
- The *hybrid* DAC uses a combination of the above techniques in a single converter. Most DAC integrated circuits are of this type due to the difficulty of getting low cost, high speed, and high precision in one device.

Example 5.29. The 4-bit binary-weighted DAC has four scaled resistances to obtain separately 1, 2, 4, and 8 V. Then the bit code 1011 produces $1 \times 8 + 0 \times 4 + 1 \times 2 + 1 \times 1 = 11\,\text{V}$.

□

Example 5.30. The 4-bit R2R Ladder DAC generates the voltage

$$V = V_0 \frac{R_0}{R} \left(\frac{D_0}{16} + \frac{D_1}{8} + \frac{D_2}{4} + \frac{D_3}{2} \right) ,$$

where $R_0 = R$, $V_0 = 10$ V is a reference voltage, and D_i takes the value 0 or 1. Then the bit code 1011 produces $V = 10(1/16 + 1/8 + 0/4 + 1/2) = 6.875\,\mathrm{V}$.
□

Transition from the discrete-time digital code $c(n)$ to the continuous-time signal $x(t)$ is basically modelled by three steps. First, the discrete database is decoded to discrete samples $x(n)$. Second, the discrete samples $x(n)$ are performed by a *sample-to-impulse* (S/I) converter as a sequence of impulses $x_s(t)$. Finally, a *reconstruction filter* is used to go from $x_s(t)$ to a continuous-time signal $x(t)$. Provided a conversion of $c(n)$ to $x(n)$, two types of the rest part of DAC may be recognized.

5.6.3.1 DAC with an Ideal Reconstruction Filter

Figure 5.25 shows the structure of the DAC utilizing an ideal LP filter. In the S/I converter, discrete samples are multiplied with the unit pulse-train that produces a sequence of impulses

$$x_s(t) = \sum_{n=-\infty}^{\infty} x(n)\delta(t - nT_s)\,. \tag{5.81}$$

The result (5.81) is then processed by a *reconstruction filter*, which is required to be an ideal LP filter with the transfer function

$$H_r(j\omega) = \begin{cases} T_s, & \text{if } |\omega| \leqslant \frac{\pi}{T_s} \\ 0, & \text{if } |\omega| > \frac{\pi}{T_s} \end{cases}. \tag{5.82}$$

Having such a filter, the DAC becomes "ideal." Because the impulse response of the ideal reconstruction filter is

$$h_r(t) = \frac{\sin \pi t/T_s}{\pi t/T_s}\,, \tag{5.83}$$

the output of the filter is a convolution

$$x(t) = x_s(t) * h_r(t) = \int_{-\infty}^{\infty} x_s(\theta)h_r(t-\theta)\mathrm{d}\theta\,. \tag{5.84}$$

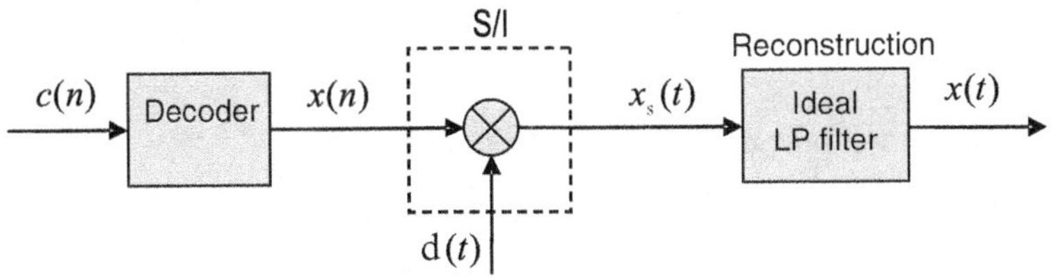

Fig. 5.25 Digital-to-analog converter with an ideal reconstruction filter.

Substituting (5.81) and (5.82) and using the sifting property of the delta function transforms (5.84) to the interpolation

$$
\begin{aligned}
x(t) &= \int_{-\infty}^{\infty} \left[\sum_{n=-\infty}^{\infty} x(n)\delta(\theta - nT_s) \right] h_\mathrm{r}(t-\theta)\mathrm{d}\theta \\
&= \sum_{n=-\infty}^{\infty} x(n) \int_{-\infty}^{\infty} \delta(\theta - nT_\mathrm{s}) h_\mathrm{r}(t-\theta)\mathrm{d}\theta \\
&= \sum_{n=-\infty}^{\infty} x(n) h_\mathrm{r}(t - nT_\mathrm{s}) \\
&= \sum_{n=-\infty}^{\infty} x(n) \frac{\sin \pi(t - nT_\mathrm{s})/T_\mathrm{s}}{\pi(t - nT_\mathrm{s})/T_\mathrm{s}}
\end{aligned} \tag{5.85}
$$

that is virtually stated by the sampling theorem (5.76).

The spectral density of the interpolated signal $x(t)$ is easily produced by using the time-shifting property of the Fourier transform to be

$$
\begin{aligned}
X(j\omega) = \mathcal{F}x(t) &= \sum_{n=-\infty}^{\infty} x(n)\mathcal{F}h_\mathrm{r}(t - nT_\mathrm{s}) \\
&= \sum_{n=-\infty}^{\infty} x(n) H_\mathrm{r}(j\omega)\mathrm{e}^{-jn\omega T_\mathrm{s}} \\
&= H_\mathrm{r}(j\omega) \sum_{n=-\infty}^{\infty} x(n)\mathrm{e}^{-jn\omega T_\mathrm{s}} = H_\mathrm{r}(j\omega) X(\mathrm{e}^{j\omega T_\mathrm{s}}) ,
\end{aligned} \tag{5.86}
$$

where $X(\mathrm{e}^{j\omega T_\mathrm{s}})$ is known in digital signal processing as the direct discrete Fourier transform of $x(n)$ that is a periodic function with period $1/T_\mathrm{s}$. Since the transfer function of the reconstruction filter is specified by (5.82), the spectral density (5.86) of the reconstructed signal becomes equivalent to

$$
X(j\omega) = \begin{cases} T_s X(e^{j\omega T_s}), & \text{if } |\omega| \leqslant \frac{\pi}{T_\mathrm{s}} \\ 0, & \text{if } |\omega| > \frac{\pi}{T_\mathrm{s}} \end{cases} . \tag{5.87}
$$

The result (5.87) means that the spectral density of a discrete signal $x(n)$ is magnitude-and frequency-scaled, as it is required by (5.73), and the ideal LP filter removes all its frequency components above the cutoff frequency π/T_s to produce the interpolated signal $x(t)$.

5.6.3.2 DAC with a Zero-order Hold

An ideal reconstruction filter cannot be realized practically, because of its infinite impulse responses. Therefore, many DACs utilize a *zero-order hold.*

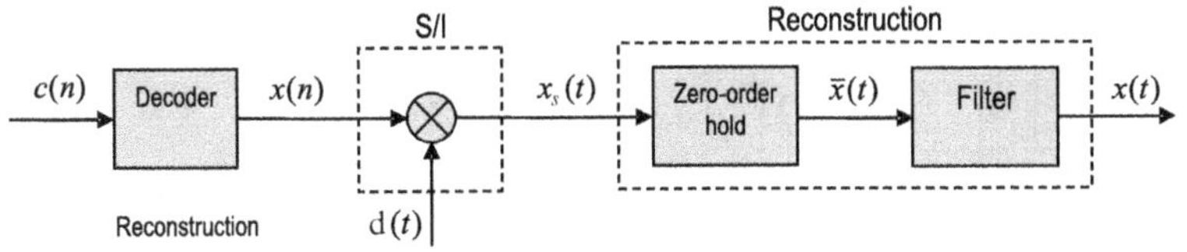

Fig. 5.26 Digital-to-analog converter utilizing a zero-order hold and compensation filter.

Figure 5.26 shows a structure of the DAC with a zero-order hold utilizing a compensation filter as a reconstruction filter.

In this DAC, conversion from the discrete samples $x(n)$ to the impulse samples $x_{\mathrm{s}}(t)$ is obtained in the same way as in Fig. 5.25 that is illustrated in Fig. 5.27a and b. Thereafter, a zero-order hold transforms $x_{\mathrm{s}}(t)$ to the staircase function $\bar{x}(t)$ (Fig. 5.27c) and the compensation filter recovers a continuous signal $x(t)$ (Fig. 5.27d). Since the postprocessing output compensation filter plays a key role here, we need to know what exactly should be its transfer function. An analysis is given below.

The impulse response and the transfer function of a zero-order hold are given respectively, by

$$h_0(t) = \begin{cases} 1, & \text{if } 0 \leqslant t \leqslant T_{\mathrm{s}} \\ 0, & \text{otherwise} \end{cases} \tag{5.88}$$

$$H_0(j\omega) = T_{\mathrm{s}} \left(\frac{\sin \omega T_{\mathrm{s}}/2}{\omega T_{\mathrm{s}}/2} \right) \mathrm{e}^{-j\omega \frac{T_2}{2}} \tag{5.89}$$

Generic relation (5.86) then gives us the spectral density of the reconstructed signal without postprocessing,

$$\begin{aligned} X_0(j\omega) &= H_0(j\omega) X(\mathrm{e}^{j\omega T_{\mathrm{s}}}) \\ &= T_{\mathrm{s}} \frac{\sin \omega T_{\mathrm{s}}/2}{\omega T_{\mathrm{s}}/2} \mathrm{e}^{-j\omega \frac{T_2}{2}} X(\mathrm{e}^{j\omega T_{\mathrm{s}}}) \,, \end{aligned} \tag{5.90}$$

which shape is shown in Fig. 5.28a. We, however, still want to obtain errorless reconstruction. Therefore, the area of distortions (shadowed area in Fig. 5.28a) must somehow be reduced to zero. Basically, it is achieved to include at the output of the DAC an auxiliary compensating filter with the transfer function

$$H_0(j\omega) = \begin{cases} \frac{\omega T_{\mathrm{s}}/2}{\sin \omega T_{\mathrm{s}}/2} e^{j\omega \frac{T_{\mathrm{s}}}{2}}, & \text{if } |\omega| \leqslant \frac{\pi}{T_{\mathrm{s}}} \\ 0, & \text{otherwise} \end{cases} \tag{5.91}$$

that is shown in Fig. 5.28b. This means that the resulting effect of a series connection of a zero-order hold (Fig. 5.28a) and a compensation filter (Fig. 5.28b) is desirable to be very close to that of an ideal LP filter. However, again we need to implement a filter, which transient range (Fig. 5.28b) is sharp and thus it is also a kind of ideal filters. To do it efficiently, the postprocessing of a staircase-function signal may be provided utilizing external resources.

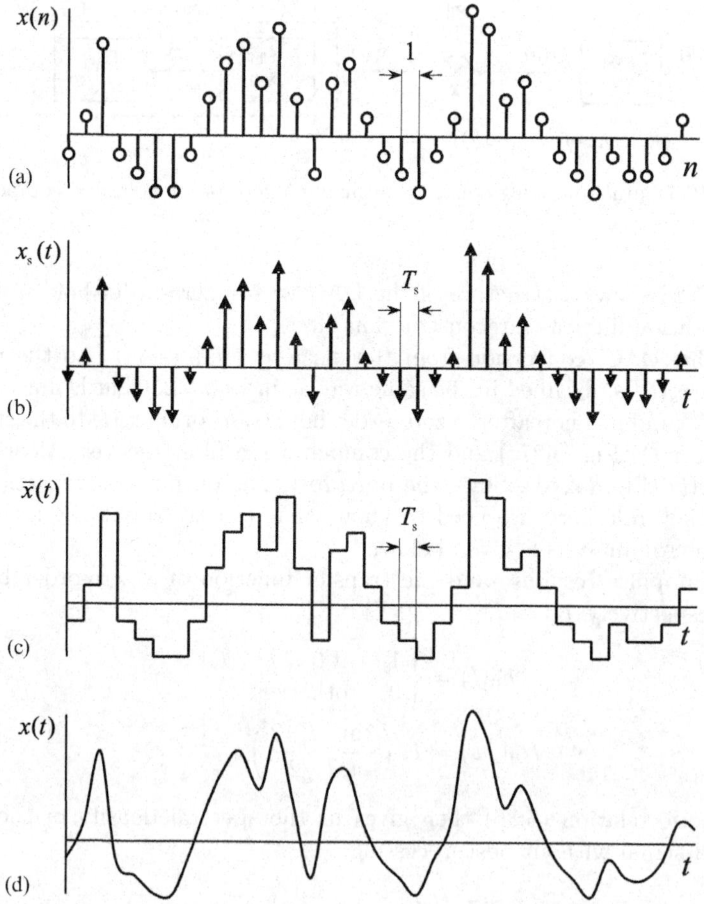

Fig. 5.27 Digital-to-analog conversion with a zero-order hold: **(a)** discrete samples; **(b)** impulse samples; **(c)** staircase-function signal; and **(d)** recovered continuous-time signal.

5.7 Summary

Bandlimited signals may be described, studied, and generated using the tool named Hilbert transform. What are the major features of these signals? First of all, *bandlimited signals are infinite in time.* Does this mean that they cannot be realized physically? Theoretically, Yes! Practically, however, no real physical process may pretend to be absolutely bandlimited with some frequency. Energy of many real bandlimited processes diminishes with frequency and approaches zero asymptotically. A concept of a bandlimited signal is thus conditional in this sense. An illustration is the Gaussian waveform that is infinite both in time and in frequency; however, practically is finite in time and frequency bandlimited.

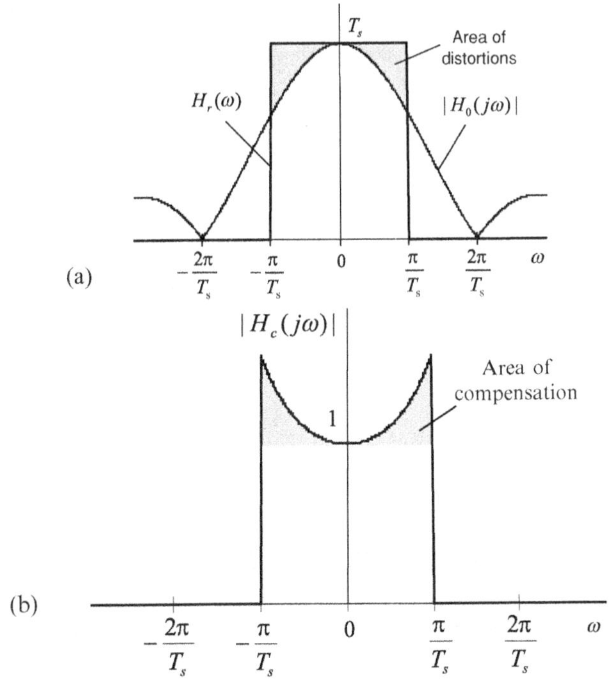

Fig. 5.28 Amplitude vs. frequency responses of the reconstruction filters: **(a)** ideal $H_r(\omega)$ and zero-order hold $H_0(j\omega)$ and **(b)** compensation $H_c(j\omega)$.

The theory of bandlimited signals postulates the following:

- Spectral width of a narrowband signal is much smaller than the carrier frequency.
- Narrowband signals are quasiharmonic; their amplitudes, informative phases, and frequencies are slowly changing in time functions.
- The Hilbert transform does not change the amplitudes of the spectral components of signals. It just alters their phases by $\pi/2$ for all negative frequencies and by $-\pi/2$ for all positive frequencies.
- The Hilbert transform couples the real and imaginary parts of an analytic signal of the spectral density of a causal signal and of the transfer function of any system.
- An analytic signal is a complex signal, which real component is a real signal and which imaginary component is the Hilbert transform of a real signal.
- Spectral density of an analytic signal exists only in the positive frequency domain.
- Any analytic signal is characterized with the envelope, total phase, and instantaneous frequency.

- Interpolation is the process to present signal samples by a continuous-time function; interpolation is associated with bandlimited signals.
- The sampling theorem establishes a correspondence between the signal boundary frequency (Nyquist frequency) and the sampling time; it also gives a unique interpolation formula in the orthogonal basis of the functions $\sin(x)/x$.
- Aliasing is a distortion of a signal interpolation if a number of samples is not enough per a time unit.
- Any real system operates with real signals; to operate with complex signals in discrete time, a computer-based digital SP part is included.
- Translation of continuous-time real signals to digital codes and vice versa is obtained using ADCs and DACs, respectively.

5.8 Problems

5.31 (Signals with bandlimited spectrum). The envelope of an ideal band-pass signal (5.2) has a peak value $y(0) = 5\,\text{V}$. Its principle lobe has zero at $|t| = 10\,\text{ms}$ Determine this signal in the frequency domain. Draw the plot of its spectral density.

5.32. Explain why the concept of an equivalent bandwidth W_{XX} is not an appropriate measure of the Nyquist rate $2W$ for all signals?

5.33. Propose a proper relation between W_{XX} and $2W$ for the following Gaussian-waveform signals avoiding aliasing distortions:

1. $x(t) = \text{e}^{(2t)^2}$
2. $y(t) = \text{e}^{(2t)^2} \cos 2\pi 5t$

5.34. The following relations are practically established between the equivalent bandwidth of a signal W_{XX} and its Nyquist rate $2W$. What are these signals? Draw plots.

1. $2W_{XX} = 2W$
2. $3W_{XX} = 2W$
3. $\infty W_{XX} = 2W$

5.35. Given the following LP signals. Define time functions of these signals:

1. $X(j\omega) = \begin{cases} X_0 \text{e}^{j\omega\tau}, & \text{if } |\omega| \leq \frac{\pi}{2\tau} \\ 0, & \text{otherwise} \end{cases}$
2. $X(j\omega) = \begin{cases} X_0(1 - \text{e}^{j\omega\tau}), & \text{if } |\omega| \leq \frac{\pi}{2\tau} \\ 0, & \text{otherwise} \end{cases}$

3. $X(j\omega) = X_0 e^{j\omega\tau - \omega^2\tau^2}$

4. $X(j\omega) = \begin{cases} X_0(a + be^{j\omega\tau}), & \text{if } |\omega| \leq \frac{\pi}{2\tau} \\ 0, & \text{otherwise} \end{cases}$

5. $X(j\omega) = \begin{cases} X_0(a + be^{-j\omega^2\tau^2}), & \text{if } |\omega| \leq \frac{\tau}{2} \\ 0, & \text{otherwise} \end{cases}$

6. $X(j\omega) = X_0 \omega\tau e^{-j\omega\tau}$

7. $X(j\omega) = \begin{cases} X_0 - |\omega|, & \text{if } |\omega| \leq X_0 \\ 0, & \text{otherwise} \end{cases}$

5.36 (Narrowband signals). Why the narrowband signal is also called a quasiharmonic signal? Give a simple graphical explanation.

5.37. A narrowband signal has a step-changed frequency:

$$y(t) = \begin{cases} A_0 \cos \omega_0 t, & \text{if } t \leqslant 0 \\ A_0 \cos(\omega_0 + \Omega)t, & \text{if } t > 0 \end{cases} .$$

Determine the complex envelope of this signal.

5.38. The following signals are given as modulated in time. Rewrite these signals in the narrowband form (5.5). Determine the envelope (5.9), informative phase (5.11), and instantaneous frequency (5.13):

1. $y(t) = A_0(1 + a \cos \Omega t) \cos (\omega_0 t + \psi_0)$
2. $y(t) = A_0 \left[u\left(t + \frac{\tau}{2}\right) - u\left(t - \frac{\tau}{2}\right)\right] \sin (\omega_0 t - \psi_0)$
3. $y(t) = A_0 e^{-(\alpha t)^2} \cos (\omega_0 t - \psi_0)$
4. $y(t) = A_0 \cos \left[\omega_0 t + b \sin(\Omega_0 t - \psi_0)\right]$
5. $y(t) = A_0 \cos \left(\omega_0 t + \frac{\alpha t^2}{2}\right)$
6. $y(t) = A_0 \cos \left[\omega_0 t + b \cos(\Omega_0 t + \psi_0)\right]$

5.39. A real signal has an asymmetric spectral density that is given in the positive frequency domain with

$$Y(j\omega) = \begin{cases} 0.5Y_0\left(1 - \frac{\omega - \omega_0}{\omega_1 - \omega_0}\right), & \text{if } \omega_0 \leqslant \omega \leqslant \omega_1 \\ 0, & \text{if } 0 \leqslant \omega < \omega_0 \text{ , } \omega_1 < \omega \end{cases}$$

Based on Example 5.1, use an identity, $\int x e^{\alpha x} dx = e^{\alpha x}\left(\frac{x}{\alpha} - \frac{1}{\alpha^2}\right)$ and write the spectral density of the complex envelope of this signal and determine its physical envelope, total phase, and instantaneous frequency. Show plots of these functions.

5.40 (Hilbert transform). Why the Hilbert transform of any signal is orthogonal to this signal? Give a simple graphical explanation.

5.41. A signal is given as a sum of harmonic signals. Define its Hilbert transform. Determine the physical envelope, total phase, and instantaneous frequency:

1. $y(t) = 2\cos\omega_0 t + \cos 2\omega_0 t$
2. $y(t) = \cos\omega_1 t + 2\sin\omega_2 t$
3. $y(t) = 2\cos\omega_1 t - \sin\omega_2 t + 2\cos(\omega_2 - \omega_1)t$

5.42. Prove the Hilbert transforms of the given signals. Show plots of the signal and its Hilbert transform.

1. $\mathcal{H}\frac{ab-t^2}{(ab-t^2)^2+t^2(a+b)^2} = \frac{t(a+b)}{(ab-t^2)^2+t^2(a+b)^2}$, a and b are real
2. $\mathcal{H}[\delta(t-a)+\delta(x+a)] = \frac{2}{\pi}\frac{a}{t^2-a^2}$, a is real
3. $\mathcal{H}\begin{cases} 1-\frac{t}{b}, & \text{if } 0 \leqslant t \leqslant b \\ 0, & \text{otherwise} \end{cases} = \frac{1}{\pi}\left[\frac{t-b}{b}\ln\left|\frac{t-b}{t}\right| + 1\right]$

5.43 (Analytic signals). Given two harmonic signals, $A_0\cos\omega_0 t$ and $A_0\sin\omega_0 t$. Write their analytic forms.

5.44. An ideal pass band signal has a value 2 V of its physical envelope at $Wt = \pi$ and a value 2 V/Hz of its spectral density at $\omega = 0$. Determine the boundary frequency W of this signal.

5.45. The Hilbert transform of an analytic signal is given as $\hat{y}_a(t) = -je^{j\omega t}$. Determine an analytic signal $y_a(t)$.

5.46. Given two analytic signals, $y_{a1}(t) = 2e^{j\omega_0 t}$ and $y_{a2}(t) = e^{2j\omega_0 t}$. Define their Hilbert transforms and determine the products: $\hat{y}_{a1}(t)y_{a2}(t)$, $y_{a1}(t)\hat{y}_{a2}(t)$, and $y_{a1}(t)y_{a2}(t)$.

5.47. AM signals with a simplest harmonic modulation and known carrier frequency ω_0 are given with their oscillograms (Fig. 5.29). Write each signal in the analytic form.

5.48. Spectral densities of analytic signals are shown in Fig. 5.30. Give a time presentation for each of these signals. Determine the physical envelope, total phase, and instantaneous frequency.

5.49 (Interpolation). Bring examples when interpolation is necessary to solve practical problems.

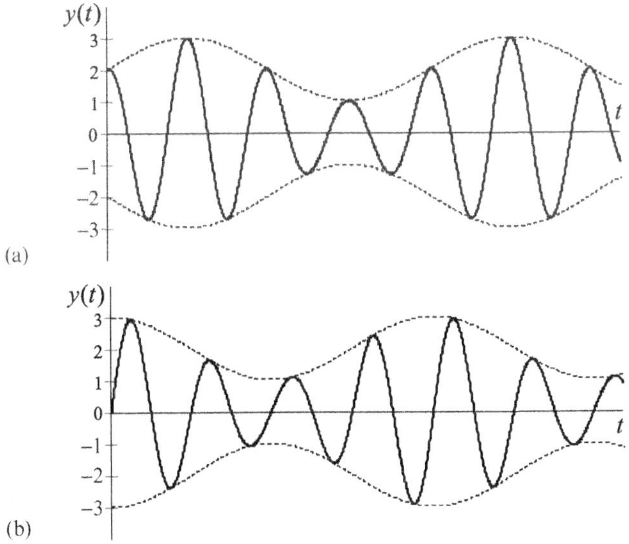

Fig. 5.29 AM signals.

5.50. A signal is measured at three points of an angular measure ϕ: $y(-\pi/2) = 0$, $y(0) = 2$, and $y(\pi/2) = 0$. Interpolate this signal with the Lagrange polynomial. Compare the interpolating curve with the cosine function $2\cos\phi$.

5.51. A rectangular pulse is given in the time interval from $t_{\min} = -1$ to $t_{\max} = 1$ (in seconds) with the function

$$x(t) = \begin{cases} 1, & \text{if } -0.5 \leqslant t \leqslant 0.5 \\ 0, & \text{otherwise} \end{cases}$$

Present this signal with equally placed samples and interpolate using the Newton form for a number of samples $n = M + 2$, where M is your personal number. Predict the interpolation function for $n \gg 1$. Compare the result with a presentation by the Fourier series. Can you see the Gibbs phenomenon in the interpolation function?

5.52 (Sampling). Argue why the sampling theorem is of practical importance? Why not just to sample a signal with a huge number of points?

5.53. A signal $y(t)$ is sampled with different rates. Spectral densities of sampled signals are shown in Fig. 5.31. Argue which spectrum corresponds to the critical sampling, undersampling, and oversampling?

5.54. The Nyquist sample rate is practically limited with $f_s \leqslant 10\text{kHz}$. What is the frequency limitation for the continuous-time harmonic signal?

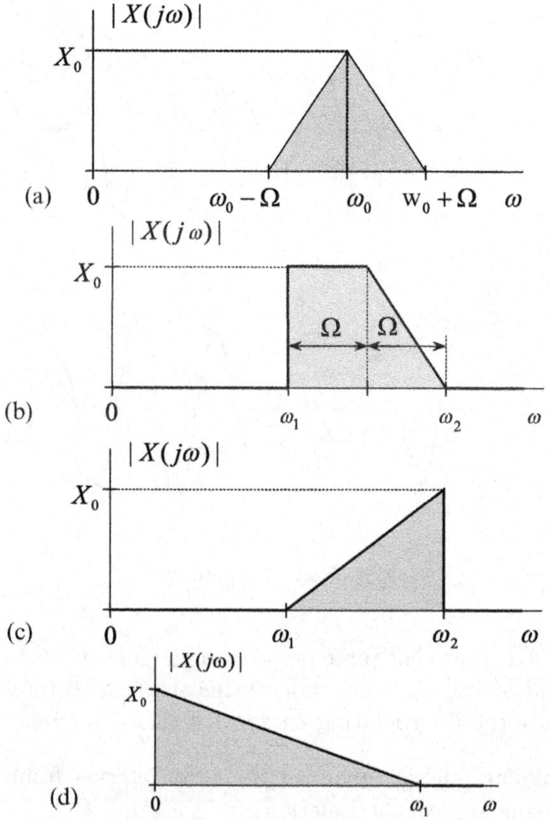

Fig. 5.30 Spectral densities of analytic signals.

5.55. The Nyquist sample rate of a signal $x(t)$ is f_s. Determine the Nyquist sample rate for the following signals:

1. $x(2t)$
2. $x(t/2)$
3. $x(3t) + x(t)$

5.56. Determine a minimum sample rate for the following bandlimited signal, which exist in the given frequency range:

1. 2 MHz $< f <$ 5 MHz
2. 900 kHz $< f <$ 1100 kHz
3. 100 MHz $< f <$ 250 MHz and 350 MHz $< f <$ 450 MHz

5.57 (Analog-to-digital conversion). Argue the necessity to provide digitization of the continuous-time signals for some applications. Why not to put an analog signal directly to the computer and obtain processing?

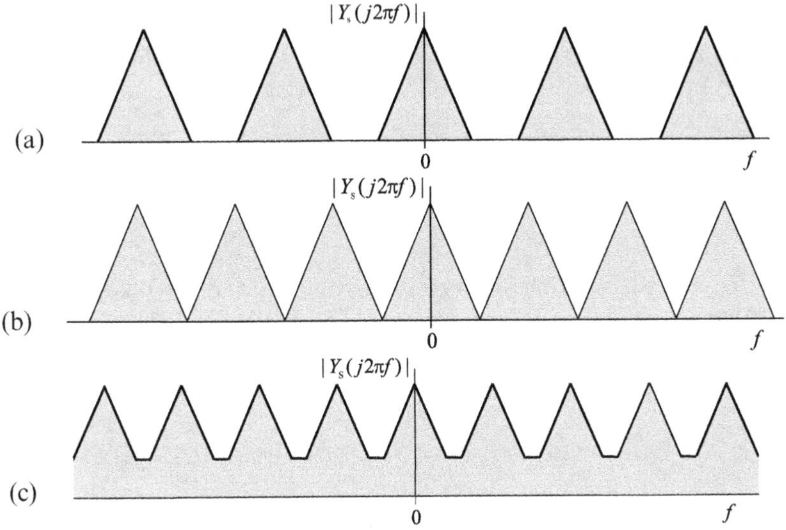

Fig. 5.31 Spectral densities of a sampled signal with different rates.

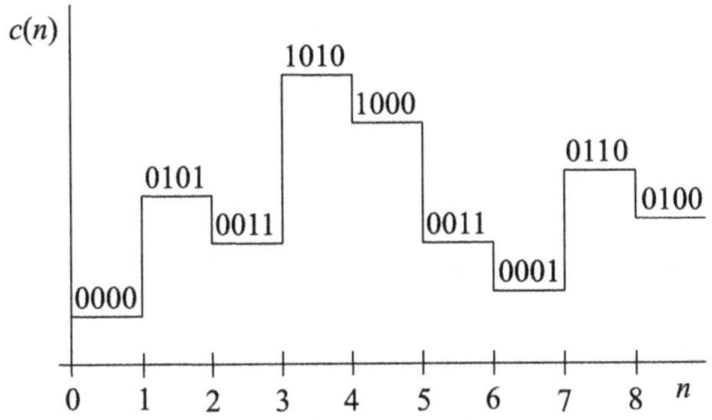

Fig. 5.32 4-bit digital code of a converted analog signal.

5.58. An analog signal is digitized with the 4-bit ADC as it is shown in Fig. 5.32. A resolution of the ADC is 2 mV and sampling is provided with a sampling time 2 ms. Restore samples of the signal. What may be its continuous-time function if a signal is (1) critically sampled and (2) undersampled?

5.59. An analog signal is measured in the range of 0–10 V. The root-mean-square noise voltage in the system is 2.1 mV. What should be a resolution of the ADC in bits to provide the ADC voltage resolution at the noise level?

5.60 (Digital-to-analog conversion). Why the conversion of digital signals is of importance? Why not to use digital codes directly to provide filtering by an analog filter?

5.61. The output voltage range of the 6-bit binary-weighted DAC is +10 to −10 and a zero code 000000 corresponds to −10 V. What is the output voltage generated by the codes 101101, 110001, and 010001?

5.62. The 6-bit R2R Ladder DAC generates the voltage

$$V = V_0 \frac{R_0}{R} \left(\frac{D_0}{64} + \frac{D_1}{32} + \frac{D_2}{16} + \frac{D_3}{8} + \frac{D_4}{4} + \frac{D_5}{2} \right),$$

where $R_0 = R$, $V_0 = 10\,\text{V}$ is a reference voltage, and D_i takes the value 0 or 1. What is the output voltage generated by the codes 101100, 011101, and 101010?

A

Tables of Fourier Series and Transform Properties

Table A.1 Properties of the continuous-time Fourier series

$$x(t) = \sum_{k=-\infty}^{\infty} C_k e^{jk\Omega t} \qquad C_k = \frac{1}{T}\int_{-T/2}^{T/2} x(t)e^{-jk\Omega t}dt$$

Property	Periodic function $x(t)$ with period $T = 2\pi/\Omega$	Fourier series C_k
Time shifting	$x(t \pm t_0)$	$C_k e^{\pm jk\Omega t_0}$
Time scaling	$x(\alpha t)$, $\alpha > 0$	C_k with period $\frac{T}{\alpha}$
Differentiation	$\frac{d}{dt}x(t)$	$jk\Omega C_k$
Integration	$\int_{-\infty}^{t} x(t)dt < \infty$	$\frac{1}{jk\Omega}C_k$
Linearity	$\sum_i \alpha_i x_i(t)$	$\sum_i \alpha_i C_{ik}$
Conjugation	$x^*(t)$	C^*_{-k}
Time reversal	$x(-t)$	C_{-k}
Modulation	$x(t)e^{jK\Omega t}$	C_{k-K}
Multiplication	$x(t)y(t)$	$\sum_{i=-\infty}^{\infty} C_{xi}C_{y(k-i)}$
Periodic convolution	$\int_T x(\theta)y(t-\theta)d\theta$	$TC_{xk}C_{yk}$
Symmetry	$x(t) = x^*(t)$ real	$\begin{cases} C_k = C^*_{-k}\,, \lvert C_k\rvert = \lvert C_{-k}\rvert\,, \\ \operatorname{Re} C_k = \operatorname{Re} C_{-k}\,, \\ \operatorname{Im} C_k = -\operatorname{Im} C_{-k}\,, \\ \arg C_k = -\arg C_{-k} \end{cases}$
	$x(t) = x^*(t) = x(-t)$ real and even	$\begin{cases} C_k = C_{-k}\,, C_k = C^*_k\,, \\ \text{real and even} \end{cases}$
	$x(t) = x^*(t) = -x(-t)$ real and odd	$\begin{cases} C_k = -C_{-k}\,, C_k = -C^*_k\,, \\ \text{imaginary and odd} \end{cases}$
Parseval's theorem	$\frac{1}{T}\int_{-T/2}^{T/2} \lvert x(t)\rvert^2 dt = \sum_{k=-\infty}^{\infty} \lvert C_k\rvert^2$	

Table A.2 Properties of the continuous-time Fourier transform

$$x(t) = \frac{1}{2\pi}\int_{-\infty}^{\infty} X(j\omega)\mathrm{e}^{j\omega t}\mathrm{d}\omega \qquad X(j\omega) = \int_{-\infty}^{\infty} x(t)\mathrm{e}^{-j\omega t}\mathrm{d}t$$

Property	Nonperiodic function $x(t)$	Fourier transform $X(j\omega)$
Time shifting	$x(t \pm t_0)$	$\mathrm{e}^{\pm\omega t_0}X(j\omega)$
Time scaling	$x(\alpha t)$	$\frac{1}{\lvert\alpha\rvert}X\left(\frac{j\omega}{\alpha}\right)$
Differentiation	$\frac{\mathrm{d}}{\mathrm{d}t}x(t)$	$j\omega X(j\omega)$
Integration	$\int_{-\infty}^{t} x(t)\mathrm{d}t$	$\frac{1}{j\omega}X(j\omega) + \pi X(j0)\delta(\omega)$
	$\int_{-\infty}^{\infty} x(t)\mathrm{d}t$	$X(j0)$
Frequency integration	$2\pi x(0)$	$\int_{-\infty}^{\infty} X(j\omega)\mathrm{d}\omega$
Linearity	$\sum_i \alpha_i x_i(t)$	$\sum_i \alpha_i X_i(j\omega)$
Conjugation	$x^*(t)$	$X^*(-j\omega)$
Time reversal	$x(-t)$	$X(-j\omega)$
Modulation	$x(t)\mathrm{e}^{j\omega_0 t}$	$X(j\omega - j\omega_0)$
Multiplication	$x(t)y(t)$	$\frac{1}{2\pi}X(j\omega) \times Y(j\omega)$
Convolution	$x(t) * y(t)$	$X(j\omega)Y(j\omega)$
Symmetry	$x(t) = x^*(t)$ real	$\begin{cases} X(j\omega) = X^*(-j\omega), \\ \lvert X(j\omega)\rvert = \lvert X(-j\omega)\rvert, \\ \mathrm{Re}\, X(j\omega) = \mathrm{Re}\, X(-j\omega), \\ \mathrm{Im}\, X(j\omega) = -\mathrm{Im}\, X(-j\omega), \\ \arg X(j\omega) = -\arg X(-j\omega) \end{cases}$
	$x(t) = x^*(t) = x(-t)$ real and even	$\begin{cases} X(j\omega) = X(-j\omega), \\ X(j\omega) = X^*(j\omega), \\ \text{real and even} \end{cases}$
	$x(t) = x^*(t) = -x(-t)$ real and odd	$\begin{cases} X(j\omega) = -X(-j\omega), \\ X(j\omega) = -X^*(j\omega), \\ \text{imaginary and odd} \end{cases}$
Rayleigh's theorem		$E_x = \int_{-\infty}^{\infty} \lvert x(t)\rvert^2\mathrm{d}t = \frac{1}{2\pi}\int_{-\infty}^{\infty} \lvert X(j\omega)\rvert^2\mathrm{d}\omega$

B

Tables of Fourier Series and Transform of Basis Signals

Table B.1 The Fourier transform and series of basic signals

Signal $x(t)$	Transform $X(j\omega)$	Series C_k
1	$2\pi\delta(\omega)$	$C_0 = 1\ ,\ C_{k\neq 0} = 0$
$\delta(t)$	1	$C_k = \frac{1}{T}$
$u(t)$	$\frac{1}{j\omega} + \pi\delta(\omega)$	—
$u(-t)$	$-\frac{1}{j\omega} + \pi\delta(\omega)$	—
$e^{j\Omega t}$	$2\pi\delta(\omega - \Omega)$	$C_1 = 1\ ,\ C_{k\neq 1} = 0$
$\sum\limits_{k=-\infty}^{\infty} C_k e^{jk\Omega t}$	$2\pi \sum\limits_{k=-\infty}^{\infty} C_k \delta(\omega - k\Omega)$	C_k
$\cos \Omega t$	$\pi[\delta(\omega - \Omega) + \delta(\omega + \Omega)]$	$C_1 = C_{-1} = \frac{1}{2},\ C_{k\neq\pm 1} = 0$
$\sin \Omega t$	$\frac{\pi}{j}[\delta(\omega - \Omega) - \delta(\omega + \Omega)]$	$C_1 = -C_{-1} = \frac{1}{2j},\ C_{k\neq\pm 1} = 0$
$\frac{1}{\alpha^2+t^2}$	$e^{-\alpha\|\omega\|}$	$\frac{1}{T} e^{-\frac{2\pi\alpha\|k\|}{T}}$
Rectangular (Fig. 2.16c)	$A\tau \frac{\sin(\omega\tau/2)}{\omega\tau/2}$	$\frac{A}{q} \frac{\sin(k\pi/q)}{k\pi/q}$
Triangular (Fig. 2.21a)	$\frac{A\tau}{2} \frac{\sin^2(\omega\tau/4)}{(\omega\tau/4)^2}$	$\frac{A}{2q} \frac{\sin^2(k\pi/2q)}{(k\pi/2q)^2}$
Trapezoidal (Fig. 2.30)	$A\tau \frac{\sin(\omega\tau/2)}{\omega\tau/2} \frac{\sin(\omega\tau_s/2)}{\omega\tau_s/2}$	$\frac{A}{q} \frac{\sin(k\pi/q)}{k\pi/q} \frac{\sin(k\pi/q_s)}{k\pi/q_s}$
Ramp (Fig. 2.34b)	$\frac{A}{j\omega}\left[\frac{\sin(\omega\tau/2)}{\omega\tau/2} e^{j\frac{\omega\tau}{2}} - 1\right]$	$\frac{A}{j2\pi k}\left[\frac{\sin(k\pi/q)}{k\pi/q} e^{j\frac{k\pi}{q}} - 1\right]$
Ramp (Fig. 2.34c)	$\frac{A}{j\omega}\left[1 - \frac{\sin(\omega\tau/2)}{\omega\tau/2} e^{-j\frac{\omega\tau}{2}}\right]$	$\frac{A}{j2\pi k}\left[1 - \frac{\sin(k\pi/q)}{k\pi/q} e^{-j\frac{k\pi}{q}}\right]$
$\frac{\sin \alpha t}{\alpha t}$	$\begin{cases} \frac{\pi}{\alpha}, & \|\omega\| < \alpha \\ 0, & \|\omega\| > \alpha \end{cases}$	$\begin{cases} \frac{\pi}{\alpha T}, & \|k\| < \frac{\alpha T}{2\pi} \\ 0, & \|k\| > \frac{\alpha T}{2\pi} \end{cases}$
$e^{-\alpha t}u(t)$, $\mathrm{Re}\,\alpha > 0$	$\frac{1}{\alpha + j\omega}$	$\frac{1}{\alpha T + j2\pi k}$
$te^{-\alpha t}u(t)$, $\mathrm{Re}\,\alpha > 0$	$\frac{1}{(\alpha + j\omega)^2}$	$\frac{T}{(\alpha T + j2\pi k)^2}$

Table B.1 The Fourier transform and series of basic signals (*Contd.*)

$\frac{t^{n-1}}{(n-1)!}e^{-\alpha t}u(t)$, $\mathrm{Re}\,\alpha > 0$	$\frac{1}{(\alpha+j\omega)^n}$	$\frac{T^{n-1}}{(\alpha T+j2\pi k)^n}$
$e^{-\alpha\|t\|}$, $\alpha > 0$	$\frac{2\alpha}{\alpha^2+\omega^2}$	$\frac{2\alpha T}{\alpha^2T^2+4\pi^2k^2}$
$e^{-\alpha^2t^2}$	$\frac{\sqrt{\pi}}{\alpha}e^{-\frac{\omega^2}{4\alpha^2}}$	$\frac{\sqrt{\pi}}{\alpha T}e^{-\frac{\pi^2k^2}{\alpha^2T^2}}$

C_k corresponds to $x(t)$ repeated with period T, τ and τ_s are durations, $q = \frac{T}{\tau}$, and $q_s = \frac{T}{\tau_s}$.

Table B.2 The Fourier transform and series of complex signals

Signal $y(t)$	Transform $Y(j\omega)$	Series C_k
Burst of N pulses with known $X(j\omega)$	$X(j\omega)\frac{\sin(\omega NT/2)}{\sin(\omega T/2)}$	$\frac{1}{T_1}X\left(j\frac{2k\pi}{T_1}\right)\frac{\sin(k\pi/q_2)}{\sin(k\pi/Nq_2)}$
Rectangular pulse-burst (Fig. 2.47)	$A\tau\frac{\sin(\omega\tau/2)}{\omega\tau/2}\frac{\sin(\omega NT/2)}{\sin(\omega T/2)}$	$\frac{A}{q_1}\frac{\sin(k\pi/q_1)}{k\pi/q_1}\frac{\sin(k\pi/q_2)}{\sin(k\pi/Nq_2)}$
Triangular pulse-burst	$\frac{A\tau}{2}\frac{\sin^2(\omega\tau/4)}{(\omega\tau/4)^2}\frac{\sin(\omega NT/2)}{\sin(\omega T/2)}$	$\frac{A}{2q_1}\frac{\sin^2(k\pi/2q_1)}{(k\pi/2q_1)^2}\frac{\sin(k\pi/q_2)}{\sin(k\pi/Nq_2)}$
Sinc-shaped pulse-burst	$\begin{cases}\frac{A\pi}{\alpha}\frac{\sin(\omega NT/2)}{\sin(\omega T/2)}, & \|\omega\| < \alpha\\ 0, & \|\omega\| > \alpha\end{cases}$	$\begin{cases}\frac{A\pi}{\alpha T_1}\frac{\sin(k\pi/q_2)}{\sin(k\pi/Nq_2)}, & \|k\| < \frac{\alpha T_1}{2\pi}\\ 0, & \|k\| > \frac{\alpha T_1}{2\pi}\end{cases}$

C_k corresponds to $y(t)$ repeated with period T_1, τ is pulse duration, T is period of pulse in the burst, T_1 is period of pulse-bursts in the train, $q_1 = \frac{T_1}{\tau}$, and $q_2 = \frac{T_1}{NT}$.

C

Tables of Hilbert Transform and Properties

Table C.1 Properties of the Hilbert transform

$$y(t) = \frac{1}{\pi}\int_{-\infty}^{\infty}\frac{\hat{y}(\theta)}{\theta - t}\mathrm{d}\theta \qquad \hat{y}(t) = \frac{1}{\pi}\int_{-\infty}^{\infty}\frac{y(\theta)}{t-\theta}\mathrm{d}\theta$$

Property	Function $y(t)$	Transform $\hat{y}(t)$
Filtering	$y(t)$ is constant	$\hat{y}(t)$ is zero
	is real	is real
	is even (odd)	is odd (even)
Causality	If $y(t)$ is causal with known transform $Y(j\omega) = Y_{\mathrm{r}}(\omega) + jY_{\mathrm{i}}(\omega)$, then:	
	$Y_{\mathrm{r}}(\omega)$	$Y_{\mathrm{i}}(\omega)$
	$Y_{\mathrm{i}}(\omega)$	$-Y_r(\omega)$
Linearity	$\sum_i \alpha_i y_i(t)$	$\sum_i \alpha_i \hat{y}_i(t)$
Time shifting	$y(t \pm \theta)$	$\hat{y}(t \pm \theta)$
Time reversal	$y(-t)$	$-\hat{y}(-t)$
Scaling	$y(at)$	$\hat{y}(at)$
	$y(-at)$	$-\hat{y}(-at)$
Multiple transform	$\hat{y}(t)$	$-y(t)$
		$\mathcal{H}^3 y(t) = \mathcal{H}^{-1} y(t)$
		$\mathcal{H}^4 y(t) = y(t)$
		$\mathcal{H}^n y(t) \overset{\mathcal{F}}{\Leftrightarrow} [-j\,\mathrm{sgn}(\omega)]^n Y(j\omega)$
Differentiation	$\frac{\mathrm{d}}{\mathrm{d}t} y(t)$	$\frac{\mathrm{d}}{\mathrm{d}t}\hat{y}(t)$
Integration	$\int_a^b y(t)\mathrm{d}t$, a and b are constants	0
Convolution	$y_1(t) * y_2(t)$	$y_1(t) * \hat{y}_2(t)$
Autocorrelation	$y(t) * y(t)$	$\widehat{y(t) * \hat{y}(t)}$
Multiplication	$ty(t)$	$t\hat{y}(t) + \frac{1}{\pi}\int_{-\infty}^{\infty} y(t)\mathrm{d}t$

Table C.2 Useful relations between $y(t)$ and its Hilbert transform $\hat{y}(t)$

Property	Relation
Orthogonality	$\int_{-\infty}^{\infty} y(t)\hat{y}(t)\mathrm{d}t = 0$
Integration	$\int_{-\infty}^{\infty} y_1(t)\hat{y}_2(t)\mathrm{d}t = \int_{-\infty}^{\infty} \hat{y}_1(t)y_2(t)\mathrm{d}t$
	$\int_{-\infty}^{\infty} y_1(t)y_2(t)\mathrm{d}t = \int_{-\infty}^{\infty} \hat{y}_1(t)\hat{y}_2(t)\mathrm{d}t$
	$\int_{-\infty}^{\infty} y^2(t)\mathrm{d}t = \int_{-\infty}^{\infty} \hat{y}^2(t)\mathrm{d}t$
Energy	$\int_{-\infty}^{\infty} \lvert y(t)\rvert^2(t)\mathrm{d}t = \int_{-\infty}^{\infty} \lvert\hat{y}(t)\rvert^2\mathrm{d}t$
Autocorrelation	$\int_{-\infty}^{\infty} y(t)y(t-\theta)\mathrm{d}t = \int_{-\infty}^{\infty} \hat{y}(t)\hat{y}(t-\theta)\mathrm{d}t$

Table C.3 The Hilbert transform of analytic signals

Property	Signal	Transform
Analytic signal	$y_\mathrm{a}(t) = y(t) + j\hat{y}(t)$	$-jy_\mathrm{a}(t) = \hat{y}(t) - jy(t)$
Multiplication	$y_{\mathrm{a}1}(t)y_{\mathrm{a}2}(t)$	$-jy_{\mathrm{a}1}(t)y_{\mathrm{a}2}(t)$
		$= \hat{y}_{\mathrm{a}1}(t)y_{\mathrm{a}2}(t)$
		$= y_{\mathrm{a}1}(t)\hat{y}_{\mathrm{a}2}(t)$
Power n	$y_\mathrm{a}^n(t)$	$-jy_\mathrm{a}^n(t)$
Real product	$\mathrm{Re}[y_{\mathrm{a}1}(t)y_{\mathrm{a}2}(t)]$	$\mathrm{Im}[y_{\mathrm{a}1}(t)y_{\mathrm{a}2}(t)]$
Imaginary product	$\mathrm{Im}[y_{\mathrm{a}1}(t)y_{\mathrm{a}2}(t)]$	$-\mathrm{Re}[y_{\mathrm{a}1}(t)y_{\mathrm{a}2}(t)]$

Table C.4 Products of the analytic signals

Product	Relation
$y_{\mathrm{a}1}(t)y_{\mathrm{a}2}(t)$	$= j\hat{y}_{\mathrm{a}1}(t)y_{\mathrm{a}2}(t) = jy_{\mathrm{a}1}(t)\hat{y}_{\mathrm{a}2}(t)$
$\hat{y}_{\mathrm{a}1}(t)y_{\mathrm{a}2}(t)$	$= -jy_{\mathrm{a}1}(t)y_{\mathrm{a}2}(t) = y_{\mathrm{a}1}(t)\hat{y}_{\mathrm{a}2}(t)$

Table C.5 The Hilbert transform of basic signals

Signal	Transform
$\delta(x)$	$\frac{1}{\pi x}$
$\delta'(x)$	$-\frac{1}{\pi x^2}$
$\cos x$	$\sin x$
$\sin x$	$-\cos x$
$\cos(ax)J_n(bx) \quad [0<b<a]$	$-\sin(ax)J_n(bx)$
$\frac{\sin x}{x}$	$\frac{\sin^2 x/2}{x/2}$
e^{jx}	$j\mathrm{e}^{jx}$
$\frac{a}{x^2+a^2} \quad [a>0]$	$\frac{x}{x^2+a^2}$
$\frac{ab-x^2}{(ab-x^2)^2+x^2(a+b)^2}$	$\frac{x(a+b)}{(ab-x^2)^2+x^2(a+b)^2}$
$\delta(x-a)-\delta(x+a)$	$\frac{2}{\pi}\frac{a}{x^2-a^2}$
$\delta(x-a)+\delta(x+a)$	$\frac{2}{\pi}\frac{x}{x^2-a^2}$
$\mathrm{e}^{-x} \quad [x>0]$	$\frac{1}{\pi}\mathrm{e}^{-x}\mathrm{Ei}(x)$
$\mathrm{sign}\ x\mathrm{e}^{-\lvert x\rvert}$	$\frac{1}{\pi}[\mathrm{e}^{-x}\ \mathrm{Ei}\ (x)+\mathrm{e}^{x}\mathrm{Ei}(-x)]$
$\mathrm{e}^{-\lvert x\rvert}$	$\frac{1}{\pi}[\mathrm{e}^{-x}\mathrm{Ei}(x)-\mathrm{e}^{x}\mathrm{Ei}(-x)]$
$\lvert x\rvert^{v-1} \quad [0<\mathrm{Re}v<1]$	$\cot\left(\frac{\pi v}{2}\right)\lvert x\rvert^{v-1}\ \mathrm{sign}\ x$
$\lvert x\rvert^{v-1}\ \mathrm{sign}\ x \quad [0<\mathrm{Re}v<1]$	$-\tan\left(\frac{\pi v}{2}\right)\lvert x\rvert^{v-1}$
$\mathrm{sign}\ x$	$-\infty$
Even rectangular pulse $u(x+\tau)-u(x-\tau)$	$\frac{1}{\pi}\ln\left\lvert\frac{x+\tau}{x-\tau}\right\rvert$
Odd rectangular pulse $-u(x+\tau)+2u(x)-u(x-\tau)$	$-\frac{1}{\pi}\left[\ln\left\lvert\frac{x+\tau}{x}\right\rvert+\ln\left\lvert\frac{x-\tau}{x}\right\rvert\right]$
Ramp pulse $\frac{1}{\tau}(\tau-x)[u(x)-u(x-\tau)]$	$\frac{1}{\pi}\left[\frac{x-\tau}{\tau}\ln\left\lvert\frac{x-\tau}{x}\right\rvert+1\right]$
Triangular pulse $\frac{1}{\tau}(\tau-\lvert x\rvert)[u(x+\tau)-u(x-\tau)]$	$\frac{1}{\pi}\left[\frac{x+\tau}{\tau}\ln\left\lvert\frac{x+\tau}{x}\right\rvert+\frac{x-\tau}{\tau}\ln\left\lvert\frac{x-\tau}{x}\right\rvert\right]$
Sawtooth pulse $\frac{1}{\tau}(\tau-\lvert x\rvert)[-u(x+\tau)+2u(x)-u(x-\tau)]$	$\frac{1}{\pi}\left[\frac{x-\tau}{\tau}\ln\left\lvert\frac{x-\tau}{x}\right\rvert-\frac{x+\tau}{\tau}\ln\left\lvert\frac{x+\tau}{x}\right\rvert+2\right]$

D

Mathematical Formulas

Limits:

- ▷ $\lim\limits_{x \to a} \frac{f(x)}{g(x)} = \lim\limits_{x \to a} \frac{\partial f(x)/\partial x}{\partial g(x)/\partial x}$ (L'Hospital's rule)

- ▷ $\lim\limits_{x \to 0} \frac{\sin x}{x} = 1$

- ▷ $\lim\limits_{x \to 0} \frac{\sin Nx}{\sin x} = N$

- ▷ $\int\limits_0^\infty \sin bx \, \mathrm{d}x = \lim\limits_{\alpha \to 1} \int\limits_0^\infty x^{\alpha-1} \sin bx \, \mathrm{d}x = \frac{\Gamma(\alpha)}{b^\alpha} \sin \frac{\alpha\pi}{2}\Big|_{\alpha=1} = \frac{1}{b}$

Trigonometric identities:

- ▷ $\mathrm{e}^{jx} = \cos x + j \sin x$ (Euler's formula)

- ▷ $\mathrm{e}^{(\alpha + jx)} = \mathrm{e}^{\alpha}(\cos x + j \sin x)$

- ▷ $\cos x = \frac{\mathrm{e}^{jx} + \mathrm{e}^{-jx}}{2}$

- ▷ $\sin x = \frac{\mathrm{e}^{jx} - \mathrm{e}^{-jx}}{2j}$

- ▷ $\cos^2 x + \sin^2 x = 1$

- ▷ $\cos^2 x - \sin^2 x = \cos 2x$

- ▷ $2 \cos x \sin x = \sin 2x$

- ▷ $\cos^2 x = \frac{1}{2}(1 + \cos 2x)$

- ▷ $\sin^2 x = \frac{1}{2}(1 - \cos 2x)$

- ▷ $\cos x \cos y = \frac{1}{2}[\cos(x+y) + \cos(x-y)]$

- ▷ $\sin x \sin y = \frac{1}{2}[\cos(x-y) - \cos(x+y)]$

▷ $\sin x \cos y = \frac{1}{2}[\sin(x+y) + \sin(x-y)]$

▷ $\cos(x \pm y) = \cos x \cos y \mp \sin x \sin y$

▷ $\sin(x \pm y) = \sin x \cos y \pm \cos x \sin y$

▷ $\cos x + \cos y = 2\cos\frac{x+y}{2}\cos\frac{x-y}{2}$

▷ $\cos x - \cos y = -2\sin\frac{x+y}{2}\sin\frac{x-y}{2}$

▷ $\sin x \pm \sin y = 2\sin\frac{x\pm y}{2}\cos\frac{x\mp y}{2}$

▷ $a\cos x + b\sin x = r\sin(x+\varphi) = r\cos(x-\psi)$,
$r = \sqrt{a^2+b^2}$, $\sin\varphi = \frac{a}{r}$, $\cos\varphi = \frac{b}{r}$, $\sin\psi = \frac{b}{r}$, $\cos\psi = \frac{a}{r}$

▷ $\frac{d}{dx}\arcsin x = \frac{1}{\sqrt{1-x^2}}$

▷ $\frac{d}{dx}\arccos x = -\frac{1}{\sqrt{1-x^2}}$

▷ $\frac{d}{dx}\arctan x = \frac{1}{1+x^2}$

▷ $\frac{d}{dx}\text{arccot}\, x = -\frac{1}{1+x^2}$

Hyperbolic identities:

▷ $\sinh x = -\sinh(-x) = \pm\sqrt{\cosh^2 x - 1} = \pm\sqrt{\frac{1}{2}(\cosh 2x - 1)} = \frac{e^x - e^{-x}}{2}$

▷ $\cosh x = \cosh(-x) = \sqrt{\sinh^2 x + 1} = \sqrt{\frac{1}{2}(\cosh 2x + 1)} = 2\cosh^2\frac{x}{2} - 1 = \frac{e^x + e^{-x}}{2}$

▷ $\tanh x = \frac{\sinh x}{\cosh x} = \frac{e^x - e^{-x}}{e^x + e^{-x}}$

▷ $\coth x = \frac{\cosh x}{\sinh x} = \frac{e^x + e^{-x}}{e^x - e^{-x}}$

▷ $\cosh^2 x - \sinh^2 x = 1$

▷ $\cosh x + \sinh x = e^x$

▷ $\cosh x - \sinh x = e^{-x}$

Exponents:

▷ $e^{\ln x} = x$

▷ $\frac{e^x}{e^y} = e^{x-y}$

▷ $e^x e^y = e^{x+y}$

▷ $(e^x)^\alpha = e^{\alpha x}$

Logarithms:

- $\ln e^x = x$
- $\ln \frac{x}{y} = \ln x - \ln y$
- $\ln \alpha x = \ln \alpha + \ln x$
- $\ln x^\alpha = \alpha \ln x$

Extension to series:

- $\sin x = x - \frac{x^3}{3!} + \frac{x^5}{5!} - \ldots + (-1)^n \frac{x^{2n+1}}{(2n+1)!} + \ldots$
- $\cos x = 1 - \frac{x^2}{2!} + \frac{x^4}{4!} - \frac{x^6}{6!} + \ldots + (-1)^n \frac{x^{2n}}{(2n)!} + \ldots$
- $e^x = 1 + x + \frac{x^2}{2!} + \frac{x^3}{3!} + \ldots + \frac{x^n}{n!} + \ldots$
- $e^{jz\cos\phi} = \sum_{n=-\infty}^{\infty} j^n J_n(z) e^{jn\phi}$

Series:

- $\sum_{n=0}^{N-1} x^n = \frac{1-x^N}{1-x}\,, \quad x \neq 1$ (by geometric progression)
- $\sum_{n=0}^{N-1} e^{\alpha n} = \frac{1-e^{\alpha N}}{1-e^{\alpha}}$
- $\sum_{n=0}^{\infty} x^n = \frac{1}{1-x}\,, \quad |x| < 1$
- $\sum_{k=1}^{\infty} \frac{\sin^2(k\pi/q)}{k^2} = \frac{\pi^2(q-1)}{2q^2}$

Indefinite integrals:

- $\int f'(x)g(x)\mathrm{d}x = f(x)g(x) - \int f(x)g'(x)\mathrm{d}x$ (integration by parts)
- $\int f(x)\mathrm{d}x = \int f[g(y)]g'(y)\mathrm{d}y \quad [x = g(y)]$ (change of variable)
- $\int \frac{dx}{x} = \ln|x|$
- $\int e^x \mathrm{d}x = e^x$
- $\int a^x \mathrm{d}x = \frac{a^x}{\ln a}$
- $\int \sin x \,\mathrm{d}x = -\cos x$
- $\int \cos x \,\mathrm{d}x = \sin x$

▷ $\int x \begin{Bmatrix} \sin x \\ \cos x \end{Bmatrix} \mathrm{d}x = \begin{Bmatrix} \sin x \\ \cos x \end{Bmatrix} \mp x \begin{Bmatrix} \cos x \\ \sin x \end{Bmatrix}$

▷ $\int \frac{\mathrm{d}x}{x(ax^r+b)} = \frac{1}{rb} \ln \left| \frac{x^r}{ax^r+b} \right|$

▷ $\int \sin^2 x \, \mathrm{d}x = -\frac{1}{4} \sin 2x + \frac{x}{2}$

▷ $\int \cos^2 x \, \mathrm{d}x = \frac{1}{4} \sin 2x + \frac{x}{2}$

▷ $\int x \mathrm{e}^{\alpha x} \mathrm{d}x = \mathrm{e}^{\alpha x} \left(\frac{x}{\alpha} - \frac{1}{\alpha^2} \right)$

▷ $\int x^2 \mathrm{e}^{\alpha x} \mathrm{d}x = \mathrm{e}^{\alpha x} \left(\frac{x^2}{\alpha} - \frac{2x}{\alpha^2} + \frac{2}{\alpha^3} \right)$

▷ $\int x^\lambda \mathrm{e}^{\alpha x} \mathrm{d}x = \frac{1}{\alpha} x^\lambda \mathrm{e}^{\alpha x} - \frac{\lambda}{\alpha} \int x^{\lambda-1} \mathrm{e}^{\alpha x} \mathrm{d}x$

▷ $\int \mathrm{e}^{-(ax^2+bx+c)} \mathrm{d}x = \frac{1}{2} \sqrt{\frac{\pi}{a}} \mathrm{e}^{\frac{b^2-4ac}{4a}} \operatorname{erf} \left(x\sqrt{a} + \frac{b}{2\sqrt{a}} \right)$

▷ $\int \frac{1}{x} \mathrm{e}^{\alpha x} \mathrm{d}x = \mathrm{Ei}(\alpha x) \quad [\alpha \neq 0]$

▷ $\int \frac{1}{\sqrt{x}} \mathrm{e}^{-\alpha x} \mathrm{d}x = \sqrt{\frac{\pi}{\alpha}} \operatorname{erf}\left(\sqrt{\alpha x}\right) \quad [\alpha > 0]$

▷ $\int \frac{\mathrm{e}^x}{x^2+\alpha^2} \mathrm{d}x = \frac{1}{\alpha} \mathrm{Im}[\mathrm{e}^{j\alpha} \, \mathrm{Ei}(x - j\alpha)]$

Definite integrals:

▷ $\int\limits_{-\infty}^{\infty} \frac{\sin \alpha x}{x} \mathrm{d}x = \pi$

▷ $\int\limits_{-\infty}^{\infty} \mathrm{e}^{-\alpha x^2} \mathrm{d}x = \sqrt{\frac{\pi}{\alpha}}$

▷ $\int\limits_{-\infty}^{\infty} x^2 \mathrm{e}^{-\alpha x^2} \mathrm{d}x = \sqrt{\pi} \alpha^{-3/2}$

▷ $\int\limits_{0}^{\infty} \frac{\sin \alpha x}{x} \mathrm{d}x = \frac{\pi}{2} \operatorname{sgn} \alpha$

▷ $\int\limits_{0}^{\infty} \frac{\sin^2 \alpha x}{x^2} \mathrm{d}x = \frac{\pi\alpha}{2}$

▷ $\int\limits_{-\infty}^{\infty} \frac{\sin^4 \alpha x}{x^2} \mathrm{d}x = \frac{\pi\alpha}{2}$

▷ $\int\limits_{0}^{\infty} \frac{\mathrm{d}x}{\alpha^2+x^2} = \frac{\pi}{2\alpha}$

▷ $\int\limits_{0}^{\infty} x^{\alpha-1} \sin bx \, \mathrm{d}x = \frac{\Gamma(\alpha)}{b^\alpha} \sin \frac{\alpha\pi}{2}$

▷ $\int\limits_0^{\infty} \sin bx \, \mathrm{d}x = \frac{1}{b}$

▷ $\int\limits_{-\infty}^{\infty} \frac{\sin x}{x(x-\alpha)} \mathrm{d}x = \frac{\pi}{\alpha}(\cos\alpha - 1)$, α is real

▷ $\int\limits_{-\infty}^{\infty} \frac{\cos(ax)}{b^2-x^2} \mathrm{d}x = \frac{\pi}{2b} \sin(ab)$, $a, b > 0$

Special functions:

▷ $\mathrm{erf}\,(x) = \frac{2}{\sqrt{\pi}} \int\limits_0^x \mathrm{e}^{-t^2} dt$ (Error function)

▷ $\mathrm{Ei}\,(x) = -\int\limits_{-x}^{\infty} \frac{\mathrm{e}^{-t}}{t} \mathrm{d}t = \int\limits_{-\infty}^{x} \frac{\mathrm{e}^{t}}{t} \mathrm{d}t$ $[x < 0]$ (Exponential-integral function)

▷ $\mathrm{Ei}\,(x) = \mathrm{e}^x \left[\frac{1}{x} + \int\limits_0^{\infty} \frac{\mathrm{e}^{-t}}{(x-t)^2} \mathrm{d}t \right]$ $[x > 0]$ (Exponential-integral function)

▷ $S(x) = \frac{2}{\sqrt{2\pi}} \int\limits_0^x \sin t^2 \mathrm{d}t = \int\limits_0^x \sin \frac{\pi t^2}{2} \mathrm{d}t$ (Fresnel integral)

▷ $C(x) = \frac{2}{\sqrt{2\pi}} \int\limits_0^x \cos t^2 \mathrm{d}t = \int\limits_0^x \cos \frac{\pi t^2}{2} \mathrm{d}t$ (Fresnel integral)

▷ $J_n(z) = \frac{1}{2\pi} \int\limits_{-\pi}^{\pi} \mathrm{e}^{-j(n\theta - z\sin\theta)} \mathrm{d}\theta$ (Bessel function of the first kind)

References

1. Anderson, J. B., Aulin, T., and Sundberg, C.-E. *Digital Phase Modulation,* New York, Plenum Press, 1986.
2. Baskakov, S. I. *Radio Circuits and Signals,* 2nd edn., Moscow, Vysshaya Shkola, 1988 (in Russian).
3. Bracewell, R. N. *The Fourier Transform and its Applications,* 3rd edn., New York, McGraw-Hill, 1999.
4. Chen, C.-T. *Signals and Systems,* 3rd edn., New York, Oxford University Press, 2004.
5. Cracknell, A. and Hayes, L. *Introduction to Remote Sensing,* London, Taylor & Francis, 1990.
6. Gonorovsky, I. S. *Radio Circuits and Signals,* 2nd edn., Moscow, Radio i sviaz, 1986 (in Russian).
7. Gradshteyn, I. S. and Ryzhik, I. M. *Tables of Integrals, Series, and Products,* San Diego, CA, Academic Press, 1980.
8. Elali, T. and Karim, M. A. *Continuous Signals and Systems with MATLAB,* Boca Raton, FL, CRC Press, 2001.
9. Frerking, M. *Digital Signal Processing in Communication Systems,* Boston, MA, Kluwer, 1994.
10. Haykin, S. and Van Veen, B. *Signals and Systems,* 2nd edn., New York, Wiley, 2002.
11. Hsu, H. P. *Signals and Systems,* New York, McGraw-Hill, 1995.
12. Komarov, I. V. and Smolskiy, S. M. *Fundamentals of Short-Range FM Radars,* Boston, MA, Artech, 2003.
13. Lathi, B. P. *Linear Systems and Signals,* New York, Oxford University Press, 2002.
14. Le Chevalier, F. *Principles of Radar and Sonar Signal Processing,* Norwood, MA, Artech, 2002.
15. Logston, T. *Understanding the Navstar GPS, GIS, and IVHS,* 2nd edn., New York, Chapman & Hall, 1995.
16. Nathanson, F. E., Reilly, J. P., and Cohen, M. N. *Radar Design Principles,* 2nd edn., Mendham, NJ, SciTech Publishing, 1999.
17. Oppenheim, A. V., Willsky, A. S., and Hamid Nawab, S., *Signals and Systems,* 2nd edn., New York, Prentice-Hall, 1994.

18. Poularikas, A. D. *The Handbook of Formulas and Tables for Signal Processing,* Boca Raton, FL, CRC Press, 1999.
19. Roberts, M. J. *Signals and Systems Analysis Using Transform Methods and MATLAB,* New York, McGraw-Hill, 2004.
20. Schwartz, M. *Mobile Wireless Communications,* Cambridge University Press, 2005.
21. Soliman, S. S. and Srinath, M. D. *Continuous and Discrete Signals and Systems,* New York, Prentice-Hall, 1990.
22. Tetley, L. and Calcutt, D. *Electronic Navigation Systems,* 3rd edn., Oxford, Elsevier, 2001.
23. Taylor, F. J. *Principles of Signals and Systems,* New-York, McGraw Hill, 1994.
24. Tikhonov, V. I. and Mironov, M. A, *Markov Processes,* 2nd edn., Moscow, Sovetskoe Radio, 1977 (in Russian).
25. Trakhtman, A. M. and Trakhtman, V. A. Tables of Hilbert transform, *Radiotekhnika*, 25, 3, 1970.
26. Vig, J. *Introduction to Quartz Frequency Standards,* Fort Monmouth, US Army SLCET-TR-92-1 (Rev. 1), 1992.
27. Shannon, C. E. A mathematical theory of communication, *The Bell System Technical Journal*, 25, 379–423, July 1948.

Index

CPSIA information can be obtained at www.ICGtesting.com
Printed in the USA
LVOW03*1525141015

458090LV00007BB/610/P